New Developments in Quantum Field Theory and Statistical Mechanics

Cargèse 1976

NATO ADVANCED STUDY INSTITUTES SERIES

A series of edited volumes comprising multifaceted studies of contemporary scientific issues by some of the best scientific minds in the world, assembled in cooperation with NATO Scientific Affairs Division.

Series B: Physics

RECENT VOLUMES IN THIS SERIES

Volume 19 – Defects and Their Structure in Nonmetallic Solids
edited by B. Henderson and A. E. Hughes

Volume 20 – Physics of Structurally Disordered Solids
edited by Shashanka S. Mitra

Volume 21 – Superconductor Applications: SQUIDs and Machines
edited by Brian B. Schwartz and Simon Foner

Volume 22 – Nuclear Magnetic Resonance in Solids
edited by Lieven Van Gerven

Volume 23 – Photon Correlation Spectroscopy and Velocimetry
edited by E. R. Pike and H. Z. Cummins

Volume 24 – Electrons in Finite and Infinite Structures
edited by P. Phariseau and L. Scheire

Volume 25 – Chemistry and Physics of One-Dimensional Metals
edited by Heimo J. Keller

Volume 26 – New Developments in Quantum Field Theory and Statistical Mechanics–Cargèse 1976
edited by Maurice Lévy and Pronob Mitter

Volume 27 – Topics in Theoretical and Experimental Gravitation Physics
Edited by V. De Sabbata and J. Weber

Volume 28 – Material Characterization Using Ion Beams
Edited by J. P. Thomas and A. Cachard

Volume 29 – Electron–Phonon Interactions and Phase Transitions
Edited by Tormod Riste

The series is published by an international board of publishers in conjunction with NATO Scientific Affairs Division

A	Life Sciences	Plenum Publishing Corporation
B	Physics	New York and London
C	Mathematical and Physical Sciences	D. Reidel Publishing Company Dordrecht and Boston
D	Behavioral and Social Sciences	Sijthoff International Publishing Company Leiden
E	Applied Sciences	Noordhoff International Publishing Leiden

New Developments in Quantum Field Theory and Statistical Mechanics

Cargèse 1976

Edited by
Maurice Lévy and Pronob Mitter
Laboratory of Theoretical Physics and High Energies
Université Pierre et Marie Curie
Paris, France

PLENUM PRESS • NEW YORK AND LONDON
Published in cooperation with NATO Scientific Affairs Division

Library of Congress Cataloging in Publication Data

Main entry under title:

New developments in quantum field theory and statistical mechanics–Cargèse 1976.

(NATO advanced study institutes series: Series B, Physics; v. 26)
1. Quantum field theory–Congresses. 2. Statistical mechanics–Congresses. I. Lévy, Maurice, 1922- II. Mitter, Pronob. III. Series.
QC174.45.A1N47 530.1'43 77-8847
ISBN 0-306-35726-7

Proceedings of the 1976 Cargèse Summer Institute held at Cargèse, France, July 12-31, 1976, sponsored in part by NATO

A Division of Plenum Publishing Corporation
227 West 17th Street, New York, N.Y. 10011

Printed in the United States of America

PREFACE

The 1976 Cargèse Summer Institute was devoted to the study of certain exciting developments in quantum field theory and critical phenomena. Its genesis occurred in 1974 as an outgrowth of many scientific discussions amongst the undersigned, who decided to form a scientific committee for the organization of the school.

On the one hand, various workers in quantum field theory were continuing to make startling progress in different directions. On the other hand, many new problems were arising from these various domains. Thus we felt that 1976 might be an appropriate occasion both to review recent developments and to encourage interactions between researchers from different backgrounds working on a common set of unsolved problems. An important aspect of the school, as it took place, was the participation of and stimulating interaction between such a broad spectrum of theorists.

The central topics of the school were chosen from the areas of solitons, phase transitions, critical behavior, the renormalization group, gauge fields and the analysis of nonrenormalizable field theories. A noteworthy feature of these topics is the interpenetration of ideas from quantum field theory and statistical mechanics whose inherent unity is seen in the functional integral formulation of quantum field theory. The actual lectures were partly in the form of tutorials designed to familiarize the participants with recent progress on the main topics of the school. Others were in the form of more specialized seminars reporting on recent research.

We hope that the cross fertilization of Cargèse 1976 will leave a more than temporary imprint on the scientific research discussed.

We wish to express our gratitude to NATO whose generous financial help made it possible to organize Cargèse 1976, as in past years. We are equally grateful to the C.N.R.S., D.G.R.S.T., C.E.A. and N.S.F. for making available travel grants. Thanks are due to the Université de Nice for making available the facilities of the Institut d'Etudes Scientifiques de Cargèse. Last but not least we wish to thank Melle Marie-France Hanseler for her excellent organizational work before, during and after the school.

E. Brezin
A. Jaffe
H. Lehmann
M. Lévy
P. K. Mitter
R. Stora
K. Symanzik

In addition to the lectures published in this volume two lecture courses by Professors L. Kadanoff and E. Brézin and seminars by Professors H. Lehmann and A. Luther formed an integral part of Cargèse 1976. The content of these lectures have appeared elsewhere, and are thus not reproduced in this volume. For the reader's convenience we give the title of these lectures/seminars and the necessary references.

FIELD THEORETIC APPROACH TO CRITICAL PHENOMENA

E. Brézin
DPhT., C.E.N. de Saclay
B.P. n°2, 91190 Gif sur Yvette, France

The content of the lectures can be found in:

E. Brézin, J.C. Le Guillou, and J. Zinn-Justin, PHASE TRANSITION AND CRITICAL PHENOMENA, Vol. 6 (C. Domb and M.S. Green, eds.), Academic Press, N. Y., 1976

APPLICATION OF RENORMALIZATION GROUP TECHNIQUES TO QUARKS AND STRINGS

Leo P. Kadanoff
Department of Physics
Brown University, Providence, R.I.

The lectures were based on material in:

"Lectures on the application of renormalization group techniques to quarks and strings," to be published in Reviews of Modern Physics.

BOSE FIELD STRUCTURE ASSOCIATED WITH A FREE MASSIVE DIRAC FIELD IN 1-SPACE DIMENSION

H. Lehmann
DESY, Hamburg

The lecture was based on the article of the same title (co-authored by J. Stehr), D.E.S.Y. preprint, 1976.

SOLUTION OF THE MASSIVE THIRRING MODEL ON A LATTICE, AND THE SU(2) MASSLESS THIRRING MODEL

A. Luther
Nordita, Copenhagen, Denmark

The lectures were based on material published in:

Phys. Rev. B, 14, 2153 (1976), and
Phys. Rev. B, 15, 403 (1977).

CONTENTS

A TUTORIAL COURSE IN CONSTRUCTIVE FIELD THEORY

James Glimm[1] Arthur Jaffe[2]

The Rockefeller University Harvard University

New York, N.Y. 10021 Cambridge, MA 02138

---

1. Supported in part by the National Science Foundation under Grant PHY76-17191.
2. Supported in part by the National Science Foundation under Grant PHY75-21212.

1.1 INTRODUCTION

Prior to 1970, the major focus of constructive field theory was the mathematical framework required to establish the existence of quantum fields [1, 2]. Since 1970, the emphasis has gradually shifted, first toward verifying physical properties of the known models, and more recently toward bringing constructive field theory closer to the mainstream of physics [3]. In fact, by 1973 it had become more or less clear that the mathematical framework developed to give the first examples in $d = 2, 3$ space-time dimensions would be adequate to study $d = 4$. However, it was also clear that to solve the ultraviolet problem in $d = 4$, it would be necessary to incorporate into mathematical physics a deeper physical understanding of the questions being studied. In particular ideas of scaling and of critical behavior may be useful to select and analyze a suitable nontrivial critical point as the first step in dealing with the $d = 4$ ultraviolet problem. An infinite scaling transformation connects the problem of removing the ultraviolet cutoff with the problem of existence of scaling behavior at the critical point.

In these lectures we describe some results for boson (generally φ^4) fields. For $d = 4$, scaling arguments (e.g. the renormalization group) indicate that the φ^4 theory is not free at high energy. Thus the study of the φ_4^4 theory is a strong coupling problem which must be approached independent of perturbation theory. See, e.g. [4, 5] for a discussion of the mathematical aspects of this program. On the other hand, asymptotic freedom suggests that $d = 4$, nonabelian gauge field theories may be free at high energy and thus amenable to perturbation theory. For this reason gauge fields should play an important role in the future of the mathematical study of field theory, and the lecture of Osterwalder provides an introduction to that topic.

1.2 e^{-tH} AS A FUNCTIONAL INTEGRAL

It is now easy to formulate the connection between Euclidean field theory (functional integrals) and quantum mechanics (Hilbert space and Hamiltonians). We give a complete mathematical presentation of this connection for boson fields in the accompanying article, "Functional Integral Methods in Quantum Field Theory" [6]. Here we give a brief summary.

Assume we are given a functional integral $d\mu(\varphi)$, i.e. a measure $d\mu(\varphi)$ on generalized functions $\varphi \in \mathscr{S}'(R^d)$. Here φ is a classical field, and given a functional $F(\varphi)$, we define the expectation $\langle F \rangle$ by

$$\langle F \rangle = \int F(\varphi) d\mu(\varphi) \ . \tag{1.1}$$

We require the normalization condition

$$\langle 1 \rangle = \int d\mu(\varphi) = 1 \ , \tag{1.2}$$

i.e. $d\mu(\varphi)$ is a probability measure. Of particular interest is the choice $F(\varphi) = \exp(i\varphi(f))$, which leads to the Fourier transform of $d\mu(\varphi)$, which we denote

$$S\{f\} = \int e^{i\varphi(f)} d\mu(\varphi) \ . \tag{1.3}$$

$S\{f\}$ is also known as the generating functional.

We now wish to define a Hilbert space $\mathcal{H}$ of quantum mechanics in $d - 1$ space dimensions, associated with $d\mu(\varphi)$. In order to define $\mathcal{H}$, the measure $d\mu$ must satisfy three simple requirements, which we expect from every such functional integral which arises in physics. The vectors in $\mathcal{H}$ (at least a dense set of vectors) will be functionals $A = A(\varphi)$, for example $A = \varphi(f)$ or $A = \exp(i\varphi(f))$, and for certain $f \in \mathscr{S}$. The fundamental formula which relates the Hamiltonian H to functional integrals is

$$\boxed{\langle A, e^{-tH} A \rangle_{\mathcal{H}} = \int \overline{\vartheta A}\, A_t \, d\mu(\varphi)} \ , \tag{1.4}$$

where ϑ denotes time inversion, $\overline{A}$ denotes complex conjugation and A_t denotes time translation. In (1.4) A plays a dual role. On the left side, A is a vector in the Hilbert space of quantum mechanics $\mathcal{H}$, while on the right side $A(\varphi)$ is a functional of the classical field. The importance of (1.4) is that any question about H can be reduced to a question about a functional integral. For example: Does H have a unique vacuum? Does H have a mass gap? Does H have bound states and what are their masses? Do the asymptotic states span $\mathcal{H}$ (asymptotic completeness)? etc. We start with a brief explanation of (1.4), see [6] for details.

Let us start by mentioning the three requirements on $S\{f\}$ or $d\mu$ needed to obtain $\mathcal{H}$. (The generalization to include spinor or tensor fields poses no fundamental difficulty.) The conditions are

(1) Invariance
(2) Reflection Positivity
(3) Regularity

(1) We require invariance of $S\{f\}$ under a one parameter group of time translations $f(\vec{x}, s) \to f_\tau(\vec{x}, t - \tau)$ where $\tau \in R$ (or $\tau = n\delta$, $n = 0, \neq 1, \pm 2, \cdots$, i, e, lattice translations) and also invariance under time inversion $f(\vec{x}, t) \to (\vartheta f)(\vec{x}, t) = f(\vec{x}, -t)$. Then $S\{f_\tau\} = S\{\vartheta f\} = S\{f\}$.

(2) Let $f^{(j)}$, $j = 1, 2, \cdots, r$ be any sequence of real functions which vanish unless $t > 0$. Then reflection positivity is the requirement $S\{f_i - \vartheta f_j\} = M_{ij}$ is a positive matrix. In other words, if $A = \sum_{j=1}^{r} c_j \exp(i\varphi(f^{(j)}))$, then

$$\int \vartheta A \, A d\mu \geq 0 . \tag{1.5}$$

This condition was discovered by Osterwalder and Schrader [7].

(3) The regularity assumption is important for a continuous time translation group, in which case $S\{f_\tau\} \to S\{f\}$ as $\tau \to 0$.

From (1-3) we obtain $\mathcal{H}$ and e^{-tH} (in the case of a continuous time translation group). In the case of discrete time translations we obtain $\mathcal{H}$ and a transfer matrix K, with K^t playing the role of e^{-tH}. In [6] we assume invariance under the full Euclidean group on R^d (rotations, translations and reflections in hyperplanes) and we obtain a stronger result, namely the construction of a full relativistic field theory.

In fact, assuming (2), we may define a scalar product on functionals $A(\varphi)$ at positive time, e.g. functionals of the form

$A(\varphi)$ in (1.5). We let

$$(1.6) \qquad \langle A, B\rangle_{\mathcal{H}} \equiv \int \overline{\vartheta A}\, B d\mu(\varphi) \,.$$

The Hilbert space $\mathcal{H}$ is the completion of these positive time $A(\varphi)$ in the scalar product (1.6), and it is these positive time functionals $A(\varphi)$ which are allowed in (1.4). Time translation is defined by

$$(1.7) \qquad A_t = \sum c_j \exp(i\varphi(f^{(j)}))_t = \sum c_j \exp(i\varphi(f_t^{(j)})) \,,$$

so for $t \geq 0$, A_t is also a positive time functional. Thus the formula

$$(1.8) \qquad \langle A, R(t)A\rangle_{\mathcal{H}} = \int \overline{\vartheta A}\, A_t d\mu$$

defines an operator $R(t)$ on $\mathcal{H}$. As a consequence of (1) - (2), it follows that

$$(1.9) \qquad 0 \leq R(t) = R(t)^* \leq I, \quad R(t+s) = R(t)R(s) \,,$$

see for example [6]. For $t = 1$, $R(1) = K$ is the self adjoint transfer matrix, and $R(t) = K^t$. In case that time is continuous, then (3) ensures that $R(t) = e^{-tH}$, where H is a positive, self adjoint Hamiltonian, $0 \leq H = H^*$. Furthermore, since the functional $A(\varphi) = 1$ satisfies $A_t = A = 1$, it follows that $R(t)1 = 1$ and 1 is a ground state for H, $H1 = 0$. The vector 1 in $\mathcal{H}$ is generally denoted by Ω, so $H\Omega = 0$.

1.3 EXAMPLES

A. Gaussian Examples

Consider the case of full Euclidean symmetry. The simplest example is the Gaussian measure $d\mu_0$ corresponding to the free field of mass $\sigma^{\frac{1}{2}}$

$$(1.10) \qquad S_0\{f\} = \int e^{i\varphi(f)} d\mu_0(\varphi) = e^{-\frac{1}{2}\langle f, (-\Delta+\sigma)^{-1} f\rangle} .$$

In other words, $d\mu_0(\varphi)$ is the unique Gaussian measure with mean zero and covariance $(-\Delta + \sigma)^{-1}$. The only question to verify is whether (1.10) satisfies reflection positivity. This fact follows from

$$(1.11) \qquad \langle \vartheta f, (-\Delta + \sigma)^{-1} f \rangle = \left(\int_0^\infty \| e^{-t\mu} (2\mu)^{-\frac{1}{2}} f(\cdot, t) \|_{L_2} dt \right)^2$$

$$\geq 0 ,$$

where f vanishes unless $t \geq 0$ and where $\mu^2 = -(\vec{\nabla})^2 + \sigma$, where $\vec{\nabla}$ is the gradient in the $\vec{x}$ directions.

We can approximate R^d by a lattice Z^d, in which case we obtain

$$(1.12) \qquad d\mu_{0,\delta} = \lim_{\Lambda \nearrow Z^d_\delta} \frac{1}{N(\delta, \Lambda)} e^{-\frac{1}{2} \sum_{(n,n)} \delta^{d-2} (\varphi(i) - \varphi(i'))^2 - \frac{1}{2}\sigma \sum_i \delta^d \varphi(i)^2} \prod_i d\varphi(i).$$

In (1.12), Λ denotes a finite subset of the lattice Z^d_δ with lattice spacing δ. The sum and product over i extend over lattice sites in Λ, while the sum $\sum_{(n,n)}$ extends over nearest neighbor lattice sites in Λ. The measure $d\varphi(i)$ is Lebesgue measure, and $N(\delta, \Lambda)$ is chosen so $\int d\mu_{0,\delta} = 1$. The limit $\Lambda \nearrow Z^d$ exists in the sense of convergence of $S_{0,\delta,\Lambda}\{f\}$.

We can rewrite (1.12) as

$$(1.13) \qquad d\mu_{0,\delta} = \lim_{\Lambda \nearrow Z^d_\delta} e^{\beta \sum_{(n,n)} \varphi(i)\varphi(i')} \prod_i d\nu_0(\varphi(i)) ,$$

where $\beta = \delta^{d-2}$ and $d\nu_0(\varphi)$ is a Gaussian. Thus $d\mu_{0,\delta}$ is a ferromagnetic, nearest neighbor spin system with inverse temperature β and a single spin distribution function $d\nu_0(\varphi)$. The ferromagnetic coupling arises from the off-diagonal part of the gradient term in (1.12).

The limit $\delta \to 0$ for (1.12) - (1.13) exists and is (1.10). (This limit, however, must be taken in the form $S_{0,\delta}\{f\} \to S\{f\}$, since the measures $d\mu_{0,\delta}$ do not converge in the usual sense of convergence of measures on a finite dimensional space.) Thus formally

$$(1.14) \qquad d\mu_0 = N^{-1}\int e^{-\frac{1}{2}\int[(\nabla\varphi(x))^2+\sigma\varphi(x)^2]dx} \prod_{x\in R^d} d\varphi(x)\,,$$

but the mathematically meaningful form of (1.14) is (1.10).

B. Non-Gaussian Examples

In the case of invariance of $d\mu$ under continuous time translations, we obtain non-Gaussian examples from non-relativistic quantum mechanics ($d = 1$) and nonlinear quantum fields ($d = 2, 3$). In the case of nonrelativistic quantum mechanics with one degree of freedom, the measure

$$(1.15) \qquad d\mu(q) = \frac{1}{N} e^{-\int_{-\infty}^{\infty} V(q(t))dt} d\mu_0(q)$$

formally satisfies (1-3). The quantum mechanics Hamiltonian is just

$$(1.16) \qquad H = -\frac{1}{2}\frac{d^2}{dq^2} + \frac{1}{2}q^2 + V(q) - E$$

$$= H_0 + V - E\,,$$

i.e., the perturbation of the harmonic oscillator Hamiltonian H_0 by the potential V. Here E is a constant chosen so $H \geq 0$, $H\Omega = 0$. The formula (1.15) is exactly the Feynman-Kac functional integral representation of the ground state of H. In fact integration of $A(q(0))$ over $d\mu(q)$ is expectation of A in the state Ω:

$$(1.17) \qquad \int A(q_0)d\mu(q) = \lim_{T\to\infty} \frac{1}{N_T}\int e^{-\int_{-T}^{T} V(q(t))dt} A(q(0))d\mu_0(q)$$

$$= \lim_{T\to\infty} \frac{\langle e^{-TH}\Omega_0, Ae^{-TH}\Omega_0\rangle}{\|e^{-TH}\Omega_0\|^2}$$

$$= \langle \Omega, A\Omega\rangle\,.$$

Here Ω_0 is the ground state of H_0 and Ω is the ground state of H.

In quantum field theory, the known examples for $d = 2, 3$ arise from interaction energy densities $P(\varphi(x)) = V(x)$ for polynomials P. For instance the $\lambda\varphi_2^4$ model has the measure

$$d\mu = \lim_{\Lambda \nearrow R^2} \frac{1}{N_\Lambda} e^{-\int_\Lambda V(x)dx} d\mu_0 \tag{1.18}$$

$$= \lim_{\Lambda \nearrow R^2} \frac{1}{N_\Lambda} e^{-\int_\Lambda :\varphi^4(x):dx} d\mu_0 .$$

The Wick ordering of V is necessary since for $d > 1$, the measure $d\mu_0$ is concentrated on certain distributions, rather than on continuous functions as in the case of $d = 1$ (e.g. Wiener measure). The main work in this approach to proving existence of nonlinear quantum fields lies in establishing the existence of measures such as (1.18).

On a lattice, the local interaction V(x) in (1.18) does not effect the nearest neighbor term of the measure (1.12)-(1.13). Rather, V(x) contributes to the distribution of a single spin $d\nu(\varphi)$. Thus a lattice $P(\varphi)_d$ model has a measure

$$d\mu_\delta = \lim_{\Lambda \nearrow Z_\delta^d} e^{\delta^{d-2} \sum_{(n, n)} \varphi(i)\varphi(i')} \prod_i d\nu(\varphi(i)) , \tag{1.19}$$

with

$$d\nu(\varphi) = \frac{1}{N} e^{-\delta^d V(\varphi)} d\nu_0(\varphi) .$$

Here N is an appropriate normalization constant. For example, the lattice $\lambda\varphi_d^4$ model has

$$d\nu(\varphi) = \frac{1}{N} e^{-\delta^d \lambda (\varphi^2 - c)^2} d\nu_0(\varphi) , \tag{1.20}$$

where

$$c = 3\int \varphi^2 d\mu_0(\varphi) = \begin{cases} O(\ln \delta^{-1}) & d = 2 \\ O(\delta^{-(d-2)}) & d > 2 \end{cases}$$

is the Wick ordering constant.

1.4 APPLICATIONS OF THE FUNCTIONAL INTEGRAL REPRESENTATION

The usefulness of formula (1.4) is that it provides a computational framework to answer questions concerning the spectrum of H. For instance consider the question: Is the ground state Ω unique? In other words, we ask whether $e^{-tH}\theta - P_\Omega\theta$ converges to zero as $t \to \infty$. Here P_Ω is the projection onto Ω,

$$(1.21) \qquad P_\Omega \theta = |\Omega\rangle\langle\Omega|\theta\rangle = \int \theta(\varphi)d\mu(\varphi) .$$

Equivalently, is there a dense set of positive time functionals $\theta(\varphi)$ such that

$$(1.22) \quad F_\theta(t) \equiv \langle\theta, e^{-tH}(1 - P_\Omega)\theta\rangle = \int \overline{\vartheta\theta}\, \theta_t d\mu - \left|\int \theta d\mu\right|^2$$

converges to zero as $t \to +\infty$? If so, then Ω is the unique ground state of H. On the other hand if there is a positive time functional θ for which $F_\theta(t) \not\to 0$, then Ω is degenerate. In terms of the measure $d\mu(\varphi)$, the uniqueness of the vacuum Ω is equivalent to ergodicity of $d\mu(\varphi)$ under the group of time translations.

If, more generally, there are exactly r normalized, orthogonal vacuum states Ω_i, $i = 1, 2, \cdots, r$, then the function

$$F(t) = \int \overline{\vartheta\theta}\; \theta_t d\mu - \sum_{i=1}^{r} \left|\int \Omega_i \theta d\mu\right|^2$$

would converge to zero as $t \to +\infty$, rather than (1.22).

We see later that in the case of $\lambda\varphi_2^4$ or $\lambda\varphi_3^4$ quantum field models, the vacuum Ω is unique for $\lambda \ll 1$. However, Ω is degenerate for $\lambda \gg 1$. The construction of $d\mu(\varphi)$ outlined above, yields for $\lambda \gg 1$ an even mixture of vacuum states and $\langle\varphi\rangle = 0$. By introducing boundary conditions to select a particular vacuum, we may obtain solutions breaking the $\varphi \to -\varphi$ symmetry, $\langle\varphi\rangle \neq 0$, and with a unique vacuum. See the lectures of Fröhlich and Spencer for further discussion of phase transitions, and in particular a discussion of continuous symmetry breaking.

A second question concerning H is the existence of a mass gap, i.e. a gap in the spectrum corresponding to massive particles. The occurrence of a gap $(0, m)$ in the spectrum is equivalent to

$$|F_{\theta}(t)| \le O(1)e^{-mt}, \tag{1.23}$$

where the constant $O(1)$ depends on θ. Thus again the spectral properties of H are reduced to asymptotic decay rates of certain functional integrals. The proof of such decay rates in models has been established by expansion methods or by using correlation inequalities as described below.

In a theory which is even (e.g. a φ^4 model in which the symmetry $\varphi \to -\varphi$ is not broken) we can decompose the Hilbert space $\mathcal{H} = \mathcal{H}_e + \mathcal{H}_o$ into subspaces even or odd under the transformation $\varphi \to -\varphi$. The vacuum lies in $\mathcal{H}_e$, while the one particle states lie in $\mathcal{H}_o$. Let m denote the bottom of the spectrum on $\mathcal{H}_o$. On $\mathcal{H}_e$, we thus expect a mass gap of magnitude m', where $m < m' \le 2m$. (Since two particle scattering states occur in $\mathcal{H}_e$, the Hamiltonian will always have spectrum throughout the interval $[2m, \infty)$.) The statement $m' = 2m$ is the statement that two particle bound states do not occur in $\mathcal{H}_e$. This is equivalent to

$$|F_{\theta}(t)| \le O(1)e^{-2mt} \tag{1.24}$$

as θ ranges over a dense set of $\mathcal{H}_e$. We discuss this further in the next section.

Finally, in order to analyze the bound states or scattering of several particles, it is useful to study kernels (e.g. exact Bethe-Salpeter kernels) which characterize the Hamiltonian for n-body processes. Such kernels have a functional integral representation, and a detailed study has been made by Spencer and Zirilli [7, 8] in the case $n = 2$. (See also [2, 3, 9]

1.5 ISING, GAUSSIAN AND SCALING LIMITS

We briefly mention the qualitative structure of the φ_d^4 lattice quantum field model of §1.3, in its dependence on the parameters δ, λ, σ. In particular, we discuss the measures $d\mu_\delta$ defined in (1.19)-(1.20); an analogous discussion could be given for $P(\varphi)$ models. See [4, 5, 10].

To begin with, consider the (λ, δ) parameter space with fixed σ. For δ fixed, we study $\lambda \to 0$ and $\lambda \to \infty$, the minimum and maximum coupling. It is clear that the $\lambda \to 0$ limit of (1.19)-(1.20) is Gaussian, in fact $d\mu_\delta \to d\mu_{0,\delta}$. (In every case we define convergence as convergence

$$(1.25) \quad S_\delta\{f\} = \int e^{i\varphi(f)} d\mu_\delta(\varphi) \to S_{0,\delta}\{f\} = \int e^{i\varphi(f)} d\mu_{0,\delta}(\varphi)$$

of generating functionals.) On the other hand, for $\lambda \to \infty$, with δ fixed, the measure $d\nu(\varphi)$ becomes concentrated at the points where $|\varphi| = c^{\frac{1}{2}}$, i.e. $\varphi = \pm c^{\frac{1}{2}}$. Since the integral of $d\mu_\delta$ is normalized to one, in this limit

$$(1.26) \qquad d\mu_\delta = \lim_{\Lambda \nearrow Z_\delta^d} e^{\beta \sum_{(n,n)} \varphi(i)\varphi(i')} \prod_i d\nu(\varphi_i)$$

where $d\nu(\varphi) = \frac{1}{N}(\delta(\varphi - c^{\frac{1}{2}}) + \delta(\varphi + c^{\frac{1}{2}}))$. In other words, $d\mu_\delta$ is an Ising model with lattice spacing δ and spin φ normalized to take the values $\pm c^{\frac{1}{2}}$. The mathematical existence of this Ising limit was established [11]. Furthermore, for $d = 2$ (or with $d = 3$ and the proper choice $\sigma = \sigma(\delta) = O(\ln \delta^{-1})$ to ensure massrrenormalization, the $\delta \to 0$ limit can be taken with λ fixed. This continuum limit yields the Euclidean φ^4 model.

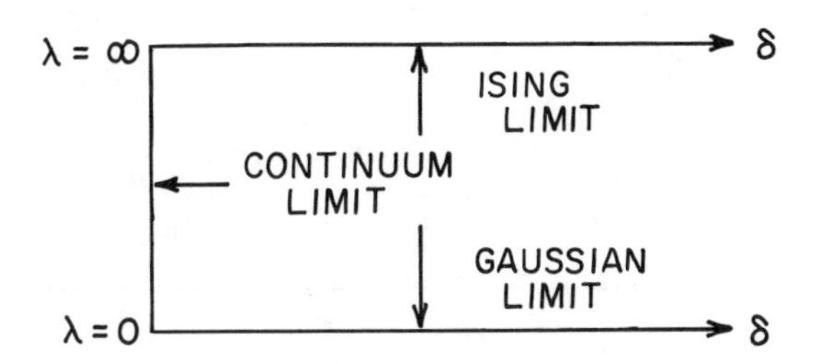

We next modify this picture slightly by fixing the mass gap m. On the lattice, m is defined as the gap in the spectrum of $-\ln K$ (K is the transfer matrix); if $\delta = 0$, m is defined as the gap in the spectrum of H. Since $m = m(\lambda, \delta, \sigma)$, we achieve this by choosing σ in such a way that we remain in the single phase region and such that m = const. The required continuity of m follows by [12, 13]. We now plot the projection of such a $\sigma = \sigma(\lambda, \delta)$ surface in the (λ, δ) plane.

One can now ask whether the $\delta \to 0$ limit of Ising models exist, and whether the $\lambda \to \infty$ limit of continuum models exists. The first is a scaling limit of the Ising model, the second is a scaling limit of the $\lambda\varphi^4$ model. In the continuum theory ($d = 2, 3$), increasing λ with σ fixed would result in a phase transition and $m \to 0$. Thus in the scaling limit, with m fixed, it follows that $\sigma(\lambda, \delta = 0) \to \infty$ as $\lambda \to \infty$, i.e. there is an infinite change of scale. Also the scaling limit is formally an infinite scaling of a $m = 0$ (critical) theory. We conjecture that both the $\lambda \to \infty$ and the $\delta \to 0$ scaling limits exist, and that they agree. (See §1.5.)

In studying this limit, it is also useful to consider curves in the (λ, δ) plane with constant unrenormalized, dimensionless charge $g_0 = \lambda\delta^{4-d}$. For $d < 4$ (the superrenormalizable case) these curves lead to the scaling limit ($\lambda = \infty$, $\delta = 0$) discussed above, i.e. strong coupling. For $d > 4$ (the nonrenormalizable case) these curves lead to $\lambda = \delta = 0$, i.e. weak coupling.

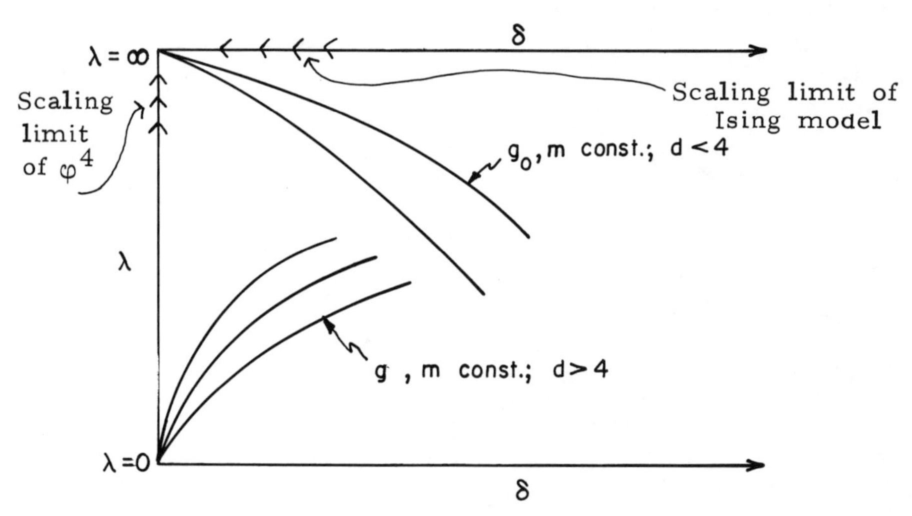

This picture leads us to conjecture that for $d > 4$, the $\delta \to 0$ limit for g_0 bounded is Gaussian (trivial), but a nontrivial theory could result with $g_0 \to \infty$ as $\delta \to 0$ (charge renormalization). See §2.5 for further discussion, and also [20, 10].

1.6 MAIN RESULTS

We sketch the main results for the φ^4 model, φ^4 lattice fields and Ising models, some of which we discuss in detail in the next two chapters. There are two main methods to derive these results: correlation inequalities (discussed in Chapter 2) and expansion techniques (discussed in Chapter 3). The correlation inequalities for the φ^4 model express in part the repulsive character of the forces in these models (in the single phase region). We obtain mathematical proofs for portions of the critical behavior in these models, as well as an initial analysis of the elementary particle and bound state problems.

A. The critical point has a conventional φ^4 structure. For $\sigma >> 0$, there is a unique phase, independent of boundary conditions, and for $\sigma << 0$ and $d \geq 2$, there are at least two pure phases, depending on the boundary conditions [14, 15, 16] The critical value $\sigma_c = \sigma_c(\lambda, \delta)$ is defined as the largest (and presumably the only) value of σ for which $m = m(\sigma) \to 0$ as $\sigma \searrow \sigma_c$. This critical σ_c exists [13] and $m(\sigma)$ is monotonic increasing for $\sigma \geq \sigma_c$ [17]. For $d \geq 3$ (and presumably for $d = 2$ also) $m(\sigma_c) = 0$, and for $\sigma = \sigma_c$ there is a unique phase and zero magnetization [13]. For $\sigma \geq \sigma_c$, the physical charge g (defined as the amputated connected four point function evaluated at zero momentum) is finite and bounded, uniformly as $\sigma \searrow \sigma_c$, and $\lambda \to \infty$. (See §2.3.) Furthermore the n-point Schwinger functions $S^{(n)}$ satisfy a Gaussian upper bound

$$0 \leq S^{(n)}(x_1 \cdots x_n) \leq \sum_{\text{pairing}} \prod_{\text{pairs}} S^{(2)}(x_i x_{i'}) ,$$

see [18, 19]. Closely related are critical exponent bounds of the form

$$\text{Gaussian exponent} \leq \varphi\text{-exponent} \leq \varphi^2\text{-exponent} .$$

For example with η the anomalous dimension of the field φ and η_E the anomalous dimension of the field $:\varphi^2(x):$,

$$0 \leq 2\eta \leq \eta_E .$$

See [20, 21, 22] for other recent exponent inequalities.

B. Particles do not form even bound states, for $\sigma \geq \sigma_c$. In the single phase region, particles exist for a.e. $m > 0$ [21]. (Presumably they exist for all $m > 0$, but at least for $d = 2, 3$, the particles should not exist for $m = 0$, i.e. $\sigma = \sigma_c$.) Here particles are poles in the two point function at Minkowsky momenta, or δ-functions in a Källen-Lehmann representation for the two point function or (at least in the Euclidean invariant case of a continuum field) an Ornstein-Zernike decay rate

$$\langle \varphi(x)\varphi(y) \rangle \sim Zr^{(1-d)/2} e^{-mr}$$

as $r = |x - y| \to \infty$. These particles do not form even bound states (with energies below the two particle continuum) [23, 24] and there are indications that they do not form odd bound states (with energies below the three particle continuum) [23, 24].

C. A heuristic interchange of the $\delta \to 0$ (continuum) limit and the $\lambda \to \infty$ (Ising) limit "shows" the identical critical point behavior [25,19]. Combining this idea with the known spectral properties of the $d = 2$ Ising suggests that for φ_2^4 continuum or lattice fields, with $\sigma < \sigma_c$, the elementary particle is actually a two soliton bound state. Furthermore in this picture the binding energy goes to zero (relative to the soliton mass) as $\sigma \nearrow \sigma_c$ and the field strength renormalization $Z \to 0$, and it probably vanishes faster than the strength of the two soliton continuum, thus suggesting that because of the solitons, the intermediate renormalization is correct, but the mass shell renormalization is incorrect, for $\sigma \nearrow \sigma_c$, $d = 2$, unless the mass shell renormalization includes all sectors, with the soliton as elementary particle. Introducing a small external field $-\mu\varphi(x)$ in the action, we have argued [26] that for $d = 2$ and $\sigma < \sigma_c$, but $\sigma \approx \sigma_c$, the limit $\mu \to 0$ introduces many bound states which coalesce to form a two soliton continuum at $\mu = 0$. For $\sigma \ll \sigma_c$ the same reasoning suggests many resonances coalescing to form continuous spectrum.

D. Within its region of convergence, the cluster expansion allows a nearly complete analysis of the field theory. To begin with, one can prove convergence of the infinite volume limit and uniqueness of the vacuum. (In a two phase region, suitable boundary conditions are required to select a pure phase.) The convergence is valid for λ complex in a sector about $\lambda = 0$ and it follows that the correlation functions are also analytic in λ and other parameters, for $\lambda \neq 0$. For the case of a $\lambda\varphi_2^4$ interaction in the region $\lambda \ll 1$, the perturbation series about $\lambda = 0$ is Borel summable to the exact solution [27]. For general $P(\varphi)_2$ interactions in the region (5.1), the perturbation series about $\lambda = 0$ for the Euclidean and Minkowsky correlation functions and for the S-matrix is also asymptotic [28, 29, 30]. The particles (whose scattering is described by the S-matrix) are also constructed from the cluster expansion [15]. Criteria (in terms of P) for existence or nonexistence of weak coupling $P(\varphi)_2$ bound states are given in [31] following earlier work of [8]. Here the cluster expansion permits the study of the exact Bethe-Salpeter equation, and shows that the low order terms (ladder approximation + $\cdots$) give the dominant effects. Up to some energy level, this analysis of the Bethe-Salpeter equation also shows asymptotic completeness [8]. Unfortunately, the energies allowed by present techniques apparently do not reach up to the two soliton threshold, in the two phase region.

2.1 CORRELATION INEQUALITIES

In this section we derive some basic correlation inequalities and state some others. In the following section we derive some consequences of these inequalities, and finally we consider the conjectured inequality $\Gamma^{(6)} \leq 0$.

For positive integers $a_1, \cdots, a_n$ define

$$\varphi_A = \varphi(x_1)^{a_1} \cdots \varphi(x_n)^{a_n} . \tag{2.1}$$

Theorem 2.1: For a lattice $(\lambda\varphi^4 - \mu\varphi)_d$ quantum field with $\lambda, \mu \geq 0$,

$$0 \leq \langle\varphi_A\rangle \tag{GKS 1}$$

$$0 \leq \langle\varphi_A\varphi_B\rangle - \langle\varphi_A\rangle\langle\varphi_B\rangle . \tag{GKS 2}$$

These are the first and second Griffiths (Griffiths, Kelly, Sherman) inequalities. Since they say that certain quantities are positive, they are preserved under limits, e.g. $\delta \to 0$ or

$\lambda \to \infty$, whenever such limits exist. The solution of the $P(\varphi)_2$ ultraviolet problem [1] was extended by Guerra, Rosen and Simon to lattice cutoffs [17] in order to prove GKS and related inequalities for these models. See [32] for φ_3^4.

Proof of GKS 1: We write on a finite lattice

$$(2.4) \quad \langle \varphi_A \rangle = \int \varphi_A d\mu_\delta = \int \varphi_A e^{\beta\Sigma\varphi(i)\varphi(i')+\mu\Sigma\varphi(i)} \prod_i d\nu(\varphi(i)) .$$

Note that $d\nu(\varphi) = d\nu(-\varphi)$, so

$$(2.5) \quad \int \varphi(i)^{a_i} d\nu(\varphi(i)) = \begin{cases} 0 \text{ if } a_i \text{ odd} \\ \text{positive if } a_i \text{ even} \end{cases}$$

The basic idea of the proof is to expand the exponentials in (2.4) in power series and then factor the resulting integrals over lattice sites. Using (2.5), we obtain a sum of products of positive terms, and hence (2.4) is positive. We then take the limit as the finite lattice increases to Z^d.

Proof of GKS 2: The basic idea is to use the technique of duplicate variables. Let φ, ψ be independent, lattice fields. Define an expectation of functionals $A = A(\varphi, \psi)$ by

$$(2.6) \quad \langle A \rangle = \int A(\varphi, \psi) d\mu(\varphi)_\delta d\mu(\psi)_\delta .$$

Here

$$d\mu(\varphi)_\delta = e^{\beta\Sigma\varphi(i)\varphi(i')} \prod_i d\nu(\varphi(i))$$

$$d\mu(\psi)_\delta = e^{\beta\Sigma\psi(i)\psi(i')} \prod_i d\nu(\psi(i)) ,$$

and for simplicity we let $\mu = 0$. Define

$$(2.7) \quad t = \varphi + \psi, \quad q = \varphi - \psi$$

as the even and odd combinations of φ, ψ, under interchange of φ and ψ. Note

(2.8) $$\langle t\rangle = \langle\varphi\rangle + \langle\psi\rangle = 2\langle\varphi\rangle ,$$

$$\langle q\rangle = 0 ,$$

$$d\mu(\varphi)_\delta d\mu(\psi)_\delta = e^{\frac{1}{2}\beta\Sigma(t(i)t(i')+q(i)q(i'))}\prod_i d\nu\left(\frac{t(i)+q(i)}{2}\right)d\nu\left(\frac{t(i)-q(i)}{2}\right) .$$

First we remark that $d\nu(t+q)d\nu(t-q)$ is even under the transformation $t \to -t$ and also under $q \to -q$. Thus

(2.9) $$\int q^\alpha t^\beta d\nu(t+q)d\nu(t-q) = \begin{cases} 0 \text{ if } \alpha \text{ or } \beta \text{ odd} \\ \text{positive if } \alpha, \beta \text{ both even} . \end{cases}$$

Following the proof of GKS 1, and using (2.9), we find that for all A, B,

(2.10) $$0 \le \langle q_A t_B\rangle .$$

To complete the proof of GKS 2, we write

(2.11) $$\langle\varphi_A\varphi_B\rangle - \langle\varphi_A\rangle\langle\varphi_B\rangle = \langle\varphi_A(\varphi_B - \psi_B)\rangle$$

$$= 2^{-|A|-|B|}\langle(t+q)_A[(t+q)_B - (t-q)_B]\rangle .$$

But $(t+q)_B(t-q)_B$ is a polynomial in t, q with positive coefficients. Hence (2.10) shows that (2.11) is positive.

The proof above follows the presentation of Sylvester [33], which we recommend for proofs of other correlation inequalities. We now state three inequalities:

Theorem 2.2: For a lattice $(\lambda\varphi^4 - \mu\varphi)_d$ quantum field theories with $\lambda, \mu \ge 0$,

(2.12) $$\langle t_A t_B\rangle - \langle t_A\rangle\langle t_B\rangle \ge 0 ,$$

(2.13) $$\langle q_A q_B\rangle - \langle q_A\rangle\langle q_B\rangle \ge 0 ,$$

(2.14) $$\langle q_A t_B\rangle - \langle q_A\rangle\langle t_B\rangle \le 0 .$$

The inequalities (2.14) were first proved for the Ising model by Lebowitz [34], and have a number of interesting consequences. We remark that two special cases of (2.14) are

(2.15) $$\langle \varphi(x_1)\varphi(x_2)\varphi(x_3)\rangle_T \le 0 , \quad \mu \ge 0 ,$$

(2.16) $$\langle \varphi(x_1)\varphi(x_2)\varphi(x_3)\varphi(x_4)\rangle_T \le 0 , \quad \mu = 0 .$$

Here $\langle \quad \rangle_T$ denotes the truncated (connected) expectation values, defined by

(2.17) $$\langle \varphi(f)^n\rangle_T = \frac{d^n}{d\alpha^n} \ln\langle e^{\alpha\varphi(f)}\rangle \Big|_{\alpha=0} ,$$

and extended to $\langle \varphi(f_1)\cdots\varphi(f_n)\rangle_T$ by multilinearity. We obtain (2.15), the Griffiths-Hearst-Sherman inequality, by expanding $\langle t_1 q_2 q_3\rangle - \langle t_1\rangle\langle q_2 q_3\rangle \le 0$. The inequality (2.16) follows from evaluating $\langle t_1 t_2 q_3 q_4\rangle - \langle t_1 t_2\rangle\langle q_3 q_4\rangle \le 0$ in case $\mu = 0$.

2.2 ABSENCE OF EVEN BOUND STATES

In a single phase, even φ^4 model, i.e. for $\sigma > \sigma_c$, we now show that the Hamiltonian, restricted to $\mathcal{H}_{even}$, has no spectrum in the interval $(0, 2m)$, i.e. two particle bound states do not exist. We remark that $\mathcal{H}_{even}$ is spanned by vectors Ω, $\varphi(f_1)\cdots\varphi(f_n)\Omega$, $n = 2, 4, \ldots$, where suppt f_j is contained in $t > 0$.

Theorem 2.3: Consider a φ^4 field or Ising model with zero external field and $\sigma > \sigma_c$, and let A and B have an even number of elements. Then

$$\langle\varphi_A\varphi_B\rangle - \langle\varphi_A\rangle\langle\varphi_B\rangle \le \sum_{\substack{A_1\subset A,\ A_1 \text{ odd} \\ B_1\subset B,\ B_1 \text{ odd}}} \langle\varphi_{A_1}\varphi_{B_1}\rangle\langle\varphi_{A-A_1}\varphi_{B-B_1}\rangle$$

Corollary 2.4: Under the hypothesis of Theorem 3.3, there are no even bound states with energy below the two particle threshold.

Proof: Let Ω be the vacuum state, unique since it is assumed that $\sigma > \sigma_c$. We write $x = x_1, \cdots, x_d$ as

$$x = (t, \vec{x})$$

with $\vec{x} \in R^{d-1}$. In particular if

$$A + s = \{(t + s, \vec{x}): t, \vec{x} \in A\},$$

then the equation (1.4), namely

$$\langle \varphi_A e^{-sH} \varphi_B \rangle = \langle \vartheta \varphi_A \varphi_{B+s} \rangle ,$$

is valid when the times in A precede the times in B. In particular we choose A to have only negative times $t \le 0$, and B, chosen as

$$B = \{(-t, \vec{x}):(t, \vec{x}) \in A\}$$

then has only positive times. With this choice of A and B, and with P_Ω the projection onto the vacuum state, we recognize

$$\langle \varphi_{A+s} \varphi_{B+s} \rangle - \langle \varphi_{A+s} \rangle \langle \varphi_{B+s} \rangle = \| e^{-sH}(I - P_\Omega) \varphi_A \Omega \|^2$$

so that the Theorem 2.3 gives a bound on the decay rates which occur in

$$e^{-sH}(I - P_\Omega).$$

For A_1 odd, $\varphi_{A_1} \Omega$ is perpendicular to the vacuum $((\Omega, \varphi_{A_1} \Omega) = \langle \varphi_{A_1} \rangle = 0)$, and so $\langle \varphi_{A_1 - s} \varphi_{B_1 + s} \rangle$ has as its slowest exponential decay rate, m, by definition the mass of the theory. Thus by definition of m,

$$\langle \varphi_{A_1 - s} \varphi_{B_1 + s} \rangle \le C_{A_1, B_1} e^{-ms}$$

for some constant C_{A_1, B_1} depending on A_1 and B_1. The same bound holds for $\langle \varphi_{(A - A_1) - s} \varphi_{(B - B_1) + s} \rangle$, and so by Theorem 2.3,

$$\|e^{-sH}(I - P_\Omega)\varphi_A\Omega\|^2 \le \text{const.}\, e^{-2ms} .$$

Thus there are no even states, except Ω, with energy below 2m, hence in particular no even bound states in this energy range.

Proof of Theorem 2.3:

$$\langle t_A q_B\rangle = \sum_{\substack{A=A_1\cup A_2\\ B=B_1\cup B_2}} (-1)^{|B_2|}\langle \varphi_{A_1}\psi_{A_2}\varphi_{B_1}\psi_{B_2}\rangle$$

$$= \sum(-1)^{|B_2|}\langle \varphi_{A_1}\varphi_{B_1}\rangle\langle\varphi_{A_2}\varphi_{B_2}\rangle$$

$$\le \langle t_A\rangle\langle t_B\rangle = \sum(-1)^{|B_2|}\langle\varphi_{A_1}\rangle\langle\varphi_{A_1}\rangle\langle\varphi_{B_1}\rangle\langle\varphi_{B_2}\rangle .$$

We drop zero terms from the right hand side (A_2 or B_2 odd), and from the left hand side (only one of A_2, B_2 odd). For the terms with B_2 even and the partition nontrivial, we combine right and left sides and eliminate from the inequality, using Theorem 2.2. The terms remaining yield Theorem 2.3.

2.3 BOUND ON g

Define the dimensionless φ^4 coupling constant by

$$g = -m^{d-4}\chi^{-4}\int\langle\varphi(x_1)\cdots\varphi(x_4)\rangle_T dx_1 dx_2 dx_3$$

where

$$\chi = \int\langle\varphi(x_1)\varphi(x_2)\rangle dx_1 = \int_0^\infty \frac{d\rho(a)}{a} .$$

By GKS 1, $\chi \ge 0$. For a massive, single phase, even φ^4 interaction $g \ge 0$ by (2.16). We now assume in addition that the proper field strength renormalization has been performed;

in the case of an isolated particle of mass m, this means $d\rho(a) = \delta(a - m^2)da + d\sigma(a)$, where inf suppt $d\sigma > m^2$. We then prove an upper bound on g.

Theorem 2.5 [35]: Under the above assumptions,

$$0 \leq g \leq \text{const.} ,$$

where the dimensionless constant is independent of all parameters (e.g. λ, σ).

We outline the proof. For details, see the original paper. We use the basic inequality GKS 2 to derive (writing 1 for $\varphi(x_1)$, etc.)

$$(2.18)\quad 0 \leq \langle 1234 \rangle - \langle 12 \rangle\langle 34 \rangle = \langle 1234 \rangle_T + \langle 13 \rangle\langle 24 \rangle + \langle 14 \rangle\langle 23 \rangle .$$

By (2.16), $\langle 1234 \rangle_T \leq 0$ and

$$0 \leq -\langle 1234 \rangle_T \leq \langle 13 \rangle\langle 24 \rangle + \langle 14 \rangle\langle 23 \rangle .$$

After symmetrization over the choices of variables,

$$(2.19)\quad -\langle 1234 \rangle_T \leq (\langle 13 \rangle\langle 24 \rangle + \langle 14 \rangle\langle 23 \rangle)^{\frac{1}{3}}(\langle 12 \rangle\langle 34 \rangle + \langle 13 \rangle\langle 24 \rangle)^{\frac{1}{3}}$$

$$\times (\langle 14 \rangle\langle 23 \rangle + \langle 12 \rangle\langle 34 \rangle)^{\frac{1}{3}} .$$

From elementary properties of the Green's function for the Poisson operator (i.e. $\ker(-\Delta + a)^{-1}(x, y)$) we find

$$\langle xy \rangle = \int_0^\infty \ker(-\Delta + a)^{-1}(x, y)d\rho(a)$$

$$\leq \text{const.}\chi |x - y|^{-d} \exp(-m|x - y|/2) .$$

Inserting this in our bound (2.19) for $-\langle 1234 \rangle_T$ gives

$$g \leq \text{const.} m^{-4} \chi^2 .$$

Since

$$\chi = \int_{m^2}^{\infty} \frac{d\rho(a)}{a} \geq m^{-2},$$

we obtain $g \leq$ const. as claimed.

Observe that the final bound does not depend on m, and hence also holds in the limit $m \to 0$. Hence the critical point (which for $d < 4$ should be an infrared stable fixed point of the renormalization group) occurs for finite g.

2.4 BOUND ON $dm^2/d\sigma$ AND PARTICLES

Here we consider a canonical, single phase φ^4 model (i.e. without field strength renormalization). We establish

$$\frac{dm^2(\sigma)}{d\sigma} \leq Z(\sigma) \tag{2.20}$$

from which our next result follows by approximation methods:

<u>Theorem 2.6</u>: (See [21].) For almost every value of m, particles exist, i.e. $Z \neq 0$.

<u>Proof of (2.20)</u>: Consider $\Gamma(p) = -S(p)^{-1}$, where S(p) is the Fourier transform of $\langle\varphi(x)\varphi(0)\rangle$. Note

$$S(p) = \frac{Z}{p^2 + m^2} + \int \frac{d\sigma(a)}{p^2 + a},$$

and

$$Z^{-1} = -(d\Gamma/dp^2)_{p^2=-m^2} . \tag{2.21}$$

Since $\Gamma = 0$ on the one particle curve $p^2 = -m^2(\sigma)$, $\nabla\Gamma$ must be orthogonal to the vector $(dm^2/d\sigma, 1)$ in the $-p^2, \sigma$ space. Thus for $p^2 = -m^2$,

$$0 = -\frac{\partial\Gamma}{\partial p^2}\frac{dm^2}{d\sigma} + \frac{\partial\Gamma}{\partial\sigma} = Z^{-1}\frac{dm^2}{d\sigma} + \frac{\partial\Gamma}{\partial\sigma} .$$

The desired inequality follows from

Theorem 2.7: Under the above assumptions,

$$\left(\frac{\partial \Gamma}{\partial \sigma}\right)_{p^2=-m^2} \geq -1 . \tag{2.22}$$

Proof: Let $\chi(p) = \int \langle \varphi(x)\varphi(0)\rangle e^{-px} dx$. Then

$$-\frac{d\chi(p)}{d\sigma} = \frac{1}{2}\iint [\langle xozz\rangle - \langle xo\rangle\langle zz\rangle] dz e^{-px} dx$$

by (2.16), (2.18),

$$\leq \iint \langle xz\rangle\langle yz\rangle e^{-p(x-z)} e^{-pz} dx dz$$

$$= \chi(p)^2 .$$

Thus

$$0 \leq \frac{d\chi(p)^{-1}}{d\sigma} \leq 1 .$$

However $\chi(p) = -\Gamma(p)|_{p^2=-m^2}$, so (2.22) is proved.

2.5 THE CONJECTURE $\Gamma^{(6)} \leq 0$

The unamputated six point vertex function is defined by

$$\Gamma^{(6)}(xxxyyy) = \langle xxxyyy\rangle_T + \int \langle xxxz\rangle_T \Gamma(zz')\langle z'yyy\rangle_T dz dz' + 9\int \langle xxyz\rangle_T \Gamma(zz')\langle z'xyy\rangle_T dz dz' . \tag{2.23}$$

The conjecture

$$\Gamma^{(6)}(xxxyyy) \leq 0 \tag{2.24}$$

has a number of interesting consequences: e.g., the absence of three particle bound states in the propagator, the existence of the scaling limit, and certain bounds on critical exponents, see [25, 5, 20].

There is some evidence for (2.24) in single phase, even φ^4 models. For example it is true in perturbation theory (i.e. for $\sigma >> 0$ or high temperature). It holds in the one dimensional Ising model [36] and numerical studies indicate that it holds for the anharmonic oscillator [37]. There is a heuristic argument that it holds near σ_c. However, some good new idea is needed to prove (2.24).

In this section we illustrate some uses of (2.24). For example, we have

Theorem 2.8: If (2.24) holds, then

$$0 \le \Gamma(x) \le e^{-3m|x|}, \quad |x| \to \infty. \tag{2.25}$$

Remark: The bound (2.25) excludes spectrum in $\Gamma(x)$ in the interval $(0, 3m)$, and hence spectrum in $d\sigma(a)$ in the interval $(m, 3m)$. Thus no three particle bound states occur in the propagator, i.e. in the states spanned by $\varphi(x)\Omega$.

Outline of Proof: We use the integration by parts formula [25]

$$\int \varphi(x)A(\varphi)d\mu(\varphi) = \langle \varphi(x)A \rangle \tag{2.26}$$

$$= \int dyS(x-y)[\langle \frac{\delta A}{\delta\varphi(y)} \rangle - \langle \mathcal{V}'(y)(I - P_1)A \rangle].$$

Here $\mathcal{V} = \lambda :\varphi^4:$ is the interaction, and $P_1 A = \int \varphi(z)\Gamma(z - z')\langle \varphi(z')A \rangle dz dz'$. From (2.26), it follows that for $x \neq 0$,

$$\Gamma(x-y) = \langle \mathcal{V}'(x)(I - P_1)\mathcal{V}'(y) \rangle = \lambda^2 \langle \varphi^3(x)(I - P_1)\varphi^3(y) \rangle, \tag{2.27}$$

see [25]. Expanding (2.27), and using (2.23),

$$\lambda^{-2}\Gamma(x-y) = 6\langle xy \rangle^3 + 9\langle xxyy \rangle_T \langle xy \rangle \tag{2.28}$$

$$- 9\int \langle xxyz \rangle_T \Gamma(zz') \langle z'xyy \rangle_T dz dz' + \Gamma^{(6)}(xxxyyy).$$

The first term in (2.28) is $o(e^{-m|x-y|})^3$, for $|x - y| \to \infty$. The second term is negative. The third term, also has a three particle decay, which can be established using the absence of two particle bound states in $\langle xxyz\rangle_T$, see [25]. Thus (2.24) results in

$$\Gamma(x - y) \le e^{-3m|x-y|}, \quad |x - y| \to \infty .$$

The positivity of $\Gamma(x - y)$ follows from the fact that it is the Fourier transform of a Herglotz function. This completes the outline of the proof.

We finish this section with the statement of another consequence of (2.24), and an Ornstein-Zernicke upper bound

$$\langle\varphi(x)\varphi(0)\rangle \le K \min(|x|, m^{-1})^{-(d-2+\eta)} e^{-m|x|} . \tag{2.29}$$

Theorem 2.8 [20]: Assume (2.24), (2.29) and $\lambda < \infty$. Then

$$\eta \le \begin{cases} .4 & d = 4 \\ .8 & d = 3 \\ 1.2 & d = 2 \end{cases} .$$

Also $\eta = 0$, $Z^{-1} < \infty$ for $d \ge 5$.

Corollary 2.9 [20]: Assume (2.24). If the $\delta \to 0$ limit of the $\lambda\varphi_d^4$ lattice field theory is Euclidean invariant for $g_0(\delta) = \lambda\delta^{4-d} \le$ const. (finite charge renormalization), then the limit is a free field for $d \ge 6$.

3. CLUSTER EXPANSIONS

3.1 THE REGION OF CONVERGENCE

The cluster expansion, in field theory as in statistical mechanics, provides almost complete information for parameter values away from critical, and it provides only limited information for parameter values near critical. In statistical mechanics, this expansion is a variant of the virial, high temperature and low temperature (Peierls' contour) expansions. These names distinguish various regions of the coupling constants, and expansion parameters. In field theory, convergence of the cluster expansion is known for the corresponding parameter values. In particular, for a two dimensional $P(\varphi)$ field theory, the cluster expansion is convergent in the following asymptotic regions [14, 15, 38, 26, 39].

$$P(\varphi) = \lambda :P_0(\varphi): + \sigma\varphi^2, \quad \lambda \to 0 \tag{3.1}$$

$$P(\varphi) = \lambda :P_0(\varphi): - \mu\varphi, \quad \mu \to \infty \tag{3.2}$$

$$P(\varphi) = \lambda :\varphi^4: + \sigma\varphi^2, \quad \sigma \to -\infty \text{ or } \lambda \to +\infty, \tag{3.3}$$

or more generally, whenever P, expanded about a suitable global minimum φ_c of $P(\varphi)$ has a dominant quadratic term. For the Yukawa$_2$ and φ_3^4 interactions, convergence of the expansion is known in the high temperature region ($\lambda \to 0$ as in (3.1)) [40-43].

3.2 THE ZEROTH ORDER EXPANSION

The expansion is adapted from the virial expansion of statistical mechanics. In the zeroth order, all couplings are removed. We divide Euclidean space time into cells (lattice cubes) and then remove the coupling between distinct cells. In the zeroth approximation, all correlations factor,

$$\langle \varphi(x_1)\cdots\varphi(x_n)\rangle_0 = \prod_{\Delta=\text{cell}} \langle \prod_{x_j\in\Delta} \varphi(x_j)\rangle_0 .$$

Consequently the long distance behavior is trivial and all states have infinite energy in the zeroth approximation.

To define the zeroth approximation, let $\partial\Delta$ be the boundary of the cell Δ.. Then formally

$$\mathcal{A}_0(\varphi) = \infty \sum_{\Delta} \int_{\partial\Delta} \varphi^2(x)dx + \mathcal{A} \tag{3.4}$$

is the action defining the zeroth approximation, if $\mathcal{A}$ is the action of the full theory. To rewrite this expression in mathematical language, we combine the $\partial\Delta$ term in $\mathcal{A}_0$ with the gradient term in $\mathfrak{a}$,

$$(3.5) \qquad \infty \sum_{\Delta} \int_{\partial\Delta} \varphi^2 dx + \int \nabla\varphi^2(x) dx = \langle \varphi - \Delta_D \varphi \rangle ,$$

where $-\Delta_D$ is the Laplace operator with zero Dirichlet boundary conditions on all cube faces $\partial\Delta$.

3.3 THE PRIMITIVE EXPANSION

Graphical expansions in statistical mechanics are generated by the identity

$$(3.6) \qquad \prod_{i<j} e^{-V(r_i - r_j)} = \prod_{i<j} [1 - e^{-V(r_i - r_j)} - 1)]$$

$$= \sum_{\text{sets of pairs}} \prod_{\text{pairs in set}} (e^{-V(r_i - r_j)} - 1) .$$

Now a pair $\{r_i, r_j\}$ is represented by the line segment connecting r_i to r_j, so that a set of pairs is a graph. Thus (3.6) is a graph expansion. Before adapting (3.6) to field theory, we modify it slightly. Let $V^{(s)}$ be a one parameter family of potentials, $0 \le s \le 1$, with

$$V^{(0)} = 0 , \quad V^{(1)} = V .$$

Then

$$e^{-V} - 1 = \int_0^1 \frac{d}{ds} e^{-V^{(s)}} ds ,$$

and introducing a parameter s_{ij} for each pair $\{r_i, r_j\}$, we have

$$(3.7) \qquad \prod_{i<j} e^{-V(r_i - r_j)} = \sum_{\Gamma} \int \prod_{\{ij\} \in \Gamma} \frac{d}{ds_{ij}} e^{-V^{(s_{ij})}} ds_{ij}$$

where Γ is a graph constructed from pairs $\{i, j\}$. In (3.7), the s_{ij} for $\{i, j\}$ not in Γ are evaluated at $s = 0$. Writing

$$s^{\Gamma} = \{s_{ij}:\{i, j\} \in \Gamma\}$$

$$s^{\Gamma^c} = \{s_{ij}:\{i, j\} \notin \Gamma\} ,$$

this formula can be expressed in the compact notation

$$\prod_{i<j} e^{-V(r_i - r_j)} = \sum_{\Gamma} \int \partial s^{\Gamma} \prod_{\{i, j\}\in\Gamma} e^{-V(s_{ij})} ds^{\Gamma} \Big|_{s^{\Gamma^c}=0} . \tag{3.8}$$

To adapt this formula to field theory, we only need to choose a multiparameter family $s^{\Gamma} = \{s^{\gamma}:\gamma \in \Gamma\}$ interpolating between the totally decoupled Laplacian Δ_D and the ordinary Laplacian Δ. This can easily be done, and we only mention that it is convenient to choose the s-dependence to be linear in the inverse operators $(-\Delta + m_0^2)^{-1}$. In field theory, (3.8) becomes

$$\langle F(\varphi)\rangle = \sum_{\Gamma} \int \partial s^{\Gamma} \langle F(\varphi)\rangle_s ds^{\Gamma} \Big|_{s^{\Gamma^c}=0} . \tag{3.9}$$

Here $F(\varphi)$ is some function of φ and $\langle \quad \rangle_s$ is the expectation defined by the interpolating "Laplacian" $\Delta(s)$. Observe that $s^{\gamma} = 0$ corresponds to Dirichlet data on γ, and no coupling, while $s^{\gamma} > 0$ allows coupling across Γ. Thus the hypercube faces $\gamma \in \Gamma$ transmit coupling from one cube to its neighbors, while the faces $\gamma \in \Gamma^c$ do not.

3.4 FACTORIZATION AND PARTIAL RESUMMATION

In graphical language, we are concerned here with the cancellation of disconnected vacuum contributions. In (3.9), the expectation $\langle\cdot\rangle$ is normalized so that $\langle 1\rangle = 1$, while $\langle\cdot\rangle_s$ is normalized by the same denominator, so that

$$\langle 1\rangle_s = \int e^{-\int P(\varphi)dx} d\varphi_{0,s} \Big/ \int e^{-\int P(\varphi)dx} d\varphi_0$$

$$= Z(s)/Z$$

where $d\varphi_{0,s}$ is the interpolating Gaussian measure defined by the parameter s. It follows that the expansion (3.9) defines disconnected graphs.

Suppose $F(\varphi)$ depends on $\varphi(x)$ only for x in some set $S \subset R^d$. In (3.9), we consider the term with Γ fixed. Because of the Dirichlet data on Γ^c, this term decouples across Γ^c. More precisely suppose Γ^c divides R^d into a certain number of connected components. Let Y be the closure of the union of the components which meet S; graphically, Y is the region connected to S by Γ. Then the expectation $\langle F(\varphi)\rangle_s \big|_{s_{\Gamma^c}=0}$ factors across ∂Y, allowing us to write it as a product of an "inside" expectation associated with Y times an "outside" expectation associated with $R^d \sim Y$.

Now let us fix Y in place of Γ and sum over all Γ which yield this fixed Y. This sum factorizes into independent sums over $\Gamma \cap Y$ and $\Gamma \sim Y$. The sum over $\Gamma \sim Y$ of the outside expectations has no constraints, and sums to a partition function $Z(\sim Y)$ for the region $\sim Y$, defined with Dirichlet data on ∂Y. (This resummation is just the reverse of the expansion (3.9), but for the region $\sim Y$ in place of R^d.) In a similar fashion we define $Z(Y)$, the partition function for the region Y, with Dirichlet data on ∂Y. Let

$$\langle F(\varphi)\rangle_Y \, Z(Y)/Z$$

denote the sum over $\Gamma \cap Y$ of the inside expectations (including the overall normalization denominator Z). Then (3.9) becomes

$$\langle F(\varphi)\rangle = \sum_{Y} \langle F(\varphi)\rangle_Y \frac{Z(Y)Z(\sim Y)}{Z} . \tag{3.10}$$

In this expression, all disconnected vacuum components have been resummed into the single factor $Z(Y)Z(\sim Y)/Z$. Somewhat lengthy estimates, including a Kirkwood-Salzburg equation, prove the convergence of (3.10) for the parameters and interactions indicated in §3.1.

3.5 TYPICAL APPLICATIONS

To prove exponential decay of correlations, i.e. a positive mass, we use the method of duplicate variables. Thus we write

$$|\langle F(\varphi)G(\varphi)\rangle - \langle F(\varphi)G(\varphi)\rangle| = 2^{-1}|\langle[F(\varphi) - F(\psi)][G(\varphi) - G(\psi)]\rangle|$$

$$\le \sum_{Y} |\langle[F(\varphi) - F(\psi)][G(\varphi) - G(\psi)]\rangle_{Y}|$$

$$\times \frac{Z(Y)Z(\sim Y)}{Z} .$$

Here we choose Y to be the closure of the union of the connected components containing the support of F alone. Howevery Y's which do not also meet the support of G then vanish, by the $\varphi \to \psi$ symmetry. The convergence proof shows that the remaining terms have the desired exponential decay rate.

To establish the upper mass gap and the isolated one particle hyperboloid, the cluster expansion can be applied to the vertex functions $\Gamma^{(n)}$ in place of the Schwinger functions considered above. Similarly a cluster expansion in the Bethe-Salpeter kernel K defined graphically by

(3.11) [diagram: double line – circle – double line] = [double line] + [double line – (K) – double line – circle – double line]

shows that K has three particle (and for even theories four particle) decay. In Minkowski momentum space, this means that K has an analytic continuation up to a neighborhood of the three or four particle threshold. Returning to (3.11), the only singularities of the four point function up to these energies come from the two particle cut. Hence asymptotic completeness up to these energies follows.

To extend these methods to the two phase region (3.3), we choose a length $L >> 1$ and write

$$\bar{\varphi} = L^{-d} \int_{L\text{-cube}} \varphi(x)dx$$

$$\sigma(\bar{\varphi}) = \operatorname{sgn} \bar{\varphi} .$$

Then σ is an Ising variable, defined on the lattice $(LZ)^d$. An expansion in Peierls contours isolates large regions of R^d in which σ is a constant. In these regions, the field φ is concentrated within a single well of the W-shaped potential, and a Gaussian approximation to this well is possible. This Gaussian approximation is controlled by a cluster expansion. Thus the final expansion is a double expansion, first in Peierls contours and then Dirichlet contours, as in (3.10).

REFERENCES

[1] J. Glimm and A. Jaffe, Quantum Field Models, in Statistical Mechanics, C. DeWitt and R. Stora (eds.) 1970 Les Houches Lectures, Gordon and Breach Science Publishers, New York, 1971.

[2] ________, Boson Quantum Field Theory, in Mathematics of Contemporary Physics, R. Streater (ed.), London Mathematical Society, Academic Press, 1972.

[3] J. Glimm, A. Jaffe and T. Spencer, The Particle Structure of the Weakly Coupled $P(\varphi)_2$ Model and Other Applications of High Temperature Expansions, Part I. Physics of Quantum Field Models, in Constructive Quantum Field Theory, G. Velo and A. S. Wightman (eds.), 1973 Erice Lectures, Springer Lecture Notes in Physics, Vol. 25, Springer-Verlag, 1973.

[4] J. Glimm and A. Jaffe, Critical Problems in Quantum Fields, presented at the International Colloquium on Mathematical Methods of Quantum Field Theory, Marseille, June 1975.

[5] ________, Critical Exponents and Renormalization in the φ^4 Scaling Limit, to appear in Quantum Dynamics: Models and Mathematics, L. Streit (ed.), Springer.

[6] ________, Functional Integral Methods in Quantum Field Theory, proceedings of the 1976 Cargèse Summer Institute.

[7] T. Spencer, The Decay of the Bethe-Salpeter Kernel in $P(\varphi)_2$ Quantum Field Models, Commun. Math. Phys. 44, 143-164 (1975).

[8] T. Spencer and F. Zirilli, Scattering States and Bound States in $\lambda P(\varphi)_2$, Commun. Math. Phys. 49, 1-16 (1976).

[9] J. Glimm and A. Jaffe, Particles and Bound States and Progress Toward Unitarity and Scaling, presented at the International Symposium on Mathematical Problems in Theoretical Physics, Kyoto, January 23-29, 1975.

[10] R. Schrader, A Constructive Approach to φ_4^4. I. Commun. Math. Phys. 49, 131-153 (1976); II. preprint; III. Commun. Math. Phys. 50, 97-102 (1976).

[11] J. Rosen, The Ising Model Limit of φ^4 Lattice Fields, preprint, 1976.

[12] J. Glimm and A. Jaffe, The φ_2^4 Quantum Field Model in the Single Phase Region: Differentiability of the Mass and Bounds on Critical Exponents, Phys. Rev. D10, 536-539 (1974).

[13] O. McBryan and J. Rosen, Existence of the Critical Point in φ^4 Field Theory, preprint 1976.

[14] J. Glimm, A. Jaffe and T. Spencer, The Particle Structure of the Weakly Coupled $P(\varphi)_2$ Model and Other Applications of High T mperature Expansions, Part II. The Cluster Expansion, in Constructive Quantum Field Theory, G. Velo and A. S. Wightman (eds.), 1973 Erice Lectures, Springer Lecture Notes in Physics, Vol. 25, Springer-Verlag, 1973.

[15] __________, The Wightman Axioms and Particle Structure in the $P(\varphi)_2$ Quantum Field Model, Ann. Math. 100, 585-632 (1974).

[16] J. Fröhlich, B. Simon and T. Spencer, Infrared Bounds, Phase Transitions and Continuous Symmetry Breaking, Commun. Math. Phys. 50, 79-95 (1976).

[17] F. Guerra, L. Rosen, B. Simon, The $P(\varphi)_2$ Euclidean Quantum Field Theory as Classical Statistical Mechanics, Ann. Math. 101, 111-259 (1975).

[18] J. Glimm and A. Jaffe, A Remark on the Existence of φ_4^4, Phys. Rev. Lett. 33, 440-442 (1974).

[19] C. Newman, preprint.

[20] J. Glimm and A. Jaffe, Particles and Scaling for Lattice Fields and Ising Models, Commun. Math. Phys. 51, 1-14 (1976).

[21] __________, Critical Exponents and Elementary Particles, Commun. Math. Phys., to appear.

[22] R. Schrader, preprint.

[23] J. Feldman, On the Absence of Bound States in the $\lambda\varphi_2^4$ Quantum Field Model without Symmetry Breaking, Canad. J. Phys. 52, 1583-1587 (1974).

[24] T. Spencer, The Absence of Even Bound States for $\lambda(\varphi^4)_2$, Commun. Math. Phys. 39, 77-79 (1974).

[25] J. Glimm and A. Jaffe, Three Particle Structure of φ^4 Interactions and the Scaling Limit, Phys. Rev. D11, 2816-2827 (1975).

[26] J. Glimm, A. Jaffe and T. Spencer, A Convergent Expansion about Mean Field Theory, Part I. The Expansion, Ann. Phys. 101, 610-630 (1976).

[27] J.-P. Eckmann, J. Magnen, and R. Sénéor, Decay Properties and Borel Summability for the Schwinger Functions in $P(\varphi)_2$ Theories, Commun. Math. Phys. 39, 251-271 (1975).

[28] J. Dimock, The $P(\varphi)_2$ Green's Functions: Asymptotic Perturbation Expansion, Helv. Phys. Acta 49, 199-216 (1976).

[29] K. Osterwalder and R. Sénéor, The Scattering Matrix is Non-Trivial for Weakly Coupled $P(\varphi)_2$ Models, Helv. Phys. Acta 49, 525-534 (1976).

[30] J.-P. Eckmann, H. Epstein and J. Fröhlich, Asymptotic Perturbation Expansion for the S-Matrix and the Definition of Time-Ordered Functions in Relativistic Quantum Field Models, Ann. de l'Inst. H. Poincaré 25, 1-34 (1976).

[31] J. Dimock and J.-P. Eckmann, Spectral Properties and Bound State Scattering for Weakly Coupled $P(\varphi)_2$ Models, to appear in Annals of Physics.

[32] Y. Park, Lattice Approximation of the $(\lambda\varphi^4 - \mu\varphi)_3$ Field Theory in a Finite Volume, J. Math. Phys. 16, 1065-1075 (1975).

[33] G. Sylvester, Continuous-Spin Ising Ferromagnets, Ph.D. thesis, M.I.T., 1976; J. Stat. Phys., to appear.

[34] J. Lebowitz, GHS and Other Inequalities, Commun. Math. Phys. 35, 87-92 (1974).

[35] J. Glimm and A. Jaffe, Absolute Bounds on Vertices and Couplings, Ann. de l'Inst. H. Poincaré 22, 1-11 (1975).

[36] J. Rosen, Mass Renormalization for Lattice $\lambda\varphi_2^4$ Fields, preprint, 1976.

[37] D. Marchesin, private communication.

[38] T. Spencer, The Mass Gap for the $P(\varphi)_2$ Quantum Field Model with a Strong External Field, Commun. Math. Phys. 39, 63-76 (1974).

[39] J. Glimm, A. Jaffe and T. Spencer, A Convergent Expansion about Mean Field Theory, Part II. Convergence of the Expansion, Annals of Phys. 101, 631-669 (1976).

[40] J. Magnen and R. Sénéor, The Infinite Volume Limit of the φ_3^4 Model, Ann. de l'Inst. H. Poincaré 24, 95-159 (1976).

[41] J. Feldman and K. Osterwalder, The Wightman Axioms and the Mass Gap for Weakly Coupled $(\varphi^4)_3$ Quantum Field Theories, Annals of Phys. 97, 80-135 (1976).

[42] J. Magnen and R. Sénéor, Wightman Axioms for the Weakly Coupled Yukawa Model in Two Dimensions, to appear in Commun. Math. Phys.

[43] A. Cooper and L. Rosen, The Weakly Coupled Yukawa_2 Field Theory: Cluster Expansion and Wightman Axioms, preprint.

FUNCTIONAL INTEGRAL METHODS IN QUANTUM FIELD THEORY

James Glimm[1]
The Rockefeller University
New York, N.Y. 10021

Arthur Jaffe[2]
Harvard University
Cambridge, MA 02138

We give a special form of the Osterwalder-Schrader axioms in terms of conditions on a functional integral $S\{f\} = \int e^{i\varphi(f)} d\mu$. This yields a simple, self-contained construction of a Hamiltonian H, a relativistic, local boson quantum field Φ, and a Feynman-Kac formula to study perturbations $H + \Phi$ of H.

1. FUNCTIONAL INTEGRALS

Functional integral methods have a long history in quantum mechanics and in quantum field theory [1]. They date from the original proposal by Feynman for writing solutions $e^{-itH}\psi$ to the Schrödinger equation, as integrals over classical particle trajectories. Kac, in the case of quantum mechanics,

---

1. Supported in part by the National Science Foundation under Grant PHY76-17191.
2. Supported in part by the National Science Foundation under Grant PHY75-21212.

and Symanzik, in the case of quantum field theory, realized that the analytic continuations $t \to -it$ of Feynman's integrals have a probabilistic interpretation, now known as the Feynman-Kac formula. In the construction of quantum fields by the authors and others (see [2] for original references) this Feynman-Kac probability method was an important tool for proving estimates, e.g. on the spectrum of the Hamiltonian or properties of fields. Its suitability as a conceptual framework was later reemphasized by Nelson, who gave sufficient conditions for the inverse analytic continuation $-it \to t$ from probability theory back to quantum mechanics. Another version of this analytic continuation was given by Osterwalder and Schrader, allowing for fermions, and with an easily verified condition, reflection positivity, replacing an as yet unproved Markov property. Functional integral reformulations of the axioms have been extensively studied [3]. Here we present a simple version of the connection between functional integrals and boson quantum field theory, including a proof of the Feynman-Kac formula and an illustration of its use to construct local quantum fields. Our treatment is self contained, except for the use of some standard results from functional analysis, functional integrals, and the theory of functions of a complex variable.

A functional integral is defined by a measure $d\mu(\varphi)$ on some space of (generalized) functions. Equivalently, one can give the Fourier transform (or characteristic function in the language of probability theory)

$$S\{f\} = \int e^{i\varphi(f)} d\mu(\varphi) . \tag{1.1}$$

For quantum field theory applications, it is convenient to integrate over $\mathscr{S}'_{\text{real}}$, the space of real, tempered distributions. Thus we consider (1.1) in the case that $\varphi \in \mathscr{S}'(R^d)_{\text{real}}$ and f is an element of the real Schwartz space $\mathscr{S}(R^d)_{\text{real}}$. The existence of a regular probability measure $d\mu(\varphi)$ on $\mathscr{S}'_{\text{real}}$ is equivalent to the existence of a functional $S\{f\}$ satisfying three properties: (i) $S\{f\}$ is continuous; (ii) $S\{0\} = 1$; (iii) $S\{f\}$ is positive definite, i.e. $S\{f_i - f_j\} = M_{ij}$ are the entries of a positive matrix. We here assume the existence of $S\{f\}$ satisfying (i) - (iii). See, for instance, [4] for properties of function space integrals.

In order to recover relativistic quantum field theory in d space-time dimensions, we now make three further assumptions on $S\{f\}$, $f \in \mathscr{S}(R^d)_{real}$. In case $d = 1$, the construction can be interpreted as a functional integral formulation of non-relativistic quantum mechanics. If the continuous space R^d is replaced by a lattice, the construction yields the transfer matrix of lattice statistical mechanics. We now restrict attention to characteristic functions $S\{f\}$ which are suitable for the construction of scalar, boson quantum fields, and give our form of the Osterwalder-Schrader axioms.

Let γ denote an element of the Euclidean group $\mathcal{G}$ on R^d (rotations, translations and reflections in hyperplanes). We single out a particular coordinate direction t which we call "time," and we write $x = (\vec{x}, t)$, $\vec{x} \in R^{d-1}$. Let $T(t) \in \mathcal{G}$ denote the subgroup of time translations, and let $\vartheta \in \mathcal{G}$ denote reflection in the $t = 0$ hyperplane

$$(1.2) \qquad (T(s)f)(\vec{x}, t) = f(\vec{x}, t + s), \quad (\vartheta f)(\vec{x}, t) = f(\vec{x}, -t) .$$

Let $\mathscr{S}(a, b)$ denote the subspace of $\mathscr{S}(R^d)$ of functions f supported in the time interval (a, b), i.e. $\text{suppt } f \subset \{(\vec{x}, t) : a < t < b\}$. Let $\mathscr{S}_+ \equiv \mathscr{S}(0, \infty)$ be the positive time subspace of $\mathscr{S}$.

Assumptions: (i) - (iii) above, and

(1) <u>Euclidean invariance</u>: $S\{f\} = S\{\gamma f\}$, $\gamma \in \mathcal{G}$.

(2) <u>Reflection positivity</u>: $S\{f_i - \vartheta f_j\}$ is a positive matrix for any sequence $f_j \in \mathscr{S}_{+,\, real}$.

(3) <u>Regularity and Boundedness</u>: $S\{f\}$ is continuous on $\mathscr{S}_{real}$ and extends to an entire analytic functional of f in the complex Schwartz space $\mathscr{S}(R^d)$. There exists a Schwartz space norm $|h|_{\mathscr{S}}$ on $\mathscr{S}(R^{d-1})$ and a $p < \infty$, so that

$$(1.3a) \qquad |S\{f\}| \le \exp(|f|_1 + |f|_p^p)$$

where

$$|f|_p^p = \int |f(\cdot, t)|_{\mathscr{S}}^p dt .$$

Furthermore, for some ζ with $1 \le \zeta < 2$ and $\zeta < p$, and for $f = h \otimes \chi_{0,t}$,

(1.3b) $$|S\{f\}| \le \exp(|f|_1 + |f|_\zeta^p) .$$

The interpretation of (1) - (3) is the following: We define the Euclidean Hilbert space $\mathcal{E}$ as the completion in $L_2(d\mu)$ of the finite linear span of exponentials $\exp(i\varphi(f))$, for $f \in \mathscr{S}(R^d)_{\text{real}}$. The inner product on $\mathcal{E}$ is

(1.4) $$\langle A, B\rangle_{\mathcal{E}} = \int \bar{A} B d\mu .$$

Let $\mathcal{E}_+ \subset \mathcal{E}$ denote the positive time subspace of $\mathcal{E}$, obtained by restricting f to $\mathscr{S}_{+,\text{real}}$. More generally, let $\mathcal{E}(a, b)$ be the subspace generated by $f \in \mathscr{S}(a, b)_{\text{real}}$.

Assumption (1) states that $\gamma \in \mathcal{G}$ is represented on $\mathcal{E}$ by a unitary transformation (which we also denote by γ), or that $d\mu$ is $\mathcal{G}$-invariant. Continuity of the representation follows by continuity of $S\{f\}$. Assumption (2) states that the bilinear form

$$b(A, B) \equiv \langle \vartheta A, B\rangle_{\mathcal{E}}$$

is positive on $\mathcal{E}_+$, namely

(1.5) $$b(A, A) = \langle \vartheta A, A\rangle_{\mathcal{E}} \ge 0, \quad A \in \mathcal{E}_+ .$$

Assumption (3) yields the required regularity.

Remark: We may replace (1.3b) by the assumption that the two point function (second moment of $d\mu$) exists and has a convergent spectral representation

$$\int \varphi(f)^2 d\mu - \left(\int \varphi(f) d\mu\right)^2 = \int_0^\infty \langle f, (-\Delta + a)^{-1} f\rangle d\rho(a) ,$$

where $d\rho(a)$ is a positive measure. In particular (1.3b) is used to exclude a δ-function singularity in the two point function.

Theorem 1.1 (Reconstruction of Quantum Mechanics): Assume (1) - (2). Then there exists a Hilbert space $\mathcal{H}$, a canonical map $W:\mathcal{E}_+ \to \mathcal{H}$ and a self adjoint Hamiltonian operator on $\mathcal{H}$ such that for $A \in \mathcal{E}_+$ and $t \ge 0$,

(1.6) $$\langle WA, e^{-tH} WA\rangle_{\mathcal{H}} = \langle \vartheta A, T(t)A\rangle_{\mathcal{E}} = b(T(t/2)A, T(t/2)A) .$$

Furthermore, $\Omega \equiv W1$ is a vacuum vector for H,

$$H \geq 0, \quad H\Omega = 0 .$$

Theorem 1.2: (Reconstruction of Quantum Fields): Assume (1) - (3). For $f \in \mathscr{S}(R^d)_{real}$, there is a self adjoint operator $\Phi(f)$ on $\mathcal{H}$, such that $\Phi(f)$, H, Ω satisfy all the Wightman axioms with the possible exception of uniqueness of the vacuum. The vacuum Ω is unique if and only if $d\mu$ is ergodic under time translations. If suppt f, suppt g are space-like separated, then $\exp(i\Phi(f))$ and $\exp(i\Phi(g))$ commute.

Formally, $\Phi(it)W = W\varphi(t)$. In more detail, the connection between Φ and φ is given by $\Phi(f) = \int \Phi(x)f(x)dx$, where

$$\Phi(x) = e^{itH}\Phi(\vec{x}, 0)e^{-itH} \tag{1.7}$$

and where for $t \geq 0$,

$$\Phi(\vec{x}, it)W = W\varphi(\vec{x}, t) . \tag{1.8}$$

Alternatively, the Wightman distributions $W_n(x_1, \cdots, x_n) \equiv \langle \Omega, \Phi(x_1)\cdots\Phi(x_n)\Omega\rangle$, analytically continued to Euclidean points $\hat{x}_j = (\vec{x}_j, it_j)$, are the moments of $d\mu$ (Schwinger functions):

$$W_n(\hat{x}_1, \cdots, \hat{x}_n) = S_n(x_1, \cdots, x_n) = \int \varphi(x_1)\cdots\varphi(x_n)d\mu . \tag{1.9}$$

We make these connections precise in what follows.

Central to our method is the Φ bound

$$\|(H + I)^{-\frac{1}{2}}\Phi(f)(H + I)^{-\frac{1}{2}}\| \leq |f|_1 \equiv \int |f(\cdot, t)|_{\mathscr{S}} dt \tag{1.10}$$

established in Theorem 5.1 and the relation

$$[iH, \Phi(f)] = -\Phi(\partial f/\partial t) . \tag{1.11}$$

Our proof of (1.10) leads to the study of the perturbed Hamiltonian $H + \Phi(h)$, where $\Phi(h) \equiv \Phi(h, t = 0)$ is the time zero field. The main technical part of our proof of (1.10) concerns the proof of the Feynman-Kac formula

$$(1.12) \qquad e^{-t(H+\Phi(h))}W = We^{-\int_0^t \varphi(h,s)ds}\, T(t)\,,$$

see Theorem 4.1. Our analysis of (1.12) is accomplished through a gradual increase in control over the domain of $\Phi(h)$, and its definition as a bilinear form perturbation of H.

In final sections we establish the locality and covariance of Φ; we also discuss the equivalence of ergodicity of $d\mu(\varphi)$ under the time translation subgroup $T(t) \subset \mathcal{G}$, with the uniqueness of the vacuum vector Ω for H.

2. RECONSTRUCTION OF QUANTUM MECHANICS

In this section we construct the Hilbert space $\mathcal{H}$ of quantum mechanics and study the transformation W from $\mathcal{E}_+$ (the positive time Euclidean space) to $\mathcal{H}$. We begin with the remark that (1.5) defines a seminorm $b(A, A)^{\frac{1}{2}}$ on $\mathcal{E}_+$. Let $\mathfrak{n}$ denote the null space of $b(A, A)$, so $b(A, A)^{\frac{1}{2}}$ defines a norm on $\mathcal{E}_+/\mathfrak{n}$. Let $\mathcal{H}$ be the Hilbert space obtained by completing $\mathcal{E}_+/\mathfrak{n}$ in this norm. We let W denote the canonical map from $\mathcal{E}_+/\mathfrak{n}$ to $\mathcal{H}$, and we denote the scalar product on $\mathcal{H}$ by

$$(2.1) \qquad \langle WA, WB\rangle_{\mathcal{H}} = \langle \vartheta A, B\rangle_{\mathcal{E}} = b(A, B)\,.$$

By the Schwarz inequality and assumption (1) ,

$$(2.2) \qquad \|WA\|_{\mathcal{H}}^2 = b(A, A) \le \|\vartheta A\|_{\mathcal{E}}\|A\|_{\mathcal{E}} = \|A\|_{\mathcal{E}}^2\,,$$

so W is a contraction.

We now consider an operator S on $\mathcal{E}_+$ with domain $\mathcal{D}(S)$, and we assume that S transforms $\mathcal{D}(S) \cap \mathfrak{n}$ into $\mathfrak{n}$, i.e.

$$(2.3) \qquad S:\mathcal{D}(S) \to \mathcal{E}_+, \quad S:\mathcal{D}(S) \cap \mathfrak{n} \to \mathfrak{n}\,.$$

We can then define an operator $\hat{S}$ on $\mathcal{H}$ with domain $W\mathcal{D}(S)$, by the equation

$$(2.4) \qquad \hat{S}W = WS\,.$$

For the remainder of this section we study the general defining equation (2.4). The simplest example arises from choosing $S = T(t)$, $t \geq 0$, $\mathcal{D}(S) = \mathcal{E}_+$. We obtain the following result, which proves Theorem 1.1.

Theorem 2.1: Assume (1) - (2). Then for $t \geq 0$, $T(t)$ satisfies (2.3) and $T(t)^\wedge = e^{-tH}$, where $H = H^* \geq 0$, and $\Omega = W1$ satisfies $H\Omega = 0$.

Proof: Clearly $T(t): \mathcal{E}_+ \to \mathcal{E}_+$. If $A \in \mathfrak{n}$, then

$$(2.5)\quad b(T(t)A, T(t)A) = \langle \vartheta T(t)A, T(t)A\rangle_{\mathcal{E}} = \langle T(-t)\vartheta A, T(t)A\rangle_{\mathcal{E}}$$

$$= \langle \vartheta A, T(2t)A\rangle_{\mathcal{E}} = b(A, T(2t)A)$$

$$\leq b(A, A)^{\frac{1}{2}} b(T(2t)A, T(2t)A)^{\frac{1}{2}} = 0 .$$

Here we have used the Schwarz inequality for the positive form $b(A, B)$. Hence $T(t): \mathfrak{n} \to \mathfrak{n}$ and $T(t)^\wedge$ is well defined. For convenience, write $R(t) = T(t)^\wedge$. We now verify four properties of $R(t)$:

(i) semigroup law: $R(t)R(s) = R(t+s)$, $s, t \geq 0$.
(ii) $R(t)$ is hermitian.
(iii) $R(t)$ is a contraction: $\|R(t)\|_{\mathcal{H}} \leq 1$.
(iv) strong continuity: $R(t) \to I$ as $t \to 0$.

These properties say that $R(t)$ is a strongly continuous, self adjoint, contraction semigroup. Thus there exists a positive, self adjoint operator H such that $R(t) = e^{-tH}$. Furthermore $T(t)1 = 1$. Thus for $\Omega \equiv W1$,

$$e^{-tH}\Omega = WT(t)1 = \Omega ,$$

or $H\Omega = 0$.

Property (i) follows from the multiplication law for $T(t)$, namely

$$R(t)R(s)W = WT(t)T(s) = WT(t+s) = R(t+s)W .$$

Property (ii) follows since for $A \in \mathcal{E}_+$,

$$\langle R(t)WA, WA\rangle = \langle WT(t)A, WA\rangle_{\mathcal{H}} = \langle \vartheta T(t)A, A\rangle_{\mathcal{E}} = \langle T(-t)\vartheta A, A\rangle_{\mathcal{E}}$$

$$= \langle \vartheta A, T(t)A\rangle_{\mathcal{E}} = \langle WA, WT(t)A\rangle_{\mathcal{H}} = \langle WA, R(t)WA\rangle_{\mathcal{H}}$$

Property (iii) is established as follows: By the Schwarz inequality and (i), (ii), for $A \in \mathcal{E}_+$ we have

$$\|R(t)WA\|_{\mathcal{H}} = \langle R(t)WA, R(t)WA\rangle_{\mathcal{H}}^{\frac{1}{2}} = \langle WA, R(2t)WA\rangle_{\mathcal{H}}^{\frac{1}{2}}$$

$$\leq \|WA\|_{\mathcal{H}}^{\frac{1}{2}}\|R(2t)WA\|_{\mathcal{H}}^{\frac{1}{2}}.$$

Continuing to apply the Schwarz inequality in this fashion, we obtain after n steps

$$\|R(t)WA\|_{\mathcal{H}} \leq \|WA\|_{\mathcal{H}}^{1-2^{-n}} \|R(2^n t)WA\|_{\mathcal{H}}^{2^{-n}}$$

$$= \|WA\|_{\mathcal{H}}^{1-2^{-n}} \|WT(2^n t)A\|_{\mathcal{H}}^{2^{-n}}$$

$$\leq \|WA\|_{\mathcal{H}}^{1-2^{-n}} \|A\|_{\mathcal{E}}^{2^{-n}},$$

using (2.2) and the unitarity of T(t). Letting $n \to \infty$ gives $\|R(t)WA\|_{\mathcal{H}} \leq \|WA\|_{\mathcal{H}}$. Since the WA are dense in $\mathcal{H}$, this proves (iii).

To establish property (iv), we note that T(t) is strongly continuous on $\mathcal{E}_+$ and W is a contraction from $\mathcal{E}_+$ to $\mathcal{H}$. Thus R(t) is strongly continuous on the dense subset of $\mathcal{H}$ of vectors WA, $A \in \mathcal{E}_+$. Since $\|R(t)\| \leq 1$, it follows that R(t) is strongly continuous on the entire Hilbert space $\mathcal{H}$, and the proof is complete.

We now generalize these methods to construct operators $\hat{S}$ on $\mathcal{H}$ from functions $k \in \mathcal{E}(0, t)$. Since $k \in L_2(d\mu)$, multiplication by k defines an operator on $\mathcal{E}_+$ with domain $\mathcal{E}_+ \cap L_\infty$.

To construct $\hat{k}$, we restrict the domain of k.

Proposition 2.2: Consider $k \in \mathcal{E}(0,t)$, $t > 0$, as a multiplication operator on $\mathcal{E}_+$ with domain $T(t)(\mathcal{E}_+ \cap L_\infty$. Then $\hat{k}$ defined by (2.4) is (densely) defined on $\mathcal{H}$.

Proof: We verify (2.3), namely if $A \in \mathcal{D}(k) \cap \mathfrak{n}$, then $b(kA, kA) = 0$. First we assume that k is bounded, and we write $A = T(t)B$, $B \in (\mathcal{E}_+ \cap L_\infty)$, so as in (2.5), $B \in \mathfrak{n}$. Then

$$\begin{aligned} b(kA, kA) &= \langle \vartheta k T(t) B, k T(t) B \rangle_{\mathcal{E}} \\ &= b(B, T(t)(\vartheta k)^{-} k T(t) B) \\ &\leq b(B, B)^{\frac{1}{2}} b(C, C)^{\frac{1}{2}} = 0 , \end{aligned}$$

where $C = T(t)(\vartheta k)^{-} k T(t) B \in \mathcal{E}_+$, so we may apply the Schwarz inequality for the form b.

In general, k is not bounded, but we replace k by

$$k_j = \begin{cases} k & \text{if } |k| \leq j \\ 0 & \text{if } |k| > j \end{cases} .$$

Then $\|k - k_j\|_{\mathcal{E}} \to 0$ as $j \to \infty$, and by (2.2) and $k_j A \in \mathfrak{n}$,

$$\begin{aligned} b(kA, kA) &= b(k - k_j)A, (k - k_j)A) \leq \|(k - k_j)A\|_{\mathcal{E}}^2 \\ &\leq \|k - k_j\|_{\mathcal{E}}^2 \|A\|_{L_\infty}^2 \to 0 , \end{aligned}$$

to complete the proof that $\hat{k}$ is defined. Since $\mathcal{D}(\hat{k}) = WT(t)(\mathcal{E}_+ \cap L_\infty)$, $\mathcal{D}(\hat{k})$ is dense.

We next give a sufficient condition on k so that $\hat{k}e^{-tH}$ is bounded. Define k_s as the time translate of k, $k_s = T(s)kT(s)^{-1}$. Define

$$M_n \equiv \| \prod_{j=1}^{n} (\vartheta k)^{-}_{(2j-1)t} \, k_{(2j-1)t} \|_{L_2(d\mu)}^{1/2n} . \tag{2.6}$$

Theorem 2.3: Let $k \in \mathcal{E}(0,t)$, $t > 0$. Then

$$\|\hat{k} e^{-tH}\|_{\mathcal{H}} \le \limsup_{n\to\infty} M_n . \tag{2.7}$$

Proof: Let $Q = e^{-tH}\hat{k}^*\hat{k}e^{-tH}$ and $A \in \mathcal{E}_+ \cap L_\infty$. Then

$$\|\hat{k}e^{-tH}WA\|_{\mathcal{H}} = \langle WA, QWA\rangle_{\mathcal{H}}^{\frac{1}{2}} \le \|WA\|_{\mathcal{H}}^{\frac{1}{2}}\|QWA\|_{\mathcal{H}}^{\frac{1}{2}} .$$

We continue to apply the Schwarz inequality and after n applications obtain

$$\|\hat{k}e^{-tH}WA\|_{\mathcal{H}} \le \|WA\|_{\mathcal{H}}^{1-2^{-n}} \|Q^{2^{n-1}} WA\|_{\mathcal{H}}^{2^{-n}} . \tag{2.8}$$

By definition $(\hat{k}e^{-tH})^*W = WT(t)\vartheta k^- = W(\vartheta k^-)_t T(t)$, so

$$\|Q^{2^n} WA\|_{\mathcal{H}} = \|W \prod_{j=1}^{2^n} (\vartheta k^-)_{(2j-1)t} k_{(2j-1)t}) T(2^{n+1}t)A\|_{\mathcal{H}}$$

$$\le \|(\prod_{j=1}^{2^n} (\vartheta k)^-_{(2j-1)t} k_{(2j-1)t}) T(2^{n+1}t)A\|_{\mathcal{E}}$$

$$\le M_r^{2r} \|T(2rt)A\|_{L_\infty}$$

$$\le M_r^{2r} \|A\|_{L_\infty} .$$

Here we have set $2^n = r$ and used the fact that time translation is a contraction on L_∞. Substituting in (2.8), we obtain

$$\|\hat{k}e^{-tH}WA\|_{\mathcal{H}} \le \|WA\|_{\mathcal{H}}^{1-1/r} M_{r/2} \|A\|_{L_\infty}^{1/r}$$

$$\le \|WA\|_{\mathcal{H}} \limsup_{r\to\infty} M_{r/2}$$

to complete the proof.

We now define

$$\mathcal{H}_\delta \equiv e^{-\delta H}\mathcal{H} .$$

Corollary 2.4: Let $\delta > 0$, $0 \le \tau < \delta/4$, $\delta' > t + \delta/2$, $\lim\sup M_n < \infty$. Then $(k_\tau)^\wedge$ as a bilinear form on $\mathcal{H}_\delta \times \mathcal{H}_{\delta'}$ is analytic in τ and extends analytically to the circle $|\tau| < \delta/8$. Also

$$(2.9) \qquad \|e^{-\delta H}\frac{d^n}{d\tau^n}(k_\tau)^\wedge e^{-\delta' H}\| \le (8/\delta)^n n! \, \|\hat{k}e^{-(t+\delta/4)H}\| .$$

Proof: Since

$$e^{-\delta H}\frac{d}{d\tau}(k_\tau)^\wedge e^{-\delta' H} = e^{-(\delta+\tau)H}[\hat{k}, H]e^{-(\delta'-\tau)H}$$

we use the relations

$$c_m = \|H^m e^{-\delta H/4}\| \le (4/\delta)^m m! ,$$

$$c_m c_{n-m} \le (4/\delta)^n n! ,$$

and the theorem to establish (2.9). The stated analyticity then follows.

3. RECONSTRUCTION OF FIELDS

In this section we apply the construction of the previous section to the case $S = \varphi(f)$. Then $S^\wedge W = WS$ defines the (analytically continued) field operator $\varphi(f)^\wedge$. From $\varphi(f)^\wedge$ we obtain the sharp time field $\Phi(h)$ as a densely defined bilinear form on $\mathcal{H}$, and such that

$$\varphi(h \otimes \alpha)^\wedge = \int e^{-sH}\Phi(h)e^{sH}\alpha(s)ds .$$

Proposition 3.1: Assume (3). Then the measure $d\mu$ has moments of all orders, and the nth moment has a density $S_n(x_1, \cdots, x_n) \in \mathcal{S}'(R^{nd})$,

$$(3.1) \qquad \int \varphi(f_1)\cdots\varphi(f_n)d\mu = \int S_n(x_1, \cdots, x_n)\prod_{i=1}^{n} f_i(x_i)dx .$$

Proof: The operators $U(t) = \exp(it\varphi(f))$, $f \in \mathcal{S}_{real}$, form a unitary group on $\mathcal{E}$. Their infinitesimal generator is the multiplication operator $\varphi(f)$. The statement that the function 1 is in the domain of $\varphi(f)^n$, for all such f, is equivalent to the statement that the moments of $d\mu$ of degree 2n exist. (The moments of odd degree are bounded by a Schwarz inequality by those of even degree.) To show that 1 is in the domain of $\varphi(f)^n$, we use the definition $(\Delta/\Delta t)^n U(t)$ of the nth difference quotient of U(t). Then

$$\|[(\Delta/\Delta t_1)^n - (\Delta/\Delta t_2)^n]U(t)1\|^2 =$$

$$[(\Delta/\Delta t_1)^n - (\Delta/\Delta t_2)^n][(\Delta/\Delta s_1)^n - (\Delta/\Delta s_2)^n]S\{(s-t)f\}\,|_{s=t} \,.$$

Hence strong convergence of the nth difference quotient for U(t)1 follows from the analyticity (and hence differentiability) of S. Let f lie in a bounded, finite dimensional set of $\mathcal{S}$. By continuity, $|S\{f\}|$ is bounded on the set. We then use the Cauchy integral formula,

$$S_n(f_1, \cdots, f_n) = \oint S\{g\} \prod_{j=1}^{n} \left(\frac{dz_j}{2\pi i z_j}\right),$$

where $g = \sum_{j=1}^{n} z_j f_j$ and the integrals extend over $|z_j| = 1$. This provides a bound on S_n establishing continuity on $\mathcal{S} \times \mathcal{S} \times \cdots \times \mathcal{S}$. The extension of S_n to $\mathcal{S}'(R^{nd})$ then follows from the nuclear theorem, to complete the proof. Observe that we have now also shown $S\{f\} = \int e^{i\varphi(f)} d\mu(\varphi)$ for complex $f \in \mathcal{S}(R^d)$.

Let Y denote the Banach space obtained by completing $\mathcal{S}(R^d)$ in the norm

$$|f| \equiv |f|_1 + |f|_p \,, \tag{3.2}$$

where

$$|f|_q^q \equiv \int |f(\cdot, t)|_{\mathcal{S}}^q dt \,.$$

Proposition 3.2: Assume (3). Then S{f} extends by continuity to an entire function on Y, and the moments S_n of $d\mu$

extend to continuous, multilinear functionals on $Y \times \cdots \times Y$.

Proof: We use Vitali's theorem to establish analyticity of $S\{f\}$ for $f \in Y$: A sequence of functions on a compact set $\kappa \subset \mathbb{C}^n$ which is uniformly bounded and pointwise convergent, converges to a function analytic on κ. The continuity of S_n on $Y \times \cdots \times Y$ follows as above from the Cauchy integral theorem.

Remark: The Schwinger functions S_n are also continuous in the norm $|f|_1 + |f|_\zeta$, by (1.3b).

We now define $Y(0,t) \subset Y$ as the subset of functions $f \in Y$ supported in the interval $(0,t)$. Let $M(f) \equiv |f|_1 + |f|_p^p$.

Proposition 3.3: Assume (1) - (3) and let $f \in Y(0,t)_{\mathrm{real}}$. Then W maps the functions $\varphi(f)^r$ and $\exp(\varphi(f))$ into operators on $\mathcal{H}$ with domain $e^{-tH}\mathcal{H}$. Furthermore,

$$\|(e^{\varphi(f)})^\wedge e^{-tH}\| \le e^{M(2f)/2} , \tag{3.3}$$

$$\|(\varphi(f)^r)^\wedge e^{-tH}\| \le \kappa(c|f|)^r r!^{\,q} , \tag{3.4}$$

where $\kappa, c < \infty$ and $q = (p-1)/p$.

Remark: As a special case of (3.4),

$$\left|\int \varphi(f)^r d\mu\right| = |\langle \Omega, (\varphi(f)^r)^\wedge \Omega\rangle| \le \kappa(c|f|)^r r!^{\,q} .$$

Proof: The operators $(\varphi^r)^\wedge$ and $(\exp\varphi)^\wedge$ are densely defined by Propositions 2.2 and 3.1. To establish (3.3), let $k = \exp(\varphi(f))$ and calculate M_n defined in (2.6). Since the supports of $\alpha_{(2j-1)t}$ and $(\vartheta\alpha)_{(2j-1)t}$ are all disjoint (for fixed t and $j = 1, 2, \ldots$) it follows by (1.3) that $M_n \le \exp(M(2f)/2)$. Then (3.3) is a consequence of Theorem 2.3.

To prove (3.4), we define for $w_j, z_j \in \mathbb{C}$,

$$g = -i \sum_{j=1}^{n} (w_j f_{(2j-1)t} + z_j (\vartheta f)_{(2j-1)t}) .$$

Then

$$M_n^{4n} = \left(\prod_{j=1}^{n} \frac{d^{2r}}{dw_j^{2r}} \frac{d^{2r}}{dz_j^{2r}}\right) S\{g\}\Big|_{g=0} . \tag{3.5}$$

We bound (3.5) using the Cauchy integral formula,

$$M_n^{4n} = (2r)!/2\pi i)^{2n} \oint S\{g\} \left(\prod_{j=1}^{n} \frac{dw_j}{w_j^{2r+1}} \frac{dz_j}{z_j^{2r+1}} \right),$$

with each integral taken over a circle of radius $4\epsilon^{-1}$ centered at the origin. The upper bound for M(g) on this circle is $2nM(4\epsilon^{-1}f)$, since the terms in g arising from different (or time reflected) values of j have disjoint supports. Thus

$$M_n \leq \exp(M(4\epsilon^{-1}f)/2)(\epsilon/4)^r(2r)!)^{\frac{1}{2}}$$

$$\leq \exp(M(4\epsilon^{-1}f)/2)\epsilon^r r!$$

To minimize M_n, first replace f by $f/|f|$, yielding $M_n \leq \exp(c\epsilon^{-p})(\epsilon|f|)^r r!$ Then choose $\epsilon = (cp/r)^{1/p}$, to obtain (3.4).

<u>Proposition 3.4:</u> There exists a unique bilinear form $\Phi(h)$ defined on the domain $\mathcal{H}_\delta \times \mathcal{H}_\delta$ (for any $\delta > 0$) and satisfying (for $f = h \otimes \alpha$),

$$\varphi(h \otimes \alpha)^\wedge e^{-tH} = \int_0^t e^{-sH} \Phi(h) e^{-(t-s)H} \alpha(s) ds \tag{3.6}$$

on $\mathcal{H}_\delta \times \mathcal{H}_\delta$. Furthermore

$$\|e^{-\delta H} \Phi(h) e^{-\delta H}\| \leq \kappa_\delta |h|_{\mathcal{S}}, \quad \|e^{-\delta H} \varphi(f)^\wedge e^{-(t+\delta)H}\| \leq \kappa_\delta |f|_1 . \tag{3.7}$$

<u>Proof:</u> By Corollary 2.4, $\hat{\varphi}(h \otimes \alpha_\tau)$ is a form on $\mathcal{H}_\delta \times \mathcal{H}_{\delta'}$ which is C_∞ in τ for $0 \leq \tau < \delta/4$, $\delta' \geq t + \delta/2$. By (2.9) and (3.4),

$$\|e^{-\delta H} \varphi(h \otimes \frac{d^n}{d\tau^n} \alpha_\tau)^\wedge e^{-\delta' H}\| \leq \kappa_{n,\delta} |h|_{\mathcal{S}} (\|\alpha\|_{L_1} + \|\alpha\|_{L_p}) . \tag{3.8}$$

We appeal to the fact that a distribution in s, all of whose derivatives have a fixed order, must be a C^∞ function of s. Thus on $\mathcal{H}_\delta \times \mathcal{H}_{\delta'}$ there exists a bilinear form $\Phi(h)$ which satisfies

$$\varphi(h \otimes \alpha)^\wedge = \int_0^t e^{-sH} \Phi(h) e^{sH} \alpha(s) ds$$

and such that

$$F(s) = \langle e^{-sH} \psi_\delta, \Phi(h) e^{sH} \psi_{\delta'} \rangle$$

is a C^∞ function.

We now improve (3.8) to eliminate $\|\alpha\|_{L_p}$. With $n = 1$ and $\alpha = \chi_{(t_1, t_2)}$ the characteristic function of $(t_1, t_2) \subset (0, t)$, $t \leq 1$, we bound

$$\|e^{-\delta H} \varphi(h \otimes (\delta_{t_1} - \delta_{t_2}))^\wedge e^{-\delta' H}\| \leq 2\kappa_{1,\delta} |h|_{\mathscr{S}} .$$

We integrate t_2 from s_1 to $s_2 = s_1 + o(\delta)$. Since

$$(s_2 - s_1)\delta_{t_1} = \chi_{(s_1, s_2)} + \int_{s_1}^{s_2} (\delta_{t_1} - \delta_{t_2}) dt_2 ,$$

we have

$$\|e^{-\delta H} \varphi(h \otimes \delta_{t_1})^\wedge e^{-\delta' H}\| \leq (\kappa_{0,\delta} o(\delta^{-1}) + 2\kappa_{1,\delta}) |h|_{\mathscr{S}} .$$

Now integration over t_1 shows that

$$\|e^{-\delta H} \varphi(h \otimes \alpha)^\wedge e^{-\delta' H}\| \leq \kappa_\delta |h|_{\mathscr{S}} \|\alpha\|_{L_1} .$$

<u>Proposition 3.5</u>: Assume (1) - (3) and let $f = h \otimes \chi_{0,t}$, where $h \in \mathscr{S}(R^{d-1})$ is real and where $\chi_{0,t}$ is the characteristic function of $(0, t)$. Then

$$\int \varphi(f)^2 d\mu = o(t), \quad t \to 0 . \tag{3.9}$$

Proof: We use the continuity of s_2 in the norm $|f|_1 + |f|_\zeta$ (cf. (1.3b)) to conclude that

$$\left|\int \varphi(f)^2 d\mu\right| \le \text{const.} \|\chi_{0,t}\|^2_{L_\zeta} = \text{const.}\ t^{2/\zeta}$$

$$= o(t), \quad t \to 0 ,$$

since $\zeta < 2$ by hypothesis.

If instead we assume

$$\int \varphi(f)^2 d\mu - \left(\int \varphi(f) d\mu\right)^2 = \int \langle f, (-\Delta + a)^{-1} f\rangle d\rho(a) , \tag{3.10}$$

then (3.10) is $o(t)$ and $\left|\int \varphi(f) d\mu\right|^2 = O(t^2)$.

4. THE FEYNMAN-KAC FORMULA

In this section we derive the Feynman-Kac formula on $\mathcal{H}$, suitable to study perturbations of H by the bilinear form $\Phi(h)$, for $h \in \mathscr{S}(R^{d-1})_{\text{real}}$. We find that

$$(e^{-\varphi(h\otimes\chi_{0t})})^\wedge e^{-tH} = e^{-tH(h)} . \tag{4.1}$$

Here $H(h)$ is a self adjoint operator, bounded from below, which satisfies

$$H(h) = H + \Phi(h) \tag{4.2}$$

as an equation for bilinear forms. Also for h real, we find that

$$\pm \Phi(h) \le |h|_{\mathscr{S}}(H + I) .$$

Theorem 4.1: Let $S\{f\}$ satisfy (1) - (3). Then the left side of (4.1) is a semigroup $S(t)$ with a self adjoint generator $H(h)$ satisfying

$$H(h) \ge -(|h|_{\mathscr{S}} + 2^{p-1} |h|^p_{\mathscr{S}}) . \tag{4.3}$$

Furthermore, (4.2) is satisfied on the domain $\mathcal{H}_\delta \times \mathcal{H}_\delta$, for any $\delta > 0$.

Proof: Let S(t) denote the left side of (4.1). By Proposition 3.3, S(t) is a bounded operator, and $\|S(t)\| \le \exp(t(|h|_{\mathscr{L}} + 2^{p-1}|h|^p_{\mathscr{L}}))$. Also $S(t+s) = S(t)S(s)$ and the $S(t) = S(t)^*$ follow from the definition (4.1). We next establish weak differentiability of S(t) at $t = 0$, on the dense domain $\mathcal{K} \times \mathcal{K}$, where $\mathcal{K} = e^{-\delta H}W(\mathcal{E}_+ \cap L_\infty)$. Weak (and strong) continuity of S(t) follow, and hence the existence of a self adjoint generator H(g) satisfying (4.3). Write $f = -h \otimes \chi_{0,t}$, so that on $\mathcal{K} \times \mathcal{K}$,

$$(4.4)\quad t^{-1}(S(t) - I) = t^{-1}(e^{-tH} - I) + t^{-1}\varphi(f)^\wedge + t^{-1}(e^{\varphi(f)} - 1 - \varphi(f))^\wedge e^{-tH}.$$

The first term on the right of (4.4) tends to $-H$. By Proposition 3.4, the second term converges to $-\Phi(h)$. The third term converges to zero, which we see as follows: Let $A \in \mathcal{E}_+ \cap L_\infty$. Then using $|e^x - 1 - x| \le \text{const.}\left[(x^2 + x^N) + \sum_{j>N/2}^{\infty} x^{2j}/(2j)!\right]$,

$$(4.5)\quad |t^{-1}\langle e^{-\delta H}WA, (e^{\varphi(f)} - 1 - \varphi(f))^\wedge e^{-(\delta+t)H}WA\rangle|$$

$$\le t^{-1}\|A\|^2_{L_\infty}\int |e^{\varphi(f)} - 1 - \varphi(f)|\,d\mu$$

$$\le \text{const.}\, t^{-1}\|A\|^2_{L_\infty}\Big[\int \varphi(f)^2 d\mu + \int \varphi(f)^N d\mu$$

$$+ \sum_{j>N/2}^{\infty} \frac{1}{(2j)!}\int \varphi(f)^{2j} d\mu\Big],$$

for any even $N \ge 2$. The first term on the right of (4.5) tends to zero (as $t \to 0$) by Proposition 3.5. Choose $N > p$, so the remaining terms in (4.5) tend to zero (as $t \to 0$) by (3.4), and the fact that $|f| = |h|_{\mathscr{L}}(t + t^{1/p}) \le O(t^{1/p})$. This completes the proof of weak differentiability, and the identity (4.2) on $\mathcal{K} \times \mathcal{K}$. By Corollary 3.4, $e^{-\delta H}\Phi(h)e^{-\delta H}$ is bounded. Hence (4.2) extends by continuity to $\mathcal{H}_\delta \times \mathcal{H}_\delta$.

Corollary 4.2: The forms $\Phi(h)$ and $H(h)$ extend by continuity to the domain $D(H^{\frac{1}{2}}) \times D(H^{\frac{1}{2}})$, and they satisfy (4.2) - (4.3). Also

$$\|(H + I)^{-\frac{1}{2}}\Phi(h)(H + I)^{-\frac{1}{2}}\| \leq |h|_{\mathscr{S}} . \tag{4.6}$$

Proof: On $\mathcal{H}_\delta \times \mathcal{H}_\delta$ we infer from (4.2) - (4.3) that

$$\pm\Phi(h) \leq (|h|_{\mathscr{S}} + 2^{p-1}|h|_{\mathscr{S}}^{p})(H + I) ,$$

or for any constant $c \neq 0$,

$$\pm\Phi(h) = \pm 2c|h|_{\mathscr{S}}\Phi(h/2c|h|_{\mathscr{S}}) \leq (1 + c^{-(p-1)})|h|_{\mathscr{S}}(H + I) .$$

Taking $c \to \infty$ gives

$$\pm\Phi(h) \leq |h|_{\mathscr{S}}(H + I) .$$

This bound now extends by continuity, as claimed.

5. COMMUTATORS AND SELF ADJOINT FIELDS

We define real time field operators $\Phi(f)$ and establish their properties as operators on $\mathcal{H}$. On $D(H^{\frac{1}{2}}) \times D(H^{\frac{1}{2}})$ the bilinear form

$$\Phi(h, t) = e^{itH}\Phi(h)e^{-itH} \tag{5.1}$$

is continuous in t. Thus we define

$$\Phi(f) = \int \Phi(f^{(t)}, t)dt \tag{5.2}$$

where $f^{(t)}(\vec{x}) \equiv f(\vec{x}, t) \in \mathscr{S}(R^{d-1})$ for $f \in \mathscr{S}(R^d)$.

Theorem 5.1: Assume $S\{f\}$ satisfies (1) - (3). Then for $f \in \mathscr{S}(R^d)_{real}$, the bilinear form (5.2) uniquely determines a self adjoint operator $\Phi(f)$ on $\mathcal{H}$, which is essentially self adjoint on any core for H. Moreover

$$\|(H + I)^{-\frac{1}{2}}\Phi(f)(H + I)^{-\frac{1}{2}}\| \leq |f|_1 = \int |f^{(t)}|_{\mathscr{S}} dt , \tag{5.3}$$

(5.4) $$\Phi(f): D(H^n) \to D(H^{n-1}),$$

and

(5.5) $$[iH, \Phi(f)] = -\Phi(\partial f/\partial t) \text{ on } D(H^2).$$

Proof: The bound (5.3) for $\Phi(f)$ as a bilinear form follows by Corollary 4.2. The relation (5.5) for $\Phi(f)$ as a bilinear form follows from (5.3). The remaining assertions follow by the commutator Theorems 5.4 - 5.6 below.

Corollary 5.2: The Wightman functions

(5.6) $$W_n(f_1, \cdots, f_n) \equiv \langle \Omega, \Phi(f_1) \cdots \Phi(f_n)\Omega \rangle$$

exist and are tempered distributions in $\mathscr{S}'(R^{nd})$.

Let $\mathscr{D} \subset \mathcal{H}$ denote the finite linear span of vectors of the form $\Phi(f_1) \cdots \Phi(f_n)\Omega$, $f_j \in \mathscr{S}(R^d)_{real}$. We establish in Theorem 5.3 that $\mathscr{D}$ is dense in $\mathcal{H}$. Note that $e^{itH}: \mathscr{D} \to \mathscr{D}$, and $\mathscr{D} \subset (\bigcap_n \mathscr{D}(H^n))$ by (5.4). Thus $\mathscr{D}$ is a core for H and, by Theorem 5.1, a core for $\Phi(f)$.

Theorem 5.3: $\mathscr{D}$ is dense in $\mathcal{H}$.

Proof: Recall that $\mathcal{H}$ is spanned by vectors of the form $W\exp(i\varphi(f)) = (\exp(i\varphi(f)))^{\wedge}\Omega$, for $f \in \mathscr{S}_+$. Using Proposition 3.3, it is no loss of generality to restrict attention to $f = \sum_{j=1}^{n} h_j \otimes \alpha_j$, with $h_j \otimes \alpha_j \in \mathscr{S}(0, t)$. We first choose α_j such that $0 < t_1 < t_2 < \cdots < t_n < t$ for $t_j \in \text{suppt } \alpha_j$, and define for real s,

(5.7) $$\theta(s) = \int e^{ist_1 H}\Phi(h_1)e^{-is(t_1-t_2)H}\Phi(s_2)\cdots\Phi(h_n)\Omega \in \mathscr{D}.$$

By Corollary 2.4, $\theta(s)$ is the boundary value of a function of s, analytic for $\operatorname{Im} s > 0$. If χ is orthogonal to $\mathscr{D}$, then $\langle \chi, \theta(s)\rangle = 0$ for s real, and hence $\langle \chi, \theta(i)\rangle = 0$, i.e.

(5.8) $$0 = \int \langle \chi, We^{-t_1 H}\varphi(h_1, 0)e^{-|t_1-t_2|H}\varphi(h, 0)\cdots \rangle \prod_{j=1}^{n} \alpha_j(t_j)dt_j.$$

By the continuity of (5.8) in α_j which follows by Proposition 3.3, (5.8) remains valid if we translate the supports of α_j to overlap. Thus

$$\langle \chi, W\varphi(f)^n \rangle = 0 = \langle \chi, (\varphi(f)^n)^\wedge \Omega \rangle .$$

Again using (3.4), we may sum the exponential series to establish $\langle \chi, (\exp(i\varphi(f)))^\wedge \Omega \rangle = 0$. Thus $\chi = 0$, and $\mathcal{D}$ is dense.

We now prove four technical results concerning commutators and self adjointness of operators. The first three are well known, see [5] for references. A result similar to Theorem 5.7 was proved independently by Driessler and Fröhlich [6].

We start with some notation used in Theorems 5.4 - 5.7. Let $H = H^* \geq 0$ be a positive, self adjoint operator, let $\mathcal{D} \subset (\bigcap_n \mathcal{D}(H^n))$ be a core of C^∞ vectors for H. Let $R(\lambda) \stackrel{n}{=} (H + (\lambda + 1)I)^{-1}$, so that $R \equiv R(0) = (H + I)^{-1}$. Let A be a bilinear form defined on the domain $\mathcal{D} \times \mathcal{D}$ and let $\delta A = [iH, A]$, again be a bilinear form on $\mathcal{D} \times \mathcal{D}$.

<u>Theorem 5.4</u>: Let $R^{\frac{1}{2}}\delta(A)R^{\frac{1}{2}}$ be bounded. Then for any positive integer n, $R^{n/2}AR^{n/2}$ is bounded if and only if AR^n is bounded. Also $\|R^{n/2}AR^{n/2} - AR^n\| \leq n\|R^{\frac{1}{2}}\delta(A)R^{\frac{1}{2}}\|$.

Assuming the theorem, then by the Riesz representation theorem we have

<u>Corollary 5.5</u>: If $R^{\frac{1}{2}}\delta(A)R^{\frac{1}{2}}$ and $R^{n/2}AR^{n/2}$ are bounded, then A uniquely determines an operator (also denoted A) with domain $\mathcal{D}$. The operator A is symmetric if the form A is real.

<u>Proof:</u> It is sufficient to study the difference

$$AR^n - R^{n/2}AR^{n/2} = [A, R^{n/2}]R^{n/2}$$

$$= \sum_{j=0}^{n-1} R^{j/2}[A, R^{\frac{1}{2}}]R^{(2n-j-1)/2} .$$

The factors $R^{j/2}$ and $R^{(2n-j-1)/2}$ are bounded by 1, while the commutator is studied through the Cauchy integral formula [7],

$$R^{\frac{1}{2}} = \pi^{-1}\int_0^\infty \lambda^{-\frac{1}{2}} R(\lambda)d\lambda .$$

Thus

$$[A, R^{\frac{1}{2}}] = -i\pi^{-1}\int_0^\infty R(\lambda)\delta(A)R(\lambda)\lambda^{-\frac{1}{2}}d\lambda ,$$

which is bounded in norm by $\|R^{\frac{1}{2}}\delta(A)R^{\frac{1}{2}}\|$, using $\pi^{-1}\int_0^\infty(\lambda + 1)^{-1}\lambda^{-\frac{1}{2}}d\lambda = 1.$

Theorem 5.6: Let A be a symmetric operator defined on $\mathcal{D}$ and suppose that $R^{\frac{1}{2}}\,\delta(A)R^{\frac{1}{2}}$ and AR^n are bounded for some $n \geq 1$. Then A is essentially self adjoint on any core for H^n.

Proof: Since AR^n is bounded, $\mathcal{D}(A^-) \supset \mathcal{D}(H^n)$ and prove essential self adjointness on $\mathcal{D}(H^n)$. Let $\theta \in \mathcal{D}(A^*)$, $\chi \in \mathcal{D}(H^n)$. Then $R(\lambda)^n\theta \in \mathcal{D}(H^n) \subset \mathcal{D}(A)$, and

$$(5.9)\quad \langle\chi, A\lambda^n R(\lambda)^n\theta\rangle = \lambda^n\langle R(\lambda)^n A\chi, \theta\rangle$$

$$= \langle A\lambda^n R(\lambda)^n\chi, \theta\rangle + \langle[\lambda^n R(\lambda)^n, A]\chi, \theta\rangle$$

$$= \langle\chi, \lambda^n R(\lambda)^n A^*\theta\rangle + \langle[\lambda^n R(\lambda)^n, A]\chi, \theta\rangle .$$

We take the limit $\lambda \to \infty$. Since $\lambda^n R(\lambda)^n \to I$ strongly, we evaluate the first term using

$$\lambda^n R(\lambda)^n A^*\theta \to A^*\theta .$$

The commutator in the second term is bounded uniformly as $\lambda \to \infty$ because

$$\|[A, \lambda^n R(\lambda)^n]\| \le \sum_{r=0}^{n-1} \lambda^n \|R(\lambda)^n [A, R(\lambda)] R(\lambda)^{n-r-1}\|$$

$$\le \sum_{r=0}^{n-1} \lambda^n \|R(\lambda)^{r+1} \delta(A) R(\lambda)^{n-r}\|$$

$$\le O(1) .$$

Thus by (5.9)

$$\lim_{\lambda} A\lambda^n R(\lambda)\theta = A^*\theta + \lim_{\lambda} [\lambda^n R(\lambda)^n, A]^*\theta . \tag{5.10}$$

We complete the proof by showing that the last term in (5.10) has a limit equal to zero. Thus $\theta \in D(A^-)$ and A is self adjoint.

Because of the uniform bound on the norm above, it is sufficient to prove that $[A, \lambda^n R(\lambda)^n]$ converges to zero on the dense set $\mathcal{D}(H^n)$. Let $\psi \in \mathcal{D}(H^n)$. Then

$$[\lambda^n R(\lambda)^n, A]^*\psi = A\lambda^n R(\lambda)^n \psi - \lambda^n R(\lambda)^n A\psi .$$

The second term on right converges to $-A\psi$. The first term, written as

$$AR^n \lambda^n R(\lambda)^n (H + I)^n \psi$$

converges to $A\psi$. This completes the proof.

Theorem 5.7: Let A, B, $\delta(A)$, $\delta(B)$ all satisfy the hypotheses of Theorem 5.6 with $n = 1$. Let AB and BA be defined on $\mathcal{D}$ and $[A, B]\mathcal{D} = 0$. Then A^- and B^- commute.

Proof: Define the bounded, self adjoint approximation $A_\lambda = \lambda R(\lambda)^{\frac{1}{2}} A R(\lambda)^{\frac{1}{2}}$. Then for $\theta \in \mathcal{D}$, $A_\lambda \theta \to A\theta$. By Theorem 5.5, $\mathcal{D}$ is a core for A, so $\exp(iA_\lambda) \to \exp(iA^-)$, strongly. We now show $[\exp(iA_\lambda), \exp(iB_\lambda)] \to 0$.

First we establish the identity

$$(5.11)\quad [A_\lambda, B_\lambda] = -i\lambda^2 R(\lambda)^{3/2}[\delta(A)R(\lambda)B - \delta(B)R(\lambda)A]R(\lambda)^{\frac{1}{2}} .$$

The calculations used to derive this identity are performed on the domain $R(\lambda)^{-\frac{1}{2}}\mathcal{D}$. This domain is dense because $\mathcal{D}$ is a core for $R(\lambda)^{-\frac{1}{2}}$ (e.g. by Theorem 5.6 with $\mathbf{A} = R(\lambda)^{-\frac{1}{2}}$ and $n = 1$). On this domain, we use the identity $R(\lambda)^{3/2}[A, B]R(\lambda)^{\frac{1}{2}} = 0$. From (5.11), and the boundedness of $R^{\frac{1}{2}}\delta(A)R^{\frac{1}{2}}$, AR, etc., we infer that

$$(5.12)\quad \|R[A_\lambda, B_\lambda]R\| \le O(\lambda^{-\frac{1}{2}}) .$$

Next we claim that

$$(5.13)\quad \|(H + I)\exp(itA_\lambda)R\| \le \exp(\kappa|t|) ,$$

where κ is a constant independent of λ. To establish (5.13), we integrate the differential inequality

$$(5.14)\quad \frac{dF(t, \mu)}{dt} \le \kappa F(t, \mu)$$

where

$$F(t;\mu) \equiv \mu^2 R e^{-itA_\lambda} R(\mu)(H + I)^2 R(\mu)e^{itA_\lambda} R .$$

The inequality (5.14) follows from the assumption $\|\delta(A)R\| \le$ const., and κ does not depend on λ, μ, t. This yields on integration

$$(5.15)\quad F(t;\mu) \le \exp(\kappa|t|)\mu^2 R(\mu)^2 \le \exp(\kappa|t|) .$$

Since $\mu R(\mu)$ is monotone increasing in μ, also $F(t;\mu)$ is monotone increasing in μ, and the limit $F(t;\mu) \nearrow F(t)$ exists [8]. Thus the range of $\exp(itA_\lambda)R$ lies in the domain of $(H + I)$ and (5.13) holds. Similar estimates hold with B replacing A.

Finally, we study the identity between bounded operators:

$$(5.16)\quad R[e^{iA_\lambda}, e^{iB_\lambda}]R = \int_0^1 ds \int_0^1 dt\, R Q(s, t)^*[A_\lambda, B_\lambda]Q(1 - s, 1 - t)R ,$$

where $Q(s,t) \equiv \exp(itA_\lambda)\exp(isB_\lambda)$ is unitary. By (5.12), (5.13), we infer $\|R[e^{iA_\lambda}, e^{iB_\lambda}]R\| \le O(\lambda^{-\frac{1}{2}})$. The left side of (5.16) converges strongly as $\lambda \to \infty$ to $R[e^{iA^-}, e^{iB^-}]R$, while the right side converges to zero. Thus A^- and B^- commute.

Note (5.16) follows from the following identity for bounded C, D:

$$[e^C, e^D] = \int_0^1 ds\left(\frac{d}{ds} e^{sC} e^D e^{(1-s)C}\right) = \int_0^1 ds\, e^{sC}[C, e^D]e^{(1-s)C},$$

so

$$[e^C, e^D] = -\int_0^1 ds \int_0^1 dt\, e^{sC} e^{tD} [C, D] e^{(1-t)D} e^{(1-s)C}.$$

6. LORENTZ COVARIANCE

The main result of this section is Lorentz covariance of Φ and Lorentz invariance of Ω. We also prove analyticity of the Schwinger functions at noncoinciding points.

Theorem 6.1: Let $d\mu$ satisfy (1) - (3). Then there exists a strongly continuous, unitary representation $U(g)$ of the inhomogeneous Lorentz group $\mathcal{L}$ on $\mathcal{H}$ such that

$$U(g)\Omega = \Omega \tag{6.1}$$

$$U(g)\Phi(f)U(g)^{-1} = \Phi(g^{-1}f)$$

for all $g \in \mathcal{L}$. In terms of $\Phi(x)$,

$$U(g)\Phi(x)U(g)^{-1} = \Phi(gx). \tag{6.2}$$

Remark: We ask whether the assumptions (1) - (2) are sufficient for the existence of $U(g)$. We use (3) in our proof that Ω is invariant under Lorentz boosts (Lorentz rotations in a (t, x) plane).

Proposition 6.2: Let $d\mu$ satisfy (1) - (2) and let $g \to V(g)$ be a strongly continuous unitary representation of a group

$\mathcal{G}_0$ on $\mathcal{E}$ such that

(6.3) $V(g)1 = 1, \quad V(g)\mathcal{E}_+ \subset \mathcal{E}_+, \quad \vartheta V(g) = V(g)\vartheta, \quad T(t)V(g) = V(g)T(t)$.

Then U(g) defined by

$$WV(g) = U(g)W, \quad g \in \mathcal{G}_0 ,$$

is a continuous unitary representation of $\mathcal{G}_0$ on $\mathcal{H}$ such that

(6.4) $$U(g)\Omega = \Omega, \quad e^{itH}U(g) = U(g)e^{itH} .$$

Proof: As in the proof of Theorem 2.1, $V(g)$ maps $\mathcal{E}_+$ and $\mathfrak{n}$ into themselves, so U(g) is defined on the domain $W\mathcal{E}_+$. Furthermore, U(g) is unitary since V(g) commutes with ϑ. In fact

$$\langle U(g)WA, WB\rangle_{\mathcal{H}} = \langle \vartheta V(g)A, B\rangle_{\mathcal{E}} = \langle V(g)\vartheta A, B\rangle_{\mathcal{E}} = \langle \vartheta A, V(g)^{-1}B\rangle_{\mathcal{E}}$$

$$= \langle WA, WV(g^{-1})B\rangle_{\mathcal{H}} = \langle WA, U(g^{-1})WB\rangle_{\mathcal{H}} ,$$

so $U(g)^* = U(g)^{-1} = U(g^{-1})$. Thus U(g) extends to a representation of $\mathcal{G}_0$ on all of $\mathcal{H}$. Since V(g) commutes with T(t), U(g) commutes with e^{-tH} and hence with e^{itH}. Strong continuity of U(g) follows from strong continuity of V(g), while $U(g)\Omega = \Omega$ follows from $V(g)1 = 1$.

We now study the distributions

(6.5) $$W_n(x_1, \cdots, x_n) = \langle \Omega, \phi(x_1)\cdots\phi(x_n)\Omega\rangle ,$$

$$W_n(\underline{h};t) = \langle \Omega, \Phi(h_1, t)\cdots \Phi(h_n, t_n)\Omega\rangle ,$$

which are densities of the Wightman functions (5.6). Here for brevity we use the notation $\underline{h} = \{h_1, \cdots, h_n\}$.

Proposition 6.3: The $W_n(\underline{h};t)$ are boundary values in $\mathscr{S}'(R^n)$ of analytic functions $W_n(\underline{h};z)$. Here $z_j = t_j + is_j$, and the $W(\underline{h};z)$ are analytic in the region $s_{j+1} - s_j > 0$, $j = 1, 2, \cdots, n-1$. Furthermore, for $t_j = 0$, and $s_1 < s_2 < \cdots < s_n$,

$$W_n(\underline{h};is) = S_n(\underline{h}, s) = \int \varphi(h_1, s_1)\cdots\varphi(h_n, s_n)d\mu \; . \tag{6.6}$$

Proof: The bound of Corollary 4.2 shows that $W_n(\underline{h};z)$ is analytic for $s_{j+1} - s_j > 0$, $j = 1, 2, \cdots, n-1$. The bound of Theorem 5.1 shows that $W_n(\underline{h};z) \to W_n(\underline{h};t)$ in $\mathscr{S}'(R^n)$ as $s_{j+1} - s_j \searrow 0$. For $t_j = 0$ and $s_{j+1} - s_j > 0$, the $W_n(\underline{h};z)$ agree with the Schwinger functions, by the definitions of §2.3. Note the Schwinger functions are already known to be analytic in $s_{j+1} - s_j$ by Corollary 2.4.

Proof of Theorem 6.1: The space-time translations and space rotations follow from Proposition 6.2. The field Φ transforms covariantly by definition. To complete the proof we construct the Lorentz boost transformations. Let us consider a pure Lorentz rotation Λ_α by hyperbolic angle α in the $(t, \vec{x}_1) \equiv (t, x)$ plane. The infinitesimal operator for this rotation on the Wightman functions is

$$L_n = \sum_{j=1}^{n} \left(t_j \frac{\partial}{\partial x_j} + \frac{\partial}{\partial t_j} \right) .$$

We show $W_n(L_n F) = 0$ for all $F \in \mathscr{S}(R^{nd})$. In particular

$$\frac{d}{d\alpha} W_n(\Lambda_\alpha F) = W_n(L_n \Lambda_\alpha F) = 0 ,$$

so each W_n is Lorentz invariant. It follows that there exists a unitary group $U(\Lambda_\alpha)$ on $\mathcal{H}$ which implements Λ_α. The Lorentz group multiplication laws for the $U(\Lambda_\alpha)$ follows from the Lorentz group multiplication laws for the Λ_α.

The Euclidean invariance of the Schwinger function S_n, in infinitesimal form, states that

$$0 = \sum_{j=1}^{n} \left(s_j \frac{\partial}{\partial x_j} - x_j \frac{\partial}{\partial s_j} \right) S_n(\vec{x};s) \; . \tag{6.7}$$

We analytically continue (6.7) to complex $s_j = \epsilon_j - it_j$ with $\epsilon_{j+1} - \epsilon_j > 0$, i.e. within the domain of analyticity of S_n. For complex s we rewrite (6.7) as

$$0 = \sum_{j=1}^{n} [(\epsilon_j - it_j)\frac{\partial}{\partial x_j} - x_j \frac{\partial}{\partial(-it_j)}] S_n(\vec{x}, \epsilon - it) . \tag{6.8}$$

We now take $\epsilon_{j+1} - \epsilon_j \to 0$ and find $L_n W_n = 0$. With test functions, this states $W_n(L_n F) = 0$, as desired.

Corollary 6.4: The energy momentum spectrum lies in the forward cone, $|\vec{P}| \leq H$. Here $\vec{P}$ is the momentum operator, the generator of space translation on $\mathcal{H}$.

Theorem 6.5: Assume (1) - (3). Let $\delta > 0$. There exists a bilinear form $\Phi(x)$ on $\mathcal{H}_\delta \times \mathcal{H}_\delta$ such that

$$e^{-\delta H} \Phi(x) e^{-\delta H} \tag{6.9}$$

is a bounded operator, analytic in x for $|\mathrm{Im}\, x| \leqslant \delta$, and such that

$$\Phi(f) = \int \Phi(x) f(x) dx . \tag{6.10}$$

Proof: Since $\vec{P}$ and H commute, we infer from Corollary 6.4 that the series

$$e^{itH - i\vec{x}\vec{P} - \delta H} = \sum_{n, m=0}^{\infty} \frac{(itH)^n}{n!} \frac{(-i\vec{x}\vec{P})^m}{m!} e^{-\delta H}$$

converges in norm for $|t| + |\vec{x}| < \delta$. Using (5.3), it follows that for $|t| + |\vec{x}| < \delta$,

$$e^{-\delta H} \Phi(f_{t, \vec{x}}) e^{-\delta H} = F(x)$$

is a real analytic function, with

$$|| D_x^n F(x) || \leq K(\epsilon) \epsilon^{-|n|} n! \, |f|_1 , \tag{6.11}$$

for $\epsilon < \delta$ and x, t real but otherwise unrestricted. Thus as in (3.8), F(0) is the integral of a bounded C^∞ function G(x), $F(0) = \int G(x) f(x) dx$, where $G(x) = e^{-\delta H} \Phi(x) e^{-\delta H}$ defines $\Phi(x)$.

Repeating the argument leading to (6.11), we obtain

$$\|D_x^n e^{-\delta H}\Phi(x)e^{-\delta H}\| \le K(\epsilon,\delta)\epsilon^{-|n|}n! , \tag{6.12}$$

where

$$K(\epsilon,\delta) = \|e^{-(\delta-\epsilon)H}\Phi(x)e^{-(\delta-\epsilon)H}\| .$$

Since $K(\epsilon, \delta)$ is independent of x (for x real), the claimed analyticity follows by (6.12).

Corollary 6.6: The Schwinger functions $S_n(x_1, \cdots, x_n)$ are real analytic functions of $x_1, \cdots, x_n$ at non-coinciding points (i.e. $x_i \neq x_j$ for all $i \neq j$).

Proof: Since the S_n are symmetric under permutation of $x_1, \cdots, x_n$, we may assume that $t_1 \le t_2 \le \cdots \le t_n$. If some times are equal, but no two points coincide, then a small Euclidean rotation produces all unequal times. The corollary now follows from the analyticity of (6.9).

Proposition 6.7: Let x, y be given, with $\vec{x} - \vec{y} \neq 0$. Let B be a subset of R^d of points $(z^0, \vec{z})$ such that $\vec{z}$ projected on the line $\vec{x} - \vec{y}$ does not lie between $\vec{x}$ and $\vec{y}$. Let $f_1, \cdots, f_n \in \mathscr{S}$ be supported in B. Then $S_{n+2}(f_1, \cdots, f_n, x, y)$ is analytic in $x - y$ for $\vec{x} - \vec{y}$ real and $|x^0 - y^0| < |\vec{x} - \vec{y}|$.

Proof: Perform a Euclidean rotation so that $\vec{x} - \vec{y}$ is the new time axis. By construction, B contains no points in the time interval between $\vec{x}$ and $\vec{y}$. The proof now follows the proof of the theorem.

Note that in the $\xi = x^0 - y^0$ plane, the proposition allows us to connect the half planes $\mathrm{Re}\,\xi > 0$ and $\mathrm{Re}\,\xi < 0$ of analyticity by a slit $|\mathrm{Im}\,\xi| < |\vec{x} - \vec{y}|$ on the imaginary ξ axis.

7. LOCALITY

Theorem 7.1: Assume (1) - (3) and let $f, g \in \mathscr{S}_{\mathrm{real}}$ have space-like separated supports. Then three forms of locality hold:

(i) $[\exp(i\Phi(f)),\ \exp(i\Phi(g))] = 0$.

(ii) $[\Phi(f),\ \Phi(g)]\mathscr{D} = 0$.

(iii) $W_{n+2}(f_1, \cdots, f, g, \cdots, f_n) = W_{n+2}(f_1, \cdots, g, f, \cdots, f_n)$

for all n and all $f_j \in \mathscr{S}$.

Given $T > 0$ and $\vec{z} \neq 0$, $\vec{z} \in R^{d-1}$, let $B = B(T, \vec{z})$ be the subset of R^d of points $(t, \vec{x})$ with $t \geq T$, $\vec{x} \cdot \vec{z} \geq \vec{z}^2$. Let

$$\mathscr{D}_B = \operatorname{span}\{W\varphi(f)^n : f \in \mathscr{S}(B)\} .$$

Proposition 7.2: $\mathscr{D}_B$ is a core of C^∞ vectors for H.

Proof: Every vector in $\mathscr{D}_B$ is in the range of e^{-TH}. Thus we need only show that $\mathscr{D}_B$ is dense. Let $\chi \perp \mathscr{D}_B$ and define

$$F(t, \vec{x}) = \langle \chi, e^{-tH + i\vec{x}\vec{P}} W\varphi(g)^n \rangle_{\mathcal{H}} \equiv \langle \chi, \psi \rangle_{\mathcal{H}}$$

where $g \in \mathscr{S}_+ \cap C_0^\infty$. For t and $\vec{x} \cdot \vec{z}$ sufficiently large, $\psi \in \mathscr{D}_B$, and $F(t, \vec{x}) = 0$. Clearly F is real analytic in t for $t > 0$, and by Corollary 6.4, F is real analytic in $\vec{x}$. Thus $F \equiv 0$, and in particular $F(0, 0) = 0$. But the estimates (3.4) allow summation of the exponential series $e^{\varphi(g)}$ and thus ensure $\chi \perp W\exp(\varphi(g))$. However the vectors $W\exp(\varphi(g))$ span $\mathcal{H}$. Thus $\chi = 0$ and $\mathscr{D}_B$ is dense.

Proposition 7.3: Let $f, g \in C_0^\infty$ have space-like separated supports. Then for some B as above,

$$[\Phi(f), \Phi(g)]\mathscr{D}_B = 0 .$$

Proof: Study the Schwinger functions

$$S_{n+2}(\vartheta f_1, \cdots, \vartheta f_r, x, y, f_{r+1}, \cdots, f_n) = S_{n+2}(\vartheta f_1, \cdots, \vartheta f_r, y, x, f_{r+1}, \cdots, f_n), \tag{7.3}$$

with f_j supported in $B(T, \vec{z})$. Choose $T, \vec{z}$ sufficiently large so that no point in $B \cup \vartheta B$ lies in the strip bounded by two hyperplanes normal to $\vec{x} - \vec{y}$ and passing through x and y respectively. Then Corollary 6.7 applies and (7.3) is analytic in $x - y$ for $\vec{x} - \vec{y}$ real and $|x^0 - y^0| < |\vec{x} - \vec{y}|$. We evaluate (7.3) at pure imaginary $x^0 = it$, $y^0 = is$. (Note we may choose one B for all $x \in$ suppt f, $y \in$ suppt g.) Then multiply by f(x)g(y) and integrate over x, y. Thus the analytic continuation of (7.3) is the statement

$$\langle \theta_1, [\Phi(f), \Phi(g)]\theta_2 \rangle_{\mathcal{H}} = 0 ,$$

where $\theta_1, \theta_2 \in \mathcal{D}_B$. Note the restriction $|t - s| < |\vec{x} - \vec{y}|$ is just the condition that f, g have space-like separated supports.

Proof of Theorem 7.1: By Proposition 7.2 and Theorems 5.4, 5.6, $\mathcal{D}_B$ is a core for $\Phi(f), \Phi(g)$ and in the domain of their product. We let $f, g \in C_{0, \text{real}}^{\infty}$, and apply Proposition 7.3. By Theorem 5.7, (with $\mathcal{D}_B$ the domain of definition) part (i) of Theorem 7.1 holds. Since the vectors in $\mathcal{D}$ are C^{∞} for all products of field operators, parts (ii) and (iii) hold. Parts (ii) and (iii) extend by continuity from $f, g \in C_0^{\infty}$ to $f, g \in \mathcal{S}$, from which part (i) follows by a second application of Theorem 5.7.

8. UNIQUENESS OF THE VACUUM

We consider the following condition on $d\mu(\varphi)$:

(4) Ergodicity: The time translation subgroup T(t) of $\mathcal{G}$ acts ergodically on the measure space $\{\mathcal{S}'(R^d)_{\text{real}}, d\mu\}$.

Theorem 8.1: Let S{f} satisfy (1) - (3). Then (4) is satisfied if and only if Ω is the unique vector (up to scalar multiples) in $\mathcal{H}$ which is invariant under time translation exp(itH).

Remark: Ergodicity of $d\mu$ is equivalent to the statement that 1 is the unique invariant vector for the unitary group T(t) acting on the Hilbert space $\mathcal{E} = L_2(\mathcal{S}', d\mu)$. This in turn is equivalent to the cluster property

$$0 = \lim_{t\to\infty} t^{-1} \int_0^t [\langle AT(s)B \rangle - \langle A \rangle \langle B \rangle] ds , \tag{8.1}$$

for A, B in a dense subspace of $\mathcal{E}$. Here $\langle \cdot \rangle$ denotes $\int \cdot \, d\mu$. In particular, exponential clustering of the Schwinger functions ensures ergodicity of $d\mu$.

Proof: For a self adjoint contraction semigroup e^{-tH} acting on a Hilbert space $\mathcal{H}$

$$\text{st.}\lim_{t\to\infty} t^{-1}\int_0^t e^{-sH}ds = P_{inv}$$

is the projection onto the subspace of invariant vectors. (Similarly, for a unitary group T(t), $P_{inv} = \text{st.}\lim t^{-1}\int_0^t T(s)ds.$) Thus (8.1) is equivalent to the statement that 1 spans the invariant subspace of T(t) in $\mathcal{E}$. Let A, B in (8.1) be finite linear combinations of functions $\exp(i\varphi(f))$, with $f \in C^{\infty}_{0,\text{real}}$. It is no loss in generality to require $A \in \mathcal{E}_-$, $B \in \mathcal{E}_+$, because of the time translation T(s) in (8.1). Then (8.1) is equivalent to the cluster property

$$(8.2)\quad 0 = \lim_{t\to\infty} t^{-1}\int_0^t [\langle W\vartheta A, e^{-tH}WB\rangle_{\mathcal{H}} - \langle W\vartheta A, \Omega\rangle_{\mathcal{H}}\langle \Omega, WB\rangle_{\mathcal{H}}]ds\ ,$$

and thus to the uniqueness of Ω as a ground state for H.

We remark that the proof of Theorem 1.2 is now complete. We also state without proof the following:

Theorem 8.2: Let $d\mu$ satisfy (1) - (3). Then $d\mu$ is ergodic under T(t) if and only if $d\mu$ is ergodic under the full Euclidean group.

Theorem 8.3: (Decomposition into pure phases) Let $d\mu$ satisfy (1) - (3) but not (4). Let Z denote the subspace of $L_\infty(d\mu)$ which is invariant under T(t). Then there exists a positive probability measure $d\zeta$ on the spectrum of Z and a decomposition

$$d\mu = \int d\zeta\, d\mu_\zeta\ ,$$

with the property that the measure $d\mu_\zeta$ satisfies (1) - (4) for almost every ζ.

REFERENCES

[1] R. P. Feynman, Rev. Mod. Phys. 20, 367-387 (1948); M. Kac, Probability and Related Topics in Physical Sciences, Interscience, 1959; K. Symanzik, Euclidean quantum field theory, in Local Quantum Field Theory, R. Jost (ed.), Academic Press, 1969; E. Nelson, J. Funct. Anal. 12, 97-112 (1973); K. Osterwalder and R. Schrader, Commun. Math. Phys. 31, 83-112 (1973) and 42, 281-305 (1975).

[2] J. Glimm and A. Jaffe, Quantum Field Models, in Statistical Mechanics, C. DeWitt and R. Stora (eds.), 1970 Les Houches Lectures, Gordon and Breach Science Publishers, New York, 1971.

[3] J. Fröhlich, Helv. Phys. Acta 47, 265-306 (1974); Adv. Math., to appear; Ann. de l'Inst H. Poincaré 21, 271-317 (1974); Ann. Phys. 97, 1-54 (1976); G. Hegerfeldt, Commun. Math. Phys. 35, 155 (1974); F. Guerra, L. Rosen and B. Simon, Ann. Math. 101, 111-259 (1975); A. Klein and L. Landau, J. Funct. Anal. 20, 44-82 (1973); A. Klein, preprint; H. Borchers and J. Yngvason, preprint; J. Challifour, preprint; D. Uhlenbrock, preprint.

[4] I. Gelfand and N. Vilenkin, Generalized Functions, Vol. 4, Academic Press, 1964.

[5] J. Glimm and A. Jaffe, Commun. Math. Phys. 44, 293-320 (1975), footnote page 297.

[6] W. Driessler and J. Fröhlich, preprint.

[7] T. Kato, Perturbation Theory for Linear Operators, Springer Verlag, 1966, formula (3.43), page 282.

[8] Ibid., Theorem 3.13, page 459.

HOW TO PARAMETRIZE THE SOLUTIONS OF LAGRANGIAN FIELD THEORIES: SYMMETRY BREAKING, AND DIMENSIONAL INTERPOLATION AND RENORMALIZATION in the $(\phi^4)_n$ MODEL

A.S. Wightman

Princeton University
Princeton, New Jersey

1. INTRODUCTION

At the present stage of development of Lagrangian quantum field theory, information about model theories is available from several sources with quite different levels of mathematical completeness. For example, there is constructive quantum field theory, the results of which are typically mathematical theorems; it undertakes to construct solutions of specific models and then to establish their main properties with complete mathematical rigor. A second example is the analysis of high-energy and low-energy behavior using the method of the renormalization group. This analysis assumes that the models studied have solutions and deduces properties of those solutions from further assumptions (e.g. differentiability in parameters and the equivalence of a redefinition of the normalization of Green's functions to a multiplicative renormalization) about the functions controlling high energy behavior. There are other sources of information; a complete list will not be given here.

The present survey is an attempt at an overview of the solutions of the Lagrangian theory of a quartically self-coupled neutral scalar field in n space-time dimensions, customarily denoted the $(\phi^4)_n$ model. From a logical point of view, the outlook arrived at is based on a patchwork of rigorous results and more or less plausible inferences. It is to be hoped that eventually all its statements will be incorporated into constructive field theory as theorems. In fact, the main point of the present survey is to focus attention on what appear to be the key obstacles in the way of such a program.

2. SOME RESULTS OF CONSTRUCTIVE FIELD THEORY [1]...[6]

Consider first the $P(\phi)_2$ model. Here the Hamiltonian is

$$H = H_0(m_0) + H_1 \tag{2.1}$$

where (formally)

$$H_0 = \int \tfrac{1}{2} :(\pi^2 + (\nabla\phi)^2 + m_0^2\,\phi^2): \,dx \tag{2.2}$$

and

$$H_1 = \int :P(\phi(x)):_{m_1} \,dx - E_0 \tag{2.3}$$

The Wick ordering in H_0 is so defined that

$$H_0(m_0) = \int \sqrt{m_0^2+\vec{k}^2}\,dk\; a^*(k)a(k) \tag{2.4}$$

$a(k)$ and $a^*(k)$ being the annihilation and creation operators of momentum k. On the other hand, for the sake of generality the Wick ordering in H_1 is defined relative to the vacuum state, $\Phi_0(m_1)$, of $H_0(m_1)$:

$$H_0(m_1)\Phi_0(m_1) = 0 . \tag{2.5}$$

The field operator ϕ is normalized so that

$$(\Psi_0,\phi(x)\Psi_0) = \langle\phi\rangle \tag{2.6}$$

where Ψ_0 is the ground state of H. E_0 is a constant chosen so that

$$H\Psi_0 = 0 \tag{2.7}$$

P is a polynomial bounded below. The constant term in P is of no significance since it is cancelled by E_0. Thus, P will be regarded as defined modulo constants. In constructive field theory, the field ϕ is usually normalized by the canonical commutation relations

$$[\phi(x),\pi(y)] = i\delta(x-y) , \qquad \pi(x) = \frac{\partial\phi}{\partial x_0} \tag{2.8}$$

and that normalization will be assumed here unless an explicit to the contrary is made. (The above description is not precise because it does not take into account the delicacies arising in the thermodynamic limit from the fact that one must use a non-Fock representation of the commutation relations. See [1] for further information.)

The theories just described may be labeled by the symbols

$$\{\phi; m_0^2, m_1^2, P, \langle\phi\rangle\} \tag{2.9}$$

However, this is a redundant parametrization because there are several groups acting on the fields and Hamiltonians which establish relations between solutions.

i) The Wick Reordering Group

These operations $R(m_1'^2 \leftarrow m_1^2)$ relate the theory with P Wick-ordered with respect to $\Phi_0(m_1^2)$ to a theory in which the interaction is Wick-ordered with respect to $\Phi_0(m_1^2)$

$$\{\phi;m_0^2,m_1^2,P,\eta\} = \{\phi:m_0^2,m_1'^2,R(m_1'^2 \leftarrow m_1^2)P,\eta\} \tag{2.10}$$

where $R(m_1'^2 \leftarrow m_1^2)$ is defined first for monomials

$$R(m_1'^2 \leftarrow m_1^2)\xi^k = \sum_{\ell=0}^{[k/2]} (-1)^\ell \frac{n!}{(n-2\ell)!2^\ell \ell!} [\delta C(m_1' \leftarrow m_1)]^\ell \xi^{k-2\ell}$$

$$\delta C(m_1' - m_1) = \frac{1}{4\pi} \ell n\left(\frac{m_1'}{m_1}\right)^2 \tag{2.11}$$

and then by linearity for polynomials. The relation (2.10) permits one to chose at one's convenience an arbitrary mass $m_1'^2$ with respect to which the interaction polynomial is Wick-ordered, at the cost of reshuffling coefficients of P of degree $< 2N$, the degree of P

ii) The Orthogonal Group of the Fields

This is the group generated by translation

$$(\{C,1\}\phi)(x) = \phi(x) + C \qquad C \in \mathbb{R}^4 \tag{2.12}$$

and reflection

$$(\{0,-1\}\phi)(x) = -\phi(x) . \tag{2.13}$$

The first yields the relation

$$\{\phi+C;m_0^2,m_1^2,P,\eta\} = \{\phi;m_0^2,m_1^2,\{C,1\}P,\eta-C\} \tag{2.14}$$

with

$$\{C,1\}P(\xi) = P(\xi+C) \tag{2.15}$$

Translation of the field permits one to reduce the problem of solving the theory with the interaction P to one in which the term of degree $2N - 1$ is absent at the cost of introducing interactions of lower degree. The reflection operation yields

$$\{-\phi;m_0^2 m_1^2,P,\eta\} = \{\phi;m_0^2,m_1^2,\{0,-1\}P,-\eta\} \quad . \tag{2.16}$$

These transformations leave the canonical commutation relations invariant.

iii) Bare Mass Dependence

This group invariance just says that the theory depends on $m_0^2 + 2\lambda_2$ rather than on m_0^2 and λ_2 separately. Here λ_2 is the coefficient of the second degree term in P

$$\{\phi;m_0^2,m_1^2,P,\eta\} = \{\phi,m_0^2 + a,m_1^2,P - 2a\xi^2,\eta\} \tag{2.17}$$

The symbol $\{\phi;m_0^2,m_1^2,P,\langle\phi\rangle\}$ contains $2N + 2$ continuous parameters: 2 masses m_0^2 and m_1^2, 2N coefficients in P and one-order parameter $\langle\phi\rangle$. The bare mass relation (2.17) reduces these by one, the re Wick-ordering by another, and the order parameter $\langle\phi\rangle$ ought to be fixed by the rest of the parameters at least up to a discrete set of possibilities. (These are the possible pure theories which will be discussed shortly.) Thus there is a 2N-1 parameter family of theories. For the simplest non-trivial case $N = 2$, these parameters may be taken as the bare mass parameter $m_0^2 + 2\lambda_2$, the coefficient, λ_4, of ϕ^4 and the coefficient, λ_1, of the linear term ϕ .

There is one other group action in which a scaling transformation of lengths and masses is combined with an appropriately chosen scaling of fields. In n dimensions of space-time it is

$$(\{\rho\}\phi)(x) = \rho^{\frac{n-2}{2}}\phi(\rho x) \tag{2.18}$$

and consequently for $n = 2$

$$\{\{\rho\}\phi;m_0^2,m_1^2,P,\eta\} = \{\phi(\rho\cdot);\frac{m_0^2}{\rho^2},m_1^2,\frac{1}{\rho^2}P,\eta\} \tag{2.19}$$

Using an appropriate choice of ρ one can display the dependence of the theory on its coupling constants in terms of dimensionless bare coupling constants

$$g_0^{(j)} = \frac{\lambda_j}{m_0^2+2\lambda_2} \quad j=1,\dots 2N \quad \text{where} \quad P(\xi) = \sum_{j=1}^{2N} \lambda_j \xi^j \tag{2.20}$$

The explicit dependence on $m_0^2 + 2\lambda_2$ is now in $\phi(\sqrt{m_0^2+2\lambda_2}\, x)$.

The qualitative behavior of the physical mass, m, of the one particle state is plotted for $N = 2$ and $\eta = 0$ in Figures 1 and 2.

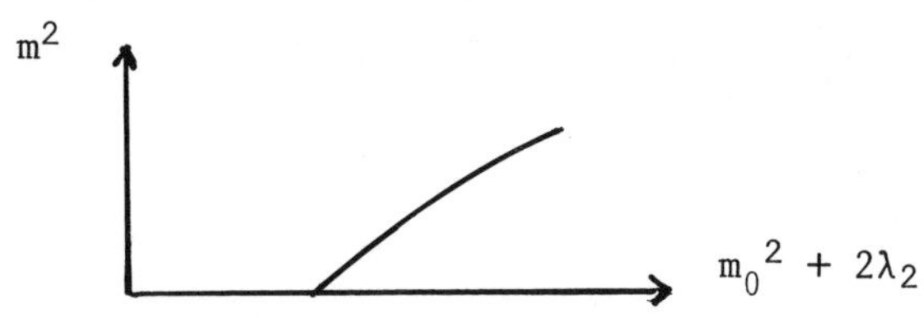

Figure 1 [4] The physical mass is monotonic in the bare mass. It vanishes at some critical value in space-time dimensions 2 and 3.

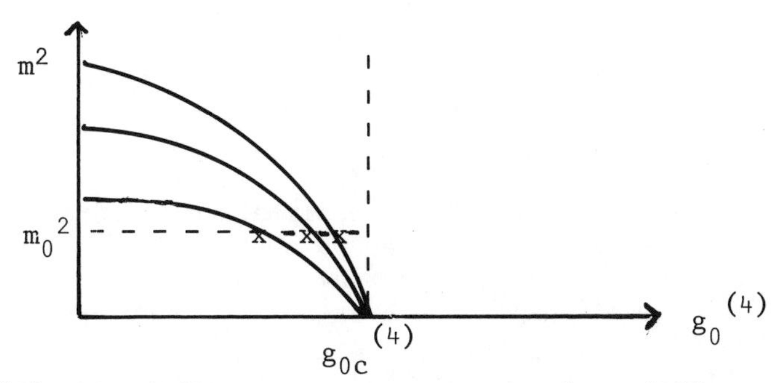

Figure 2 [5] The different curves correspond to different values of the bare mass. All curves come to zero at the critical value $g_{0c}^{(4)}$.

The striking feature of these curves is the occurrence of the parameters where the physical mass vanishes. This suggests that a phase transition has taken place and indeed one of the most remarkable recent results of constructive field theory is the proof that for sufficiently large $g_0^{(4)}$ and $g^{(1)} = 0$ there are two phases differing from one another by the sign of the order parameter $\langle\phi\rangle$, which is non-vanishing [7]. The asymptotic behavior of m as a function of $g_0^{(4)}$ has been determined rigorously; it increases. The most plausible behavior in the neighborhood of the critical point is a $g_{0c}^{(4)}$ is a cusp, but that has not yet been proved. The fact that the critical value, $g_{0c}^{(4)}$, is independent of the bare mass, $m_0^2+2\lambda_2$, is a consequence of the scaling law (2.19): zero physical mass is characterized by power law rather than exponential decay of the two-point function. By choosing $\rho^2 = m_0^2+ 2\lambda_2$ in (2.19) one sees that the theory may be regarded as a theory of the field $\phi(\rho x)$ with bare mass 1, and interaction $g_0^{(4)}:\phi^4(\rho x):_{m1}$.

Equally striking is the corresponding behavior of the renormalized coupling constant, $g^{(4)}$, shown in Figure 3. It corresponds to what Glimm and Jaffe have called critical point dominance.

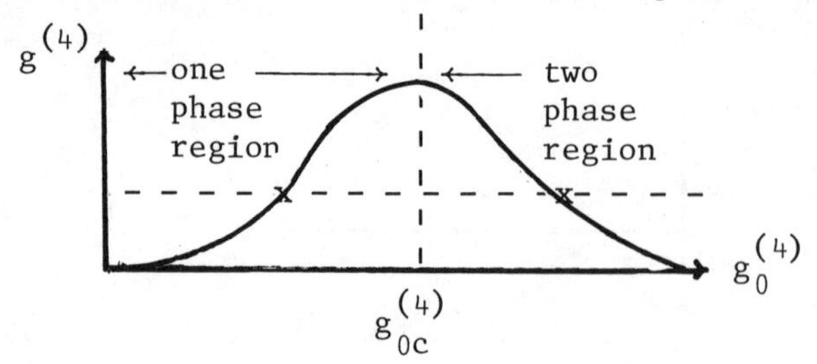

Figure 3 [3][8] The renormalized coupling constant $g^{(4)}$ as a function of unrenormalized coupling constant $g^{(4)}$ is defined by

$$g^{(4)} = -m^{-2}\,\Gamma^{(4)}(0)$$

where $\Gamma^{(4)}(p)$ is the one particle irreducible four vertex.

Figures 2 and 3 bring out two qualitative features of the solutions when they are regarded as parametrized by the physical one particle mass, m, and the renormalized coupling constant, $g^{(4)}$. For arbitrarily small $g^{(4)}$ there exist both a symmetrical solution, having an asymptotic series at $g^{(4)} = 0$ given by perturbation theory, and two broken symmetry solutions with asymptotic series at $g^{(4)} = 0$ given not by perturbation theory but by [9]. (The expansion begins with inverse powers of g.) The two values of $g_0{}^{(4)}$ corresponding to a single value of $g^{(4)}$ are indicated by the interactions of the horizontal dashed line with the curve in Figure 3. The second remarkable feature is indicated by the horizontal dashed line in Figure 2: In order to hold the physical one-particle mass constant as $g_0{}^{(4)} \to g_{0c}{}^{(4)}$ from below, one is forced to make $m_0{}^2 \to +\infty$ and $\lambda^{(4)} \to +\infty$. Thus, the limiting theory obtained in this way has an infinite mass and charge renormalization. There is a one-parameter family of such theories labeled by m, which scale into each other. The limiting theory when $m \to 0$ is a scale invariant theory much discussed in the theory of the renormalization group. On the other hand, for the same critical value, $g_{0c}{}^{(4)}$, of the bare coupling constant there is a one parameter family of superrenormalizable solutions, labeled by m_0, all having physical mass zero. It is conjectured that the limit of these theories as $m_0 \to 0$ is the same scaling limit theory.

The qualitative features of the solutions of $(\phi^4)_2$ displayed in Figures 1-3 persist in $(\phi^4)_3$ with one exception. In $(\phi^4)_2$ one can take $\lambda^{(2)} = \lambda^{(1)} = 0$ and with $\lambda^{(4)}/m_0{}^2$ sufficiently large a phase transition takes place. This comes about because changes in the definition of Wick-ordering of $:\phi^4:$ give rise to quadratic terms with negative coefficients that can be made arbitrarily large and negative. On the other hand, in three space-time dimensions the analogous terms are infinite and have to be mass renormalized away. Thus, in three space-time dimensions it is a $\lambda^{(4)}:\phi^4: + \lambda^{(2)}:\phi^2: +$ mass counterterms theory with $\lambda^{(2)}$ sufficiently negative that has the phase transition [10].

The final result of constructive field theory that will be mentioned is the Glimm-Jaffe bound on $g^{(4)}$. It says that for a solution of $(\phi^4)_n$ theory for which the ϕ field two-point Schwinger function requires no subtractions

$$\hat{S}^{(2)}(p) = \int_0^\infty \frac{d\rho(a^2)}{p^2+a^2} \qquad \int \frac{d\rho(a^2)}{a^2} < \infty$$

and for which $\phi \to -\phi$ is a symmetry

$$g^{(4)} \lesssim \frac{C}{\left[\int \left(\frac{m^2}{a^2}\right) d\rho(a^2)\right]^2}$$

where C is a numerical constant independent of all parameters of the theory except n. When the field ϕ is normalized so that the residue of $\hat{S}^{(2)}(p)$ at the single-particle pole is 1

$$\int \left(\frac{m^2}{a^2}\right) d\rho(a) \gtrsim 1$$

so $g^{(4)} \lesssim C$. This result is extraordinary not only in the simplicity of its proof but also in its consequences for the theory of the renormalization group, as will be discussed below.

3. RENORMALIZATION GROUP ANALYSIS OF $(\phi^4)_n$

The crucial quantity for the asymptotic high and low energy behavior of Schwinger functions is the function β. It has several possible definitions depending on the definition of the renormalization procedure. However, all of them yield the same qualitative behavior displayed in Figure 4.

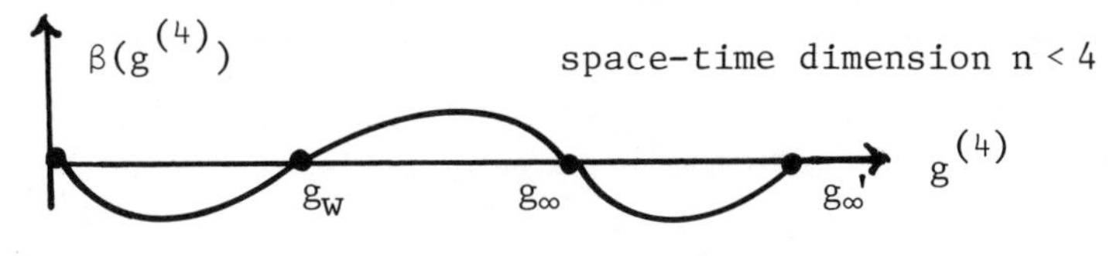

Figure 4 The zeros $g^{(4)}= 0$, g_∞ are ultraviolet stable, the zeros g_W and g_∞' are infrared stable. The subscript w stands for Wilson[11].

As far as the existing analysis based on the renormalization group goes, g_W, g_∞, g_∞' could all be infinite. However, from constructive field theory the evidence is that $g_W < \infty$; it is in fact believed to be the maximum of the curve of Figure 3. All the superrenormalizable solutions of $(\phi^4)_n$ obtained in constructive field theory have coupling constants satisfying $0 \lesssim g^{(4)} \lesssim g_W$ and have free field ultraviolet behavior. On the other hand, if there are solutions with $g_W < g^{(4)}$ and if $g_\infty, g_\infty' < \infty$, then the solutions with $g_W < g^{(4)} < g_\infty'$ will have an ultraviolet behavior determined by the theory at the ultraviolet stable zero $g^{(4)} = g_\infty$. Drastic changes in ultraviolet behavior may be expected when $g^{(4)}$ crosses g_W and g_∞' , while drastic changes in infrared behavior are expected when $g^{(4)}$ crosses g_∞ .

Glimm and Jaffe's study of the scaling limit theories with $m > 0$ and $g^{(4)} = g_W$ is an indication that $g_\infty < \infty$ and that the ultraviolet behavior of the theory with $g^{(4)} = g_\infty$ is renormalizable in the sense that it requires only infinite charge mass and field strength renormalization. (There are other definitions according to which these theories are non-renormalizable.) However, to make this indication completely convincing, two additional results would be necessary. First, their discussion assumes the celebrated inequality $\Gamma^{(6)} \lesssim 0$ which has not yet been proved. Second, one needs the existence of a solution not only exactly at $g^{(4)} = g_W$ but for some $g^{(4)}$ strictly greater than g_W in order to conclude that all theories in the interval $g_W < g^{(4)} < g'_\infty$ are renormalizable in the above sense.

The renormalization group analysis throws light on another aspect of this situation: its dependence on space-time dimension, n . It suggests that there are interpolations of the basic functions of $(\phi^4)_n$ for non-integral n whose behavior is summarized in Figure 5. The ε-expansions yield the asymptotic behavior of the curve $g_W(n)$ near $n = 4$, $g^{(4)} = 0$. If one takes Figure 5 seriously, one comes to regard the problems of solving $(\phi^4)_3$ and $(\phi^4)_2$ with $g_W < g^{(4)} < g_\infty'$ as related to that of $(\phi^4)_4$ with $0 < g^{(4)} < g_\infty'$. This provides further support for the idea that these solutions should be renormalizable but not superrenormalizable.

It is natural to ask how the Glimm-Jaffe bound fits into this picture. The simplest plausible conjecture is that theories for $g^{(4)} > g_\infty'$ violate the assumption $\int \frac{d\rho(a^2)}{a^2} < \infty$ and therefore that the Glimm-Jaffe bound is essentially an upper bound for g'_∞.

Of course, pessimists who believe that $(\phi^4)_4$ has no non-trivial solutions will note that the argument that the Glimm-Jaffe bound is an estimate for g_∞' (namely, that only at an infrared stable zero can a drastic change in ultraviolet behavior occur) would apply equally well to g_W . Then there would be no solutions of $(\phi^4)_n$ except superrenormalizable ones and the isolated family constructed by Glimm and Jaffe at $g^{(4)} = g_W$.

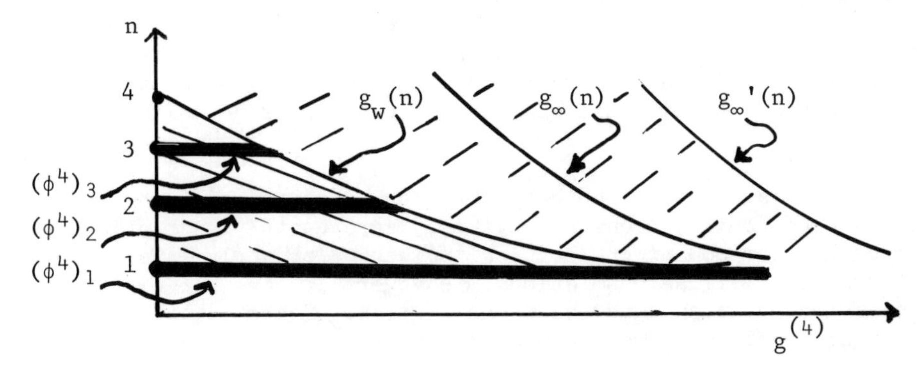

Figure 5. Diagram illustrating the solutions of $(\phi^4)_n$ in their dependence on space-time dimension n and renormalized coupling constant $g^{(4)}$. The superrenormalizable solutions obtained in constructive field theory are indicated by the heavy horizontal lines, their putative dimensional interpolations by \\\\. The expected renormalizable but not superrenormalizable solutions are indicated by ////.

The above discussion has all been carried out in terms of the symmetrical solutions of $(\phi^4)_n$. Presumably there are also broken-symmetry solutions running over the same range of m and $g^{(4)}$.

It seems to me that the conjectures summarized in Figure 5 provide a reasonable set of working hypotheses in the present state of knowledge. I have nothing to offer as a guess about what happens to the infrared behavior at $g^{(4)} = g \rightarrow g^{(4)} = g_\infty$.

4. PARAMETRIZATION OF $(\phi^4)_n$ THEORIES

If the conjectures about the behavior of the solutions of $(\phi^4)_n$ described above are accepted, one can see that there is a strong analogy between the phase diagrams of macroscopic systems in satistical mechanics and the parametrization of the solutions of $(\phi^4)_n$, an incomplete analogy however.

The superrenormalizable solutions are parametrized by $\{\phi, m_0^2, m_1^2, \Sigma_{j=1}^4 \lambda_j \xi^j, \langle\phi\rangle\}$ which we say could be taken without loss of generality as $\{\phi_0; m_0^2 + 2\lambda_2, m_1^2, \Sigma_{j=1, j\neq 2,3}^4 g_0(j)\, \xi^j, \langle\phi\rangle\}$. The use of these parameters has the advantage that the one and two phase regions are smoothly parametrized. The analogue of the bare (dimensional) coupling constant $\lambda^{(4)}$ is $1/kT$ and the analogue of $\lambda^{(1)}$ is $B\mu/kT$ where B is a magnetic field and μ a magnetic moment. One has a unique phase for weak coupling (high temperature) and multiple phases for strong coupling (low temperature). However, the bare parameters do not provide a satisfactory parametrization in the neighborhood of the critical point and they are, of course, totally useless for the non-superrenormalizable solutions.

The replacement of the bare coupling constant, $g_0^{(4)}$, by the renormalized coupling constant, $g^{(4)}$, improves the description in the meighborhood of the critical point and properly labels the non-superrenormalizable solutions. However, it does not distinguish between the one and two phase regions. There is, therefore, not a good analogy between $(m^2 g^{(4)})$ and $1/kT$.

REFERENCES

[1] Constructive Quantum Field Theory Lecture Notes in Physics #25 Ed. G. Velo and A.S. Wightman, Springer-Verlag Berlin 1973

[2] R. Baumel, Princeton Thesis, unpublished

[3] J. Glimm and A. Jaffe, Ann. Inst. Henri Poincaré XXI (1974)27-41

[4] O. McBryan and J. Rosen Existence of the Critical Point in ϕ^4 Field Theory Comm. in Math. Phys., to appear

[5] J. Glimm and A. Jaffe, Phys. Rev. D11 (1975) 2816

[6] J. Glimm and A. Jaffe, Ann. Inst. Henri Poincaré XXII(1975)109-122

[7] J.Glimm, A.Jaffe, and T.Spencer, Comm. in Math. Phys. 45 (1975) 203-216

[8] J. Glimm and A. Jaffe, Ann Inst. Henri Poincaré XXII (1975)97-107

[9] J. Glimm, A. Jaffe, and T. Spencer A Convergent Expansion about Mean Field Theory I,II, to appear

[10] S.J. Chang, Phys. Rev. D13 (1976) 2778

[11] K.G. Wilson, Phys. Rev. D7 (1973) 232

[12] B.Simon, J. Fröhlich, and T. Spencer, Phys. Rev. Lett. 36 (1976) 804-806

PHASE TRANSITIONS IN STATISTICAL MECHANICS AND QUANTUM FIELD THEORY

J. Fröhlich[*§]
Department of Mathematics
Princeton University
Princeton, N. J. 08540

T. Spencer[*+]
Department of Mathematics
Rockefeller University
New York, N. Y. 10021

1. Introduction

These lectures are a survey of some mathematically rigorous results concerning phase transitions and symmetry breaking in statistical mechanics and quantum field theory. This subject has a rather long history: The first results on phase transitions were obtained by R. Peierls in 1936, [Pe]. He showed that the Ising model in two or more dimensions has spontaneous magnetization at low temperatures. His argument was later reformulated in various ways and applied to many model systems. See the references in [Gr]. We shall apply a variant of the Peierls argument [GJS3] to prove that there is a phase transition in the two dimensional, anisotropic $(\vec{\phi} \cdot \vec{\phi})^2$ quantum field model in two dimensions and we explain how, in this model, a phase transition gives rise to soliton sectors. We show that the mass gap on the soliton sector is bounded below by the "surface tension" of this model.

In two dimensional models with continuous internal symmetries and short range interactions, (e.g. the classical or quantum Heisenberg model) there is no spontaneous magnetization and no symmetry breaking; see [MW, DS]. It seems to be very difficult to successfully apply the Peierls argument to models with <u>continuous</u> internal symmetries. However a second technique

* A. P. Sloan Foundation Fellows.
+ Supported in part by NSF under contract NSF-PHY-7617191
§ Supported in part by NSF under contract NSF-MCS-75-11864

which is related to spin wave analysis enables us to prove existence of phase transitions and symmetry breaking for such models in three of more dimensions. We apply this method [FSS] to the classical Heisenberg - and the isotropic $(\vec{\phi}\cdot\vec{\phi})^2_3$ quantum field model as well as to models with no internal symmetry whatsoever. Our lectures will concentrate only on these two methods.

The basic estimate for this second technique is an upper bound on the two point correlation function in momentum space (denoted $F(p)$)

$$0 \leq p^2 F(p) \leq \text{Const. } \beta^{-1}, \tag{1.1}$$

where β^{-1} is proportional to the temperature. We prove this estimate for classical systems on a cubic lattice with nearest neighbor ferromagnetic pair interactions and for canonical Bose quantum field theories. An analogue of this estimate — with F replaced by the Bogoliubov two point function — has been established for the quantum Heisenberg model in [DLS]. This together with other estimates enables them to prove a phase transition for this model. For the N-component, isotropic $(\vec{\phi}\cdot\vec{\phi})^2_3$-model in the multiple phase region one can conclude the existence of $N-1$ Goldstone bosons (zero mass one particle states) [ES]. As another important corollary of estimate (1.1) and of correlation inequalities we obtain the existence of <u>massless</u> $(\phi^4)_3$ field theories without long range order (i.e. without broken symmetries); this is based on [MR, GJ1]. The existence of a scaling limit for these theories is an open problem; but see [GJ2] and the contributions of Glimm and Jaffe to these proceedings for some rigorous results concerning this question.

Note that in quantum field theory inequality (1.1) (with $\beta = 1$) follows from the Källen-Lehmann representation and the canonical commutation relations which are generally valid in theories in two or three space-time dimensions. Finally we remark that (1.1) immediately implies a bound on the critical exponent η:

$$\text{If} \qquad \hat{F}^c(x) \approx \frac{\text{Const.}}{|x|^{d-2+\eta}}, \text{ as } |x| \to \infty \tag{1.2}$$

then $\eta \geq 0$. Here $\hat{F}^c$ is the connected two point function. (For some new results in $\nu = 2$ dimensions see Section 6).

2. <u>Infrared bounds</u>

We consider models of classical statistical mechanics with

ferromagnetic nearest neighbor interactions on a lattice $\mathbb{Z}^\nu$. With each site $j \in \mathbb{Z}^\nu$ we associate a spin vector $\sigma_j = (\sigma_j^1, \ldots \sigma_j^N)$ and we assign a (single spin) distribution $d\mu(\sigma_j)$ which is a positive, finite measure with the property that

$$\int e^{a|\sigma|^2} \, d\mu(\sigma) < \infty , \tag{2.1}$$

for all a. The interaction in a periodic box Λ is defined by

$$H_\Lambda(\sigma) = -\sum_{\substack{|i-j|=1 \\ i,j \in \Lambda}}^{\Lambda} \sigma_i \cdot \sigma_j + \sum_{j \in \Lambda} h \cdot \sigma_j \tag{2.2}$$

with spins at opposite faces of the box Λ identified. Note that in $\sum^\Lambda$ each nearest neighbor pair only occurs once. For a function $A(\sigma)$ of the spins in Λ we define

$$\langle A(\sigma) \rangle_\Lambda (\beta,h) = Z_\Lambda^{-1} \int A(\sigma) e^{-\beta H_\Lambda(\sigma)} \prod_{i \in \Lambda} d\mu(\sigma_i) \tag{2.3}$$

where

$$Z_\Lambda = Z_\Lambda(\beta,h) = \int e^{-\beta H_\Lambda(\sigma)} \prod_{i \in \Lambda} d\mu(\sigma_i) \tag{2.4}$$

is the partition function. For

$$d\mu(\sigma) = \delta(|\sigma|^2 - 1) d^N\sigma \tag{2.5}$$

equation (2.3) defines the finite volume expectation of the Ising model $(N = 1)$ or the classical Heisenberg model $(N = 3)$. In the case where

$$d\mu(\sigma) = \exp[-\lambda|\sigma|^4 + \gamma|\sigma|^2] d^N\sigma \tag{2.6}$$

we obtain the lattice approximation of the $(\vec{\phi} \cdot \vec{\phi})^2$-Euclidean field theory, [GRS]. The continuum field theory can be constructed by taking the lattice spacing to zero, [GRS, Pa].

We shall study the infinite volume expectations obtained by taking the limit in (2.3) as $\Lambda \to \mathbb{Z}^\nu$. Such a limit can be taken by passing to subsequences and using a standard compactness argument. For the case where $|h| > 0$ and $N = 1,2,3$ all limiting expectations of the models defined by (2.2) — (2.4) and (2.5) or (2.6) constructed by this procedure coincide, [F1]. For suitable boundary conditions and $N = 1$ or 2 the expectations of products of σ_i's can be shown, for these models, to converge, as $\Lambda \to \mathbb{Z}^\nu$ (by inclusion), by using correlation inequalities; see [DN].

To formulate the principal result of this section one considers the two point correlation $\langle\sigma_o \cdot \sigma_x\rangle(\beta)$ in momentum space. Let

$$F(p) = \sum_{x \in \mathbb{Z}^\nu} e^{ix\, p}[\langle\sigma_o \cdot \sigma_x\rangle(\beta, h)] \tag{2.7}$$

and let

$$\Delta(p) = 2 \sum_{i=1}^{\nu} (1 - \cos p_i) \tag{2.8}$$

which is the momentum space representation for minus the finite difference Laplacean. The vector p lies in the Brillouin zone $|p_i| \leq \pi$.

Theorem 2.1, [FSS]. Suppose $d\mu$ satisfies (2.1). Then there is a constant $c' \geq 0$ (depending on $d\mu$, β, h) such that

$$0 \leq F(p) - c'\delta(p) \leq N[\beta\Delta(p)]^{-1}. \tag{2.9}$$

We shall sketch the proof of this result in §5. Note that for $p \neq 0$ the right side of (2.9) is independent of μ and h. In three or more dimensions

$$\int_{|p_i| \leq \pi} \Delta(p)^{-1}\, d^\nu p < \infty \quad , \tag{2.10}$$

hence (by the Riemann-Lebesgue lemma)

$$\int_{|p_i| \leq \pi} (F(p) - c'\delta(p)) e^{ix\cdot p} d^\nu p \to 0, \tag{2.11}$$

as $|x| \to \infty$. This means that $\langle\sigma_o \cdot \sigma_x\rangle \to c \equiv (2\pi)^{-\nu} c'$.

If

$$\langle\sigma\rangle^2 \neq c \tag{2.12}$$

we say the system has long range order.

3. Phase transitions in three or more dimensions

In this section we apply Theorem 2.1 and (2.10)-(2.12) to prove the existence of phase transitions for the models introduced in Section 2 in $\nu \geq 3$ dimensions.

Theorem 3.1, [FSS]. If

$$\beta > \beta_o \equiv (2\pi)^{-\nu} N \int \Delta(p)^{-1} d^\nu p, \tag{3.1}$$

$\nu \geq 3$, then the N-component classical rotator with single spin distribution $d\mu$ given by (2.5) has long range order at $h = 0$. For $N = 1,2,3$ there is spontaneous magnetization, i.e.

$$\lim_{h^1 \downarrow 0} \langle\sigma_o^1\rangle(\beta,h) \neq 0 \tag{3.2}$$

Proof. By Theorem 2.1

$$\begin{aligned}\langle\sigma_o \cdot \sigma_o\rangle - c = 1 - c &= (2\pi)^{-\nu} \int (F(p) - c'\delta(p)) d^\nu p \\ &\leq \beta^{-1}(2\pi)^{-\nu} N \int \Delta(p)^{-1} d^\nu p = \beta^{-1}\beta_o.\end{aligned} \tag{3.3}$$

Now if $\beta > \beta_o$ $c \neq 0$. But $\langle\sigma\rangle = 0$, for $h = 0$, so our first assertion follows. When $N = 1,2,3$ the Lee-Yang Theorem holds and implies the absence of long range order, for $h \neq 0$. Hence $c = \langle\sigma_o^1\rangle(\beta,h^1)^2$ and $0 < (1 - \beta_o/\beta) \leq c = \langle\sigma_o^1\rangle(\beta,h^1)^2$, provided $h^1 \neq 0$. Thus (3.2) follows for $\beta > \beta_o$.

Theorem 3.1 states that β_o is a lower bound for the critical temperature. This lower bound agrees with that obtained by high temperature expansions to with in 14% for the 3 dimensional Ising model and 9% for the classical Heisenberg model. As $N \to \infty$ the partition function for the N-component model (with β scaled to βN) approaches that of the spherical model [K, St]. Since $\beta_o(N=1)$ is the critical temperature for the spherical model we expect our lower bound to be asymptotic for large N. Theorem 3.1 has been extended to quantum mechanical (x-y)-and Heisenberg models by Dyson, Lieb and Simon, [DLS]. Let

$$H_\Lambda = (\mp) \sum_{|i-j|=1}^{\Lambda} \sigma_i \cdot \sigma_j$$

be the Hamiltonian of the isotropic quantum mechanical Heisenberg (anti-) ferromagnet in a finite, periodic box $\Lambda \subset \mathbb{Z}^\nu$. Here $\sum^\Lambda$ is the sum over all nearest neighbor pairs in Λ, and $\sigma = (\sigma^x, \sigma^y, \sigma^z)$ are a representation of the Pauli spin matrices. Note that $\sigma \cdot \sigma = S(S+1)$, where S takes the values $1/2, 1, 3/2, \ldots$. The two point correlation function is defined by

$$\lim_{\Lambda \uparrow \mathbb{Z}^\nu} (\mathrm{tr}\, e^{-\beta H_\Lambda})^{-1} \, \mathrm{tr}[e^{-\beta H_\Lambda} \sigma_o \cdot \sigma_x] \; .$$

Dyson, Lieb and Simon have recently proven — among other interesting results on quantum lattice systems — the following

Theorem 3.2, [DLS].
(1) For the isotropic Heisenberg ferromagnet in $\nu \geq 3$ dimensions there exists a finite constant $\beta_c(S) > 0$ such that, for all $\beta > \beta_c(S)$, there is spontaneous magnetization, for all $S \geq 1/2$.
(2) For the anti-ferromagnet long range order occurs, provided $S \geq 1$.

The main ideas of the proof of (1) are as follows: Let

$$(\sigma_o, \sigma_x)_\Lambda = (\mathrm{Tr}\, e^{-\beta H_\Lambda})^{-1} \, \mathrm{Tr}\Big(\int_o^1 dt\, e^{-\beta t H_\Lambda} \sigma_o \times e^{-\beta(1-t) H_\Lambda} \sigma_x\Big)$$

denote the Bogoliubov (or Duhamel) two point function for the system in the box Λ, and let (σ_o, σ_x) denote some infinite volume limit of $\{(\sigma_o, \sigma_x)_\Lambda\}$. Then using a quantum analogue of the grad σ -bounds discussed in Section 5, one can show that the Fourier transform $\Phi(p)$ of (σ_o, σ_x) satisfies

$$\Phi(p) \leq c' \delta_o(p) + 3(\beta \Delta(p))^{-1}. \tag{3.4}$$

On the other hand one can derive a lower bound

$$\Phi(p) \geq F^c(p) \, f\Big(\frac{\Delta(p) S^2}{2F^c(p)}\Big), \tag{3.5}$$

where $F^c(p)$ is the Fourier transform of the connected two point correlation function, and f is the convex, decreasing function defined on $\mathbb{R}^+$ by the equation

$$f(x \tanh x) = \frac{1}{x} \tanh x.$$

From inequalities (3.4) and (3.5) one obtains the upper bound

$$F^c(p) \leq \sqrt{\frac{3}{2}}\, S \coth(\beta S \sqrt{\frac{2}{3}} \frac{1}{2} \Delta(p)),$$

hence

$$\begin{aligned} F(p) &\leq (2\pi)^\nu c\delta_o(p) + \sqrt{\frac{3}{2}}\, S \coth(\beta S \sqrt{\frac{2}{3}} \frac{1}{2} \Delta(p)) \\ &\approx (2\pi)^\nu c\delta_o(p) + \frac{3}{\beta\Delta(p)} \quad , \end{aligned} \tag{3.6}$$

for small $|p|$. (Note that, for small $|p|$, the upper bound agrees with the one found for the classical ferromagnet). Hence

$$\begin{aligned} S(S+1) &= \langle \sigma_o \cdot \sigma_o \rangle = (2\pi)^{-\nu} \int F(p) d^\nu p \\ &\leq c + \sqrt{\frac{3}{2}}\, S (2\pi)^{-\nu} \int \coth(\beta S \sqrt{\frac{2}{3}} \frac{1}{2} \Delta(p)) d^\nu p. \end{aligned} \tag{3.7}$$

One now defines $\beta_c(S)$ as the largest value for which $c = 0$ on the r.h.s. of (3.7) is consistent with the lower bound $S(S+1)$. Then $\beta_c(S)$ is obviously the solution of the equation

$$S+1 = \sqrt{\frac{3}{2}} (2\pi)^{-\nu} \int_{|p_i| \leq \pi} \coth(\beta_c S \sqrt{\frac{2}{3}} \frac{1}{2} \Delta(p)) d^\nu p. \tag{3.8}$$

One can show that

$$\beta_o = \lim_{S\to\infty} \frac{\beta_c(S)}{S^2}$$

is the transition temperature of the spherical model.

Even though the basic ideas of the proofs of inequalities (3.4)-(3.7) are quite simple and the strategy is similar to the one employed in the classical case the details are lengthy. We therefore refer the reader to [DLS]. (We thank F. Dyson, E. Lieb and B. Simon for informing us about their results and proofs prior to publication).

4. Phase transitions in models without internal symmetries

Phase transitions need not arise from symmetry breaking. To illustrate this fact consider an arbitrary measure $d\mu$ on $\mathbb{R}^1$ such that

$$\int_0^\infty d\mu > 0 \text{ and } \int_{-\infty}^0 d\mu > 0 \tag{4.1}$$

Theorem 4,[FSS]. For $\nu \geq 2$ there exists a β_o such that for all $\beta > \beta_o$ the magnetization

$$M(h, \beta) = \langle\sigma_o\rangle(\beta, h)$$

is discontinuous in h.

Proof. We make two simplifying assumptions: (a) $\nu \geq 3$ and (b) $\int_{-\varepsilon}^{+\varepsilon} d\mu = 0$. The case $\nu = 2$ requires different techniques; see [F3]. Assumption (b) immediately assures us that

$$\langle\sigma_o^2\rangle \geq \varepsilon^2.$$

Now suppose $M(h)$ is continuous. By condition (4.1) it is easy to see that

$$\lim_{h\to\infty} M(h) \geq \varepsilon \text{ and } \lim_{h\to-\infty} M(h) \leq -\varepsilon$$

Thus by the continuity of M there is a value of h for which $M(h) = 0$. Since there can be no long range order when M is continuous [Gu] we have $M(h)^2 = c$. But, for large β, $c \neq 0$, because

$$\varepsilon - c \leq \langle\sigma_o^2\rangle(\beta) - c \leq \beta^{-1}\beta_o.$$

Thus we conclude $M(h)$ is discontinuous.

Remarks: 1) Work on related questions has previously been done by Pirogov and Sinai [PS1] who have used Peierls-type arguments.

2) Theorem 4.1 extends to multiple component spin - and lattice gas models in $\nu \geq 2$ dimensions. Infrared bounds (Theorem 1), the Peierls argument (see Section 7) and low temperature expansions supply tools that are powerful enough to give detailed information on systems with very complicated phase diagrams. See

e.g. [PS2].

5. Gradient σ - bounds and pressure bounds

In this section we explain two key estimates which underly our results.

Let

$$\sigma(h) = \sum_j \sigma_j \cdot h(j)$$

$$(\partial_\alpha h)(j) = h(j+e_\alpha)-h(j)$$

$$(\partial^*_\alpha h)(j) = h(j-e_\alpha)-h(j)$$

where e_α is the vector $(e_\alpha)_{\alpha'} = \delta_{\alpha,\alpha'}$. Notice that $-\Delta = \sum_{\alpha=1}^{\nu} \partial^*_\alpha \partial_\alpha$.

Theorem 5.1. Let h_α take values in $\mathbb{R}^N$. Then

$$<\exp \sum_{\alpha=1}^{\nu} \sigma(\partial_\alpha h_\alpha)> \le \exp[(2\beta)^{-1} \sum_\alpha \| h_\alpha \|_2^2], \tag{5.1}$$

where $\| h_\alpha \|_2^2 = \sum_j |h_\alpha(j)|^2$. To recover the infrared bound (2.9) we set

$$h_\alpha = \partial^*_\alpha(-\Delta)^{-1/2} f, \quad f \in \text{Range } \Delta$$
$$= 0 \qquad , \quad f = \text{Const.}$$

If we subtract 1 from both sides of (5.1) and scale h we obtain

$$<[\sigma(\sum_{\alpha=1}^{\nu} \partial_\alpha h_\alpha)]^2>_\Lambda = <\sigma(f) \cdot \sigma(-\Delta f)>_\Lambda$$
$$\le \beta^{-1} \sum_j |f(j)|^2 , \tag{5.2}$$

and, for $f(x) = e^{ix\cdot p}|\Lambda|^{-1/2}$, (5.2) implies

$$\Delta(p)F(p) \le \beta^{-1} N.$$

which immediately yields (2.9). Another corollary of Theorem 5.1 is

$$\langle e^{\sigma(f) - \langle\sigma(f)\rangle}\rangle \leq e^{\frac{1}{2\beta}\langle f,\Delta^{-1}f\rangle} \tag{5.3}$$

which holds whenever $\langle \cdot \rangle$ is a pure (i.e. a clustering) state.

The second main estimate bounds expectations in terms of pressures. The pressure is defined by

$$p(\beta,h) = \lim_{\Lambda\to\mathbb{Z}^\nu} \frac{1}{|\Lambda|} \log \int e^{-H_\Lambda(\sigma)} \prod_{j\in\Lambda} d\mu(\sigma_j).$$

Consider a function $A(\sigma)$ which depends only on the variables $\{\sigma_j : j \in S\}$, where

$$S = \{j: |j_\alpha| \leq r/2, \ \alpha = 1,\ldots,\nu\} .$$

Assume that A is real and invariant under the reflections θ_α about $\{j_\alpha = 0\}$, for all $\alpha = 1,\ldots,\nu$. (Here θ_α is defined by

$$\theta_\alpha(j) = (j_1,\ldots,-j_\alpha,\ldots,j_\nu)).$$

For the purposes of this section it is important to define our lattice to be $\mathbb{Z}^\nu + (1/2,\ldots,1/2)$,

$$\Lambda = \{j \in \mathbb{Z}^\nu + (1/2,\ldots,1/2) : |j_\alpha| < \ell_\alpha\}.$$

Let $\tau_{\vec{n}}$ be translation by $\vec{n}$ and define a generalized pressure by

$$p(\beta,h,A) = \lim_{\Lambda\to\mathbb{Z}^\nu} \frac{1}{|\Lambda|} \log \Big| \int e^{-H_\Lambda(\sigma)} \prod_{\substack{\tau_{\vec{n}}S\subset\Lambda \\ \vec{n}\in\vec{r}\mathbb{Z}^\nu}} \tau_{\vec{n}}(A) \tag{5.4}$$

$$\times \prod_{j\in\Lambda} d\mu(\sigma_j) \Big|$$

Theorem 5.2. Let $\{A_{\vec{n}} : \vec{n} \in \vec{r}\,\mathbb{Z}^\nu\}$ be a family of real functions localized in S and invariant under the reflections θ_α, $\alpha = 1,\ldots,\nu$. Then

$$\Big|\langle \prod_{\vec{n}\in\vec{r}\mathbb{Z}^\nu} \tau_{\vec{n}}(A_{\vec{n}})\rangle\Big| \leq \exp[(\prod_{\alpha=1}^{\nu} r_\alpha)\sum_{\vec{n}}(p(A_{\vec{n}})-p)] \tag{5.5}$$

The proof of this theorem relies on the Schwarz inequality and Osterwalder-Schrader positivity:

$$\langle\overline{(\theta_\alpha B)}\,B\rangle \geq 0, \tag{5.6}$$

whenever B depends only on the variables $\{\sigma_j : j_\alpha > 0\}$, (for some $\alpha = 1,\ldots,\nu$). By (5.6) $\langle\overline{(\theta_\alpha B_1)}B_2\rangle$ defines an inner product on the space of such functions. The Schwarz inequality with respect to this inner product then states that if B_1 and B_2 depend only on $\{\sigma_j : j_\alpha > 0\}$ then

$$\langle\overline{(\theta_\alpha B_1)}B_2\rangle \leq \langle\overline{(\theta_\alpha B_1)}B_1\rangle^{1/2} \langle\overline{(\theta_\alpha B_2)}B_2\rangle^{1/2} \tag{5.7}$$

For example if $B_1 = \sigma_x$, $B_2 = \sigma_x^3$ with $x = (|a|,1/2,1/2)$ then

$$\langle\sigma_{-x} \quad \sigma_x^3\rangle \leq \langle\sigma_{-x} \quad \sigma_x\rangle^{1/2} \langle\sigma_{-x}^3 \quad \sigma_x^3\rangle^{1/2}$$

We now verify (5.6). We may set $\alpha = 1$ and define

$$\sigma^+ = \{\sigma_j : 1 < j_1 < \ell_1\}$$
$$\sigma^- = \{\sigma_j : -\ell_1 < j_1 < -1\}$$
$$s^+ = \{\sigma_j : j_1 = \pm 1/2\}, \quad \text{and}$$
$$t^\pm = \{\sigma_j : j_1 = \pm\ell_1\}.$$

The following diagram explains these definitions.

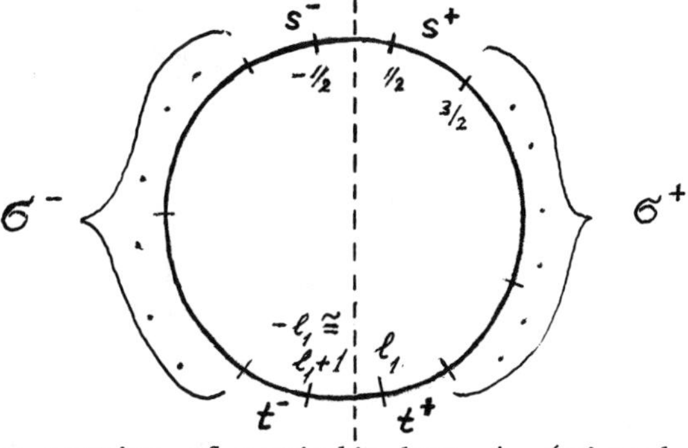

Fig. 1: Cross section of periodic box Λ (viewed as a torus); θ_1 reflects about the dashed line. By translation invariance the dashed line can be placed arbitrarily, i.e. rotated.

If we integrate $e^{-H_\Lambda}\overline{(\theta_\alpha B)}B$ over the σ^+ and σ^- variables we obtain a function

$$e^{\beta(s^+\cdot s^- + t^+\cdot t^-)} F(s^+,t^+)F^*(s^-,t^-),$$

$$\text{with} \quad F^*(s,t) = \overline{F(s,t)}$$

The expectation (5.6) then takes the form

$$\int d\mu(s^+)d\mu(s^-)\, d\mu(t^+)d\mu(t^-)\; e^{\beta(s^+\cdot s^- + t^+\cdot t^-)} F(s^+,t^+) \\ F^*(s^-,t^-).$$

where

$$d\mu(s^+) = \prod_{\sigma_j \in s^+} d\mu(\sigma_j)$$

The above integral is positive as a consequence of the following lemma when we set $x = (s^+,t^+), y = (s^-,t^-)$.

Lemma 5.3.

$$0 \leq \int \overline{F(x)}\, e^{-(\beta/2)\,(x-y)^2} F(y)d\mu(x)\, d\mu(y) \equiv \| F \|_\beta^2 \tag{5.8}$$

and, for all real vectors h,

$$\left| \int \overline{F(x)}\; e^{-\beta/2(x-y+h)^2} G(y)d\mu(x)d\mu(y) \right| \leq \|F\|_\beta \|G\|_\beta \tag{5.9}$$

Proof. First let $d\mu(x) = \mu(x)dx$. By Fourier transform the left side of (5.9) equals

$$\int (\check{\mu F})(p)(\check{\mu G})(p)\; e^{-\frac{1}{2\beta}p^2} e^{iph}\, dp \tag{5.10}$$

If we set $h = 0$ and $G = F$, (5.10) is positive and (5.8) follows. For $h \neq 0$, we use $|\exp(ihp)| = 1$; then (5.9) follows from the Schwarz inequality. The general case follows by a limiting argument.

Proof of Theorem 5.1. The proof relies on the preceeding lemma and the identity

$$e^{-\beta/2(x-y-h/\beta)^2} = e^{-h^2/2\beta}\, e^{-\beta/2\,(x-y)^2} e^{(x-y)\cdot h}.$$

Let

$$e^{-h_{+-}\cdot(s^+ - s^-)} = \prod_{\{j:j_1=1/2\}} e^{-h_1(j)(\sigma_{j-e_1} - \sigma_j)}$$

$$e^{-h'_{+-}(t^+ - t^-)} = \prod_{\{j:j_1=\ell_1+1\}} e^{-h_1(j)(\sigma_{j-e_1} - \sigma_j)}$$

If we integrate out the σ^+ and σ^- variables we see that

$$Z_\Lambda \langle \exp \sum_\alpha^\nu \sigma(\partial_\alpha h_\alpha)\rangle_\Lambda = Z_\Lambda \langle \exp \sum_\alpha^\nu (\partial^*_\alpha\sigma)(h_\alpha)\rangle_\Lambda \tag{5.11}$$

$$= e^{1/2\beta(\|h_{+-}\|^2 + \|h'_{+-}\|^2)}$$

$$\times \int \prod_{\varepsilon=\pm} d\mu(s^\varepsilon)d\mu(t^\varepsilon)\; F_1(s^+,t^+)K_{h,h'}(s,t)F_2(s^-,t^-)$$

where

$$K_{hh'}(s,t) = \exp-(\beta/2)[(s^+-s^- + \frac{h_{+-}}{\beta})^2 + (t^+ - t^- + \frac{h'_{+-}}{\beta})^2]$$

We can eliminate the h_{+-}, h'_{+-} gradient terms by applying Lemma 5.3 which proves that (5.11) is bounded by

$$e^{+1/2\beta(\|h_{+-}\|^2 + \|h'_{+-}\|^2)} \|F_1\|_\beta \; \|F_2\|_\beta \tag{5.12}$$

The norm $\|F_i\|^2_\beta$, $i = 1,2$, again has the form

$$Z_\Lambda \langle \exp \sum_\alpha^\nu \sigma(\partial_\alpha h_\alpha)\rangle_\Lambda$$

with the h_{+-}, h'_{+-} coupling set equal to zero. Next we note that

$$\|h_{+-}\|_2^2 + \|h'_{+-}\|_2^2 = \sum_{\{j:j_1=1/2,-\ell_1\}} |h_1(j)|^2$$

We now rotate the axis of reflection, (the dashed line in fig. 1) by ± 1, i.e. we shift the origin in the torus Λ. We may then apply the previous estimate to <u>both factors</u> on the r.h.s. of (5.12). Continuing in this manner and applying then the same procedure in the 2-, 3-,..., ν-direction we finally obtain (5.1)

Remarks: The above proof of Theorem 5.1 resembles an elementary proof of the Hölder inequality

$$|\mathrm{Tr}(A_1,\ldots,A_{2n})| \leq \prod_{j=1}^{2n} \{\mathrm{Tr}(A_j^* A_j)^n\}^{1/2n} \qquad (5.13)$$

without use of complex interpolation. A different proof of Theorem 5.1 based on the transfer matrix formalism and the Hölder inequality (of the form (5.13)) has been given in [FSS].

Proof of Theorem 5.2: For simplicity we only prove an upper bound on $\langle A\rangle_\Lambda$, where A is localized in S and is real and reflection invariant. Let $B = \tau_{\vec{n}_1}(A)$, $\vec{n}_1 = (r/2,0,\ldots,0)$, so that B depends only on $\{\sigma_j : j_1 \geq 0\}$. Then by (5.6) with $B_1=B$ and $B_2 = 1$ we have

$$\langle A\rangle_\Lambda = \langle B\rangle_\Lambda \leq \langle(\theta_1 B)B\rangle_\Lambda^{1/2} = \langle A(\tau_{2\vec{n}_1} A)\rangle_\Lambda$$

We translate $A(\tau_{2\vec{n}_1} A)$, so we can again apply the Schwarz inequality. By iteration we see that

$$\langle A\rangle_\Lambda \leq \langle \Pi'(\tau_{\vec{n}} A)\rangle_\Lambda^{r_1/\ell_1}$$

where the product Π' ranges over vectors $\vec{n} = (nr_1,0,\ldots,0)$, $|nr_1| \leq \ell$. By applying the inequality in the other directions we obtain the bound

$$\langle A\rangle_\Lambda \leq \langle \prod_{\tau_{\vec{n}} S \subset \Lambda} (\tau_{\vec{n}} A)\rangle_\Lambda^{\Pi r_\alpha/|\Lambda|}$$

The proof follows by taking the limit $\Lambda \uparrow \mathbb{Z}^\nu$.

6. Absence of long range order in two dimensional systems and the Goldstone theorem

In two dimensions the integral $\int \Delta(p)^{-1} d^2p$ is infinite, and so we can not conclude the existence of long range order or spontaneous magnetization by the methods discussed so far. For models with continuous internal symmetries this is no accident. The following result is related to a theorem of Mermin and Wagner [MW] (see also [Me, DS]) which excludes spontaneous magnetization for the quantum and classical Heisenberg model (more generally for models with continuous internal symmetries) in two dimensions.

Theorem 6.1, [McS]. For β sufficiently large the two point function of the two dimensional plane rotator model satisfies

$$0 \leq \langle \sigma_o \cdot \sigma_x \rangle(\beta) \leq C|x|^{-[(2\pi+\varepsilon)\beta]^{-1}} \tag{6.1}$$

where $\varepsilon \to 0$, as $\beta \uparrow \infty$.

Remark: By correlation inequalities the two point correlation function for the classical Heisenberg model is dominated by the one of the plane rotator; see [D, KPV]. The proof of the theorem is motivated by the calculation

$$Z^{-1} \int_{\infty}^{+\infty} e^{\frac{\beta}{2}\sigma\Delta\sigma} e^{i(\sigma_o - \sigma_x)} d\sigma = e^{-\frac{1}{\beta 2\pi}\log|x|}$$

Proof. We write the expectation in terms of angle variables $\{\theta_i\}$ defined so that $\sigma_i = (\cos\theta_i, \sin\theta_i)$. The two point function then takes the form

$$Z^{-1} \int_{-\pi}^{+\pi} e^{i(\theta_o - \theta_x)} \exp[\beta \sum_{|i-j|=1} \cos(\theta_i - \theta_j)]\pi d\theta_i .$$

The bound follows by a complex translation $\sigma_j \to \sigma_j + ia_j$, where

$$a_j = \beta^{-1}\langle \delta_j, \Delta^{-1}(\delta_o - \delta_x)\rangle_{\ell^2} .$$

This means we deform the paths of integration and use the periodicity of the cos to cancel the contribution from the latteral contours. From the fundamental solution of the two dimensional Laplacean

$$(a_o - a_x) \geq \frac{\beta^{-1}}{\pi} \log|x| ,$$

for large $|x|$, but $|a_j - a_{j'}| \leq$ Const. β^{-1}, provided $|j-j'| = 1$. If we make the above substitutions and use the identity $\cos(\theta + ia) = \cos\theta\cosh a - i\sin\theta\sinh a$ we get the upper bound

$$Z^{-1}\int_{-\pi}^{+\pi} \exp[\beta \sum_{|i-j|=1} \cos(\theta_i - \theta_j)\cosh(a_i - a_j)]\pi d\theta_j \; e^{-(a_o - a_x)}.$$

We express the above sum in the form

$$\sum_{|i-j|=1} \cos(\theta_i - \theta_j) + \cos(\theta_i - \theta_j)[\cosh(a_i - a_j) - 1]$$

$$= \sum_{|i-j|=1} \cos(\theta_i - \theta_j) + \sum_{|i-j|=1} \frac{(a_i - a_j)^2}{2-\varepsilon},$$

where ε can be chosen small for large β. Since

$$\beta \sum_{|i-j|=1} \frac{(a_i - a_j)^2}{2-\varepsilon} = \frac{-\beta}{2-\varepsilon} <a\Delta a> = \frac{a_o - a_x}{2-\varepsilon},$$

we see that

$$<\cos(\theta_o - \theta_\alpha)>(\beta) \leq e^{-(a_o - a_x)/(2-\varepsilon)}$$
$$\leq e^{-[(2\pi + \varepsilon)\beta]^{-1}\log|x|}$$

which completes the proof.

Next we prove a lower bound on the two point function of the classical Heisenberg models. More precisely we derive an upper bound on the exponential decay rate (mass gap) of the two point function.

Theorem 6.2. Let $d\mu(\sigma) = \delta(|\sigma| - 1)d^N\sigma$, $N \geq 2$, $\nu = 2$ and $h = 0$. Then there is a constant $c > 0$ independent of β and N such that, for β sufficiently large,

$$<\sigma_o \cdot \sigma_x>(\beta) \geq \text{Const.}\{\exp -e^{-\frac{\beta}{cN}}|x|\}.$$

Proof. Let $F(p)$ be the Fourier transform of the two point function and set $p = (p_1, p_2)$. Then, by using a transfer matrix formalism and the spectral theorem, one sees that

$$F(p) = c\delta_o(p) + \int_m^\infty \frac{d\rho(a,p_2)}{2(1-\cos p_1)+a}, \tag{6.2}$$

for some $m \geq 0$. Since $N \geq 2$, the previous result gives $c = 0$, so that m is the exponential decay rate of $\langle\sigma_o \cdot \sigma_x\rangle(\beta)$. Using a transfer matrix in the 2-direction and setting $p_1 = 0$ we see that

$$\int d\rho(a,0) \geq \int d\rho(a,p_2)$$

One then shows that $\int d\rho(a,0)$ is expressible as a double commutator of σ with the transfer matrix T; (the argument is similar to the one used to bound the spectral measure $d\rho(a)$ in the Källen-Lehmann representation of the two point function:

$$-\langle[\phi(x),[\phi(y),H]]\rangle = \delta(x-y)\int_m^\infty d\rho(a),$$

where H is the Hamiltonian). This double commutator is bounded in Section 2 and the appendix of [FSS]. We obtain

$$\int d\rho(a,p_2=0) \leq \frac{\text{Const. } N}{\beta} \tag{6.3}$$

Next we use the symmetry of F under the substitution $p_1 \to p_2$, $p_2 \to p_1$, apply (6.2) and (6.3) and use $p_1^2 + p_2^2 \leq 2p_1^2$, for $|p_1| \geq |p_2|$. This yields

$$0 \leq F(p) \leq \frac{\text{Const. } N}{\beta} \frac{1}{p_1^2+p_2^2+m^2} \tag{6.4}$$

Since $(2\pi)^{-\nu}\int F(p)d^2p = \langle\sigma_o \cdot \sigma_o\rangle = 1$, integration of (6.4) yields

$$1 = (2\pi)^{-\nu}\int F(p)d^2p \leq \frac{cN}{\beta}\log m^{-1},$$

for some constant c and β sufficiently large. Hence $m \leq e^{-\frac{\beta}{cN}}$. Since m is the exponential decay rate of the two point function, the proof is complete.

Remarks: This result indicates that the decay rate m is monotone increasing in N and monotone decreasing in β which is in agreement with heuristic predictions. Lower bounds on $\langle\sigma_o \cdot \sigma_o\rangle$ established in [FSS] can be used to extend Theorem 6.2 to very general, classical ferromagnetic spin systems that satisfy Osterwalder-Schrader positivity. Moreover the result also holds in $O(N)$-invariant $(\vec{\phi}\cdot\vec{\phi})^2$ - field theories, where

$$m \leq O(e^{-\text{Const.}\gamma}),$$

and γ is minus the bare mass squared; see [KL].

The last result of this section concerns a lattice version of the Euclidean form [Sy] of the Goldstone theorem [G] which is useful, since it does not require the full structure of relativistic quantum field theory. Let Ω be some finite region in $\mathbb{Z}^\nu$ containing 0, and $\partial\Omega$ all sites in Ω with a nearest neighbor in Ω^c. Let $\vec{\nabla}$ denote the finite difference gradient. We define a "current density"

$$\vec{J}^{\alpha}_x = \sigma^1_x(\vec{\nabla}\sigma^\alpha)_x - \sigma^\alpha_x(\vec{\nabla}\sigma^1)_x, \tag{6.5}$$

$\alpha = 2,\ldots,N$. The interaction $H_\Lambda(\sigma)$ is as in (2.2), and $\langle - \rangle \equiv \langle - \rangle(\beta,h)$ denotes an infinite volume limit of the expectations defined in (2.3).

Theorem 6.3. Let $d\mu$ be some $O(N)$-invariant single spin distribution and suppose that

$$\langle\sigma^\alpha_o\rangle = \delta_{1\alpha}\, M, \quad (M \neq 0).$$

Then

$$\sum_{x\in\partial\Omega} |\langle \vec{J}^\alpha_x\, \sigma^\alpha_o\rangle| \geq \beta^{-1}M, \quad \text{for all} \quad \alpha > 1. \tag{6.6}$$

Remarks: Using Osterwalder-Schrader positivity and a Schwarz inequality one shows

$$|\langle\vec{J}^\alpha_x\, \sigma^\alpha_o\rangle| \leq \langle\vec{J}^\alpha_{x'}\cdot\vec{J}^\alpha_o\rangle^{1/2}\langle\sigma^\alpha_{x''}\,\sigma^\alpha_o\rangle^{1/2},$$

where $|x'| = |x''| = O(|x|)$. Taking $\Omega \to \mathbb{Z}^\nu$ we conclude that

$$|\langle\vec{J}_x\cdot\vec{J}^\alpha_o\rangle\langle\sigma^\alpha_x\,\sigma^\alpha_o\rangle| \geq O(|x|^{-2d+2}),$$

as $|x| \to \infty$, and, as a consequence that $<\sigma_x^\alpha \sigma_o^\alpha>$ is bounded below by a power law decay, whenever $M \neq 0$.

If the lattice spacing can be taken to 0 and yields a Euclidean invariant limit (convergence of the lattice approximation to a Euclidean field theory) then Theorem 6.3 yields the Euclidean version of the Goldstone theorem [Sy] and, for $M \neq 0$, proves the existence of N-1 Goldstone bosons; (see also [ES]). In the case of a two dimensional field theory it proves $M = 0$, and a slightly more general version of Theorem 6.3 yields the absence of symmetry breaking for such theories, (e.g. for the isotropic $(\vec{\phi}\cdot\vec{\phi})_2^2$ -theory; see also [ES, Co]).

Proof of Theorem 6.3: This theorem is an easy corollary of (the finite difference version of) Green's theorem and the following

Lemma. If $d\mu$ is O(N)-invariant then

$$<\sigma_o^1> = -\beta \sum_{x\in\Omega} <(\vec{\nabla}\cdot\vec{J}^\alpha)_x \, \sigma_o^\alpha> ,$$

for all $\alpha > 1$.

Proof: By (6.5)

$$\sum_{x\in\Omega} <(\vec{\nabla}\cdot\vec{J}^\alpha)_x \, \sigma_o^\alpha> = \sum_{x\in\Omega} <\{\sigma_x^1(\Delta\,\sigma^\alpha)_x - \sigma_x^\alpha(\Delta\sigma^1)_x\}\sigma_o^\alpha> \tag{6.7}$$

We temporarily assume that

$$d\mu(\sigma) = e^{-\beta\nu\sigma^2 - V(\sigma)} d^N\sigma, \tag{6.8}$$

where V is some O(N)-invariant, once continuously differentiable function on $\mathbb{R}^N$, and e^{-V} is $d^N\sigma$-integrable. In this case we may apply the standard integration by parts formula ("field equation") in the variables $\{\sigma_x^\alpha\}$ $\{\sigma_x^1\}$ to derive an expression for the r.h.s. of (6.7):

$$\beta<\sigma_x^1(\Delta\sigma^\alpha)_x \quad \sigma_o^\alpha> = -<\sigma_x^1>\delta_{xo} + <\sigma_x^1 \frac{\partial V}{\partial\sigma_x^\alpha} \sigma_o^\alpha>,$$

and

$$\beta<\sigma_x^\alpha(\Delta\sigma^1)_x \quad \sigma_o^\alpha> = <\sigma_x^\alpha \frac{\partial V}{\partial\sigma_x^1} \sigma_o^\alpha>,$$

Since V is $O(n)$-invariant

$$\sigma_x^1 \frac{\partial V}{\partial \sigma_x^\alpha} = \sigma_x^\alpha \frac{\partial V}{\partial \sigma_x^1} ,$$

so that

$$\langle \{\sigma_x^1 (\Delta\sigma^\alpha)_x - \sigma_x^\alpha (\Delta\sigma^1)_x\} \sigma_o^\alpha \rangle = -\frac{1}{\beta} \langle \sigma_x^1 \rangle \, \delta_{xo}$$

This proves the lemma under the assumption (6.8). The general case then follows from a limiting argument.

7. Phase Transitions and Solitons in the Anisotropic $\lambda(\vec{\phi}\cdot\vec{\phi})_2^2$ - Field Theory

In this section we consider a relativistic quantum field model in two space-time dimensions which exhibits a phase transition and a degenerate vacuum, as some coupling constant is varied. The vacuum degeneracy is accompanied by the spontaneous breaking of a discrete symmetry and implies existence of quantum solitons (states with non-zero topological charge).

The model describes a neutral, scalar Bose field $\vec{\phi} = (\phi_1, \phi_2)$ with a quartic self-interaction. The $O(2)$-symmetry is explicitly broken by assigning different bare masses to ϕ_1 and ϕ_2. (An $O(2)$ - symmetry is of course not broken in two space-time dimensions; see Theorem 6.3).

In Subsection 7.1 we establish the existence of a phase transition for this model by constructing a vacuum state which violates the cluster decomposition properties. In Subsection 7.2 we outline the construction of soliton states and derive a lower bound for the mass gap in the soliton sectors.

In the following $\vec{\phi}$ denotes the relativistic or the Euclidean field. The free Euclidean field with bare mass 1 is determined by its Euclidean two point function (Euclidean propagator):

$$\langle \phi_i(x)\phi_j(y) \rangle^o = \delta_{ij} (2\pi)^{-2} \int \frac{e^{i(x-y)p}}{p^2+1} \, d^2p. \tag{7.1}$$

This is the kernel of the operator $\delta_{ij}(-\Delta+1)^{-1}$, where Δ is the Laplacean.

We also consider the free Euclidean field in a periodic box $\Lambda = [-L_1,L_1]\times[-L_2,L_2]$. Its two point function is the kernel of the operator $\delta_{ij}(-\Delta_\Lambda+1)^{-1}$, where Δ_Λ is the Laplacean with periodic boundary conditions at $\partial\Lambda$. The Euclidean vacuum expectation value for the free field in the box Λ is denoted by $\langle - \rangle^o_\Lambda$; $\delta_{ij}(-\Delta_\Lambda+1)^{-1}$ is called its covariance.

An interaction in the region Λ is introduced by means of a Euclidean action

$$V(\Lambda) = \int_\Lambda [\lambda :(\vec{\phi}\cdot\vec{\phi})^2:(x) - \frac{\sigma_1}{2} :\phi_1^2:(x) - \frac{\sigma_2}{2} :\phi_2^2:(x)]d^2x \tag{7.2}$$

Here :-: indicates normal ordering which, for convenience, is done with respect to bare mass 1; the coupling constant λ of the quartic term is positive; the coupling constants σ_1 and σ_2 satisfy $0 \leq \sigma_2 < \sigma_1$. The interacting Euclidean field in the box Λ is determined by cutoff Euclidean Green's functions:

$$\langle \prod_{i=1}^n \phi_{j_i}(f_i)\rangle_\Lambda(\lambda,\vec{\sigma}) \equiv [\langle e^{-V(\Lambda)}\rangle^o_\Lambda]^{-1} \langle \prod_{i=1}^n \phi_{j_i}(f_i)e^{-V(\Lambda)}\rangle^o_\Lambda , \tag{7.3}$$

where $\vec{\sigma} = (\sigma_1,\sigma_2)$, and $f_1,\ldots,f_n$ are arbitrary functions in $C_o^\infty(\Lambda)$. The cutoff vacuum energy density ("pressure") for the system in the box Λ is given by

$$p_\Lambda(\lambda,\vec{\sigma}) \equiv \frac{1}{|\Lambda|} \log \langle e^{-V(\Lambda)}\rangle^o_\Lambda . \tag{7.4}$$

It has been shown in [GRS2] that

$$p(\lambda,\vec{\sigma}) \equiv \lim_{\Lambda\to\mathbb{R}^2} p_\Lambda(\lambda,\vec{\sigma}) \quad \text{exists.} \tag{7.5}$$

In [F1] a Lee-Yang theorem and correlation inequalities due to Dunlop and Newman [DN] have been combined with the (large external field) cluster expansion of [Spl] to construct Euclidean invariant infinite volume limits of the cutoff Green's functions defined in (7.3).

For all n and arbitrary C_o^∞ functions $f_1,\ldots,f_n$

$$\langle \prod_{i=1}^{n} \phi_{j_i}(f_i) \rangle (\lambda, \vec{\sigma}) \equiv \text{"}\lim_{\Lambda \to \mathbb{R}^2}\text{"} \langle \prod_{i=1}^{n} \phi_{j_i}(f_i) \rangle_\Lambda (\lambda, \vec{\sigma}) \qquad (7.6)$$

exists. (The quotation marks indicate that there are really two limits to be taken; see [F1] for more details).

Using standard estimates (exponential $\vec{\phi}$-bounds; see Theorem 5.2) and the Osterwalder-Schrader reconstruction theorem [OS] one can show that the distributions $\{\langle \prod_{i=1}^{n} \phi_{j_i}(x_i) \rangle (\lambda, \vec{\sigma})\}_{n=0}^{\infty}$ are the Euclidean Green's functions of a quantum field theory satisfying all Wightman axioms, with the possible exception of uniqueness of the vacuum; (i.e. these distributions are the analytic continuations of the Wightman distributions, $\{\omega(\prod_{i=1}^{n} \phi_{j_i}(x_i))(\lambda, \vec{\sigma})\}_{n=0}^{\infty}$, to the Euclidean region).

7.1 Existence of phase transitions

Let $\omega = \omega(\lambda, \vec{\sigma})$ denote the physical vacuum state of the theory. The following result asserts that, for a certain choice of the bare couplings λ and $\vec{\sigma}$, ω is degenerate (i.e. violates the cluster properties).

Theorem 7.1. Given $\lambda > 0$ and $\alpha > 0$ (arbitrarily small), there exists a finite constant σ_c such that, with $\vec{\sigma} \equiv (\sigma(1+\alpha), \sigma)$, the vacuum ω is at least two fold degenerate, for all $\sigma > \sigma_c$, and there are at least two clustering (pure phase) vacuum states ω_+ and ω_- which break the $\phi_1 \to -\phi_1$ symmetry of the dynamics.

Remarks.

1. It follows from the cluster expansion [GJS1] that, for σ sufficiently small, the physical vacuum $\omega(\lambda, \vec{\sigma})$ is unique, and the theory has a positive mass gap (and one particle states, [GJS2, Sp2]).
2. Using renormal ordering and scaling one can show that Theorem 7.1 may be reformulated as follows:

Given $\alpha > 0$, choose $\sigma < 1$ such that $1 < \sigma(1+\alpha) \leq 2$; σ is then kept fixed.

Theorem 7.1'. For this choice of $\vec{\sigma}$ there exists $\lambda_c > 0$ such that, for all $0 < \lambda < \lambda_c$, $\omega(\lambda, \vec{\sigma})$ is at least two-fold degenerate.

This form of Theorem 7.1 is technically more convenient. Since $\vec{\sigma}$ is now kept fixed we set $\langle \cdot \rangle (\lambda, \vec{\sigma}) \equiv \langle \cdot \rangle (\lambda)$, $p(\lambda, \vec{\sigma}) \equiv p(\lambda)$. The proof of Theorem 7.1'-which occupies the remainder of Section 7.1 - is based on a version of the Peierls argument due to [GJS3].

Step 1: Reduction of the problem

From [OS] and [F4] we know that Theorem 7.1(') holds if we

can show that the Euclidean expectation $\langle - \rangle(\lambda)$ has <u>long range order</u> (i.e. violates the Euclidean version of the cluster properties). This problem, which is solved below, is similar to proving long range order for the symmetric state of the two dimensional Ising model at sufficiently low temperatures. We now make this analogy more precise.

Step 2: Spin variables and the Peierls argument

In order to formulate the Peierls argument in a field theoretic context we must introduce what corresponds to the spin variables $\{\sigma_i\}_{i \in \mathbb{Z}^2}$ of the Ising model. We cover $\mathbb{R}^2$ with a grid of mesh 1 (parallel to the coordinate axes). A unit square of this grid is specified by the coordinates $j = (j_1, j_2) \in \mathbb{Z}^2$ of its center; $\phi(j)$ denotes the integral of $\phi_1(x)$ over the square centered at j. Let $\chi_\pm$ be the characteristic function of the intervals $[0,\infty)$, $(-\infty,0]$, respectively. We define

$$\begin{aligned} \chi_\pm(j) &= \chi_\pm(\phi(j)), \quad \text{and} \\ \sigma_j &= \chi_+(j) - \chi_-(j) \end{aligned} \tag{7.7}$$

Since σ_j takes the values ± 1 and $\sigma_j^2 = 1$, it is the analogue of an Ising spin. The symmetric state of the Ising model has the property that $\langle \sigma_i \rangle = 0$. Therefore this state has long range order if e.g. $\langle \sigma_j \sigma_o \rangle \not\to 0$, as $|j| \to \infty$. By construction - see (7.3) and (7.6) - the expectation $\langle - \rangle(\lambda)$ has the property that

$$\langle \chi_+(i) \rangle(\lambda) = \langle \chi_-(i) \rangle(\lambda) = 1/2, \tag{7.8}$$

$$\text{hence} \quad \langle \sigma_i \rangle(\lambda) = 0, \text{ for all } i.$$

For $j = (j_o, 0)$, $j_o \in \mathbb{Z}$, it follows from Osterwalder-Schrader positivity that $\langle \sigma_j \sigma_o \rangle(\lambda)$ is positive and monotone decreasing in $|j_o|$. Therefore long range order is equivalent to showing

$$\langle \sigma_j \sigma_o \rangle(\lambda) \geq \delta^2 > 0, \quad \text{for all } j_o. \tag{7.9}$$

Osterwalder-Schrader positivity implies that, for $j = (j_o, 0)$,

$$\langle \chi_\pm(j) \chi_\pm(0) \rangle(\lambda) \geq \langle \chi_\pm(0) \rangle(\lambda)^2 = 1/4 \tag{7.10}$$

Using (7.7) we therefore conclude that (7.9) holds if

$$\langle \chi_+(j)\ \chi_-(0)\rangle(\lambda) \leq 1/4 - \delta^2 , \tag{7.11}$$

for all j_o, or (by translation invariance)

$$\langle \chi_+(-j)\ \chi_-(j)\rangle(\lambda) \leq 1/4 - \delta^2, \tag{7.12}$$

for all $j_o > 0$. By results of [F4], (7.9) implies that there exist at least two clustering (pure phase) states $\langle\!-\!\!\rangle_\pm(\lambda)$ and some $0 < \rho \leq 1/2$ such that

$$M \equiv \langle\sigma_i\rangle_+ (\lambda) = -\langle\sigma_i\rangle_-(\lambda) \geq \delta ,$$

and

$$\langle\!-\!\!\rangle (\lambda) = \rho\langle\!-\!\!\rangle_+(\lambda) + \rho\langle\!-\!\!\rangle_-(\lambda); \tag{7.13}$$

$$+ \text{ positive}.$$

M corresponds to the spontaneous magnetization.

We are left with proving (7.12). Given some $j = (j_o,0)$, let $\tilde{\Lambda} \subset \mathbb{Z}^2$ be a large square containing $\pm j$. Since $\chi_+(k) + \chi_-(k) = 1$, for all $k \in \mathbb{Z}^2$, we have

$$\langle\chi_+(-j)\chi_-(j)\rangle(\lambda) = \tag{7.14}$$

$$\langle \prod_{\substack{k\in\tilde{\Lambda}\\ k\neq\pm j}} [\, \chi_+(k) + \chi_-(k)]\chi_+(-j)\chi_-(j)\rangle(\lambda)$$

We now expand the product on the r.h.s. of (7.14).

Defintion: A configuration c is a function on $\tilde{\Lambda}$ with values in $\{+,-\}$; $\sum_c^{\tilde{\Lambda}}$ denotes a sum over all configurations c such that $c(-j) = +,\ c(j) = -$. Then, by (7.14)

$$\langle\chi_+(-j)\chi_-(j)\rangle(\lambda) = \sum_c^{\tilde{\Lambda}} \langle \prod_{k\in\tilde{\Lambda}} \chi_{c(k)}(k)\rangle(\lambda) \tag{7.15}$$

Since $0 \leq \chi_\pm (k) \leq 1$, all terms on the r.h.s. of (7.15) are positive and are increased by omitting some of the factors in $\prod_{k\in\tilde{\Lambda}}$.

Definition: A contour γ is a connected line in $\mathbb{R}^2$ consisting of sides of unit squares in the grid covering $\mathbb{R}^2$ which has the property that it decomposes $\mathbb{R}^2$ into precisely two disjoint subsets $\Omega_1 \equiv \Omega_1(\gamma) \ni -j$ and $\Omega_2 \equiv \mathbb{R}^2 \setminus \Omega_1 \ni j$. Given a contour γ, $N(\gamma)$ denotes the collection of coordinates of all unit squares with one side in γ. We set $\gamma_{\tilde{\Lambda}} = \gamma \cap \tilde{\Lambda}^{\text{int.}}$ and call it a <u>truncated contour</u>.

Since $c(-j) = +$, $c(j) = -$, we may associate with each configuration c a truncated contour $\gamma(c) \subset \tilde{\Lambda}$ with the property that

(1) for all $(k,k') \in N(\gamma(c))$ with $k \in \Omega_1$, $k' \in \Omega_2$, $c(k) = +$, $c(k') = -$; (7.16)

(2) there exists a connected set $\Omega_c \subset \Omega_1(\gamma(c))$ such that $-j \in \Omega_c$, $c(k) = +$, for all $k \in \Omega_c^c$ and $\partial\Omega_c \supset \gamma(c)$. (7.17)

We now do a resummation on the r.h.s. of (7.15) which gives

$$\langle\chi_+(-j)\chi_-(j)\rangle(\lambda)$$

$$= \sum_{\gamma_{\tilde{\Lambda}}} \sum_{\{c:\gamma(c) = \gamma_{\tilde{\Lambda}}\}} \langle \prod_{k\in\tilde{\Lambda}} \chi_{c(k)}(k)\rangle\ (\lambda) \tag{7.18}$$

$$\leq \sum_{\gamma_{\tilde{\Lambda}}} \langle \prod_{(k,k') \in N(\gamma_{\tilde{\Lambda}})} \chi_+(k)\chi_-(k')\rangle\ (\lambda) \tag{7.19}$$

(7.19) follows from (7.18) by omitting the restriction that $c(-j) = +$, $c(j) = -$ and using that $\chi_+(k) + \chi_-(k) = 1$.

<u>Lemma 7.2</u>: ("Peierls argument") Assume that

$$\langle \prod_{(k,k') \in N(\gamma)} \chi_+(k)\chi_-(k')\rangle\ (\lambda) \leq e^{-K|\gamma|}, \tag{7.20}$$

for some $K > \log 3$; (here $|\gamma|$ denotes the length of γ and is equal to the number of pairs in $N(\gamma)$). Then

$$\begin{aligned} &\langle\chi_+(-j)\chi_-(j)\rangle(\lambda) \\ &\leq \sum_{n=2}^{\infty} \sum_{\{\gamma:|\gamma|=2n\}} \langle \prod_{(k,k') \in N(\gamma)} \chi_+(k)\chi_-(k')\rangle(\lambda) \\ &\leq \sum_{n=2}^{\infty} 2n\ 3^{2n-2}\ e^{-2nK} \end{aligned} \tag{7.21}$$

<u>Proof</u>: By a rather straightforward limiting argument based on (7.20) (see e.g. [F3]) it can be shown that inequality (7.19)

remains true in the limit where $\tilde{\Lambda} = \mathbb{Z}^2$, and

$$\lim_{\tilde{\Lambda} \to \mathbb{Z}^2} \sum_{\gamma_{\tilde{\Lambda}}} \langle \prod_{(k,k') \in N(\gamma_{\tilde{\Lambda}})} \chi_+(k)\chi_-(k') \rangle(\lambda)$$

$$= \sum_{\{\gamma : |\gamma| < \infty\}} \langle \prod_{(k,k') \in N(\gamma)} \chi_+(k)\chi_-(k') \rangle(\lambda) < \infty .$$

If γ is a contour of finite length it must be closed and hence its length is even. The smallest closed contours (two in number) have length 4. Therefore

$$\sum_{\{\gamma : |\gamma| < \infty\}} \langle \prod_{(k,k') \in N(\gamma)} \chi_+(k)\chi_-(k') \rangle(\lambda)$$

$$= \sum_{n=2}^{\infty} \sum_{\{\gamma : |\gamma| = 2n\}} \langle \prod_{(k,k') \in N(\gamma)} \chi_+(k)\chi_-(k') \rangle(\lambda) .$$

Next, the number of contours of length $2n$ separating j from $-j$ is bounded above by

$$\text{card}\{\gamma : |\gamma| = 2n\} \leq 2n\, 3^{2n-2} \tag{7.22}$$

This is a well known estimate which follows from a standard argument; (see e.g. [GJS3, F3]). The lemma now follows from (7.19), (7.20) and (7.22).

Remark: Lemma 7.2 shows that

$$\langle \chi_+(-j)\chi_-(j) \rangle(\lambda) \to 0,$$

as $K = K(\lambda) \to \infty$, uniformly in j. Therefore there exists some K_o such that if inequality (7.20) is true for some $K > K_o$ then

$$\langle \chi_+(-j)\chi_-(j) \rangle(\lambda) \leq \frac{1}{4} - \delta^2, \tag{7.23}$$

for some $\delta = \delta(K) > 0$. We are thus left with proving that

$$K = K(\lambda) > K_o, \quad \text{if} \quad 0 < \lambda < \lambda_c. \tag{7.24}$$

We shall show that $K(\lambda) \to \infty$, as $\lambda \searrow 0$.

Step 3: Estimating the statistical weight of contours

We now turn to the proof of (7.20) and (7.24) and show that $K(\lambda) \to \infty$, as $\lambda \searrow 0$. The basic idea behind this proof is to reduce (7.20) and (7.24) to a "thermodynamic estimate", namely estimates on pressures.

Let $N_1(\gamma)$ be some maximal subset of $N(\gamma)$ with the property that if (k,k') and (ℓ,ℓ') are two different pairs in $N_1(\gamma)$ then $k \neq \ell$ and $k' \neq \ell'$. Obviously

$$\text{card}(N_1(\gamma)) \geq \frac{1}{4}\,\text{card}(N(\gamma)) = \frac{|\gamma|}{4}. \tag{7.25}$$

Since $0 \leq \chi_\pm \leq 1$,

$$\Big\langle \prod_{(k,k') \in N(\gamma)} \chi_+(k)\chi_-(k') \Big\rangle(\lambda)$$

$$\leq \Big\langle \prod_{(k,k') \in N_1(\gamma)} \chi_+(k)\chi_-(k') \Big\rangle(\lambda). \tag{7.26}$$

Let $\chi_\pm^1(x)$ be the characteristic function of the set $\{x: \pm x \geq J\}$, for some finite, positive J (chosen later) and $\chi_\pm^2 = \chi_\pm - \chi_\pm^1$. Then

$$\begin{aligned} \chi_+(k)\chi_-(k') &= \chi_+^1(k)\chi_-^1(k') + \chi_+^1(k)\chi_-^2(k') \\ &\quad + \chi_+^2(k)\chi_-^1(k') + \chi_+^2(k)\chi_-^2(k') \end{aligned} \tag{7.27}$$

We insert this identity into the r.h.s. of (7.26) and expand. The resulting expectations are then bounded above by means of the following elementary

Lemma 7.3:

$$0 \leq \chi_+^1(k)\chi_-^1(k') \leq F_1(k) \equiv e^{-2J} e^{\phi(k) - \phi(k')}$$

$$0 \leq \chi_+^1(k)\chi_-^2(k') \leq F_2(k) \equiv e^{\frac{\beta J^2}{2}\left(1 - \frac{1}{J^2}\phi(k')^2\right)}$$

$$0 \le \chi_+^2(k)\chi_-^1(k') \le F_3(k) \equiv e^{\frac{\beta J^2}{2}\left(1-\frac{1}{J^2}\phi(k)^2\right)}$$

$$0 \le \chi_+^2(k)\ \chi_-^2(k') \le F_4(k) \equiv F_2(k)\,F_3(k),$$

for arbitrary $\beta > 0$; (here $\phi \equiv \phi_1$).

Proof: One verifies easily that, when x and y are real variables,

$$\chi_+^1(x)\chi_-^1(y) \le e^{-2J}\ e^{x-y},$$

$$\chi_+^1(x)\chi_-^2(y) \le \chi_-^2(y) \le e^{\frac{\beta J^2}{2}\left(1-\frac{1}{J^2}y^2\right)}$$

etc.. The lemma now follows from the selfadjointness of $\phi(k)$ and $\phi(k')$. We define $I_\gamma \equiv \{k:(k,k') \in N_1(\gamma)\}$ and note that card $(I_\gamma) \ge |\gamma|/4$. From (7.26), (7.27) and Lemma 7.3 we then obtain

$$\begin{aligned} &< \prod_{(k,k') \in N(\gamma)} \chi_+(k)\chi_-(k')>(\lambda) \\ &\le \sum_{\{p(k)=1,\ldots,4\}} < \prod_{k \in I_\gamma} F_{p(k)}(k)>(\lambda). \end{aligned} \tag{7.28}$$

The r.h.s. of (7.28) can be estimated by means of (a continuum limit version of) Theorems 5.1 and 5.2.

Definition: Let $\Lambda = [-L_1,L_1]\times[-L_2,L_2]$, where L_1 and L_2 are positive integers. We define

$$p_\Lambda^{(\beta)}(\lambda) = \frac{1}{|\Lambda|}\log<e^{-\beta\sum_{j\in\Lambda\cap\mathbb{Z}^2}:\phi(j)^2:}\ e^{-V(\Lambda)}>_\Lambda^o,\ \beta \ge 0\,.$$

One can show that

$$p^{(\beta)}(\lambda) \equiv \lim_{L_1,L_2\to\infty} p_\Lambda^{(\beta)}(\lambda)$$ exists; see [F3]. Next

we define "weights"

$$W_1 \equiv 2J - 4/3, \quad W_2 = W_3 = \frac{1}{2} W_4 = -\frac{\beta J^2}{2} + \frac{\beta}{2} \langle \phi(0)^2 \rangle^o + p(\lambda) - p^{(\beta)}(\lambda) \tag{7.29}$$

Lemma 7.4.

$$\langle \prod_{k \in I_\gamma} F_{p(k)}(k) \rangle(\lambda) \le e^{-\sum_{k \in I_\gamma} W_{p(k)}}$$

Proof. First we note that

$$e^{\phi(k) - \phi(k')} = e^{(\partial_\alpha \phi)(h_\alpha)}, \quad \text{where} \tag{7.30}$$

$\alpha = \alpha(k,k')$ is the direction of $\pm(k - k')$, h_α is a function supported in the union of the unit squares centered at k and k', and $\|h_\alpha\|_2^2 = \frac{2}{3}$. Next

$$e^{\frac{\beta J^2}{2}\left(1 - \frac{1}{J^2}\phi(k)^2\right)} = e^{\frac{\beta J^2}{2} - \frac{\beta}{2}\langle \phi(k^2) \rangle^o} \, e^{-\frac{\beta}{2} :\phi(k)^2:}, \tag{7.31}$$

(and, of course, $\langle \phi(k)^2 \rangle^o = \langle \phi(0)^2 \rangle^o$). We insert (7.30) and (7.31) into the r.h.s. of (7.28) and apply the Schwarz inequality. This gives

$$\langle \prod_{(k,k') \in N(\gamma)} \chi_+(k)\chi_-(k') \rangle(\lambda)$$

$$\le \sum_{\{p(k)=1,\ldots,4\}} \langle \prod_{\substack{k \in I_\gamma \\ p(k)=1}} e^{-4J} e^{2\partial_{\alpha(k,k')}\phi(h_{\alpha(k,k')})} \rangle^{1/2}$$

$$\times \langle \prod_{\substack{k \in I_\gamma \\ p(k)=2,3,4}} F_{p(k)}(k)^2 \rangle^{1/2}.$$

To the first factor on the r.h.s. we apply Theorem 5.1, to the second factor (7.31) and Theorem 5.2. This completes the proof. More details concerning Lemma 7.4 may be found in [F3].

Finally we must choose J and β. A convenient choice for β is

$$\beta = \sigma(1+\alpha)/2. \tag{7.32}$$

We claim that $p^{(\sigma(1+\alpha)/2)}(\lambda)$ is bounded uniformly in λ on any interval of the form $[0,\lambda_o]$, $(0<\lambda_o<\infty)$, whereas (7.33)

$$p(\lambda) \to +\infty, \quad \text{as} \quad \lambda \searrow 0. \tag{7.34}$$

This shows that, for any $J<\infty$,

$$W_i \to +\infty, \quad \text{as} \quad \lambda \searrow 0, \quad i = 2,3,4.$$

Therefore, given an arbitrary $K \varepsilon (o,\infty)$, there exists $\lambda(K) > 0$ such that

$$< \prod_{(k,k') \varepsilon N(\gamma)} \chi_+(k)\chi_-(k')>(\lambda) \leq e^{-K|\gamma|},$$

for all $0 < \lambda \leq \lambda(K)$. By Lemma 7.2 and (7.23) the proof of Theorem 7.1 is now complete, up to the verification of (7.33) and (7.34).

Step 4: Pressure estimates

Lemma 7.5. Estimates (7.33) and (7.34) hold, and

$$<\exp[\frac{\sigma(1+\alpha)}{2}:\phi_1^2:(\Omega) + \frac{\sigma}{2}:\phi_2^2:(\Omega)]>^{\lambda} \to \infty,$$

as $\lambda \searrow 0$.

Proof. We start with proving (7.34), as this estimate is somewhat simpler. Let $\langle\text{—}\rangle^{\lambda}$ denote the expectation constructed in (7.6) in the case where $\sigma_1 = \sigma_2 = 0$. Let Ω be a large square in $\mathbb{R}^2$ and set

$$F_\sigma(\Omega) = \frac{\sigma(1+\alpha)}{2}: \phi_1^2:(\Omega) + \frac{\sigma}{2}:\phi_2^2:(\Omega).$$

By Theorem 5.2,

$$\langle e^{F_\sigma(\Omega)} \rangle^\lambda \leq e^{|\Omega|[p(\lambda,\vec{\sigma})-p(\lambda,0)]} .$$

Standard estimates show that $p(\lambda,0)$ is bounded uniformly in λ on any interval $[0,\lambda_o]$, $\lambda_o < \infty$. Therefore (7.34) follows if we can show that

$$\langle e^{F_\sigma(\Omega)} \rangle^\lambda \to +\infty , \text{ as } \lambda \searrow 0 .$$

Now

$$\langle e^{F_\sigma(\Omega)} \rangle^\lambda = 2\langle \cos h[F_\sigma(\Omega)]\rangle^\lambda - \langle e^{-F_\sigma(\Omega)} \rangle^\lambda$$

$$\geq 2\langle \cos h[F_\sigma(\Omega)]\rangle^\lambda$$

$$- e^{|\Omega|[p(\lambda,-\vec{\sigma})-p(\lambda,0)]} ,$$

and the inequality follows from Theorem 5.2. Again, $p(\lambda,-\vec{\sigma})$ is bounded uniformly in λ on any interval $[0,\lambda_o]$, for fixed $\sigma > 0$, $\alpha > 0$. Hence it suffices to show that

$$\langle \cos h\, F_\sigma(\Omega)\rangle^\lambda = \sum_{n=0}^{\infty} \frac{1}{(2n)!} \langle F_\sigma(\Omega)^{2n}\rangle^\lambda \tag{7.35}$$

tends to $+\infty$, as $\lambda \searrow 0$. But

$$\langle F_\sigma(\Omega)^{2n}\rangle^\lambda \longrightarrow \langle F_\sigma(\Omega)^{2n}\rangle^o , \tag{7.36}$$

as $\lambda \searrow 0$; for all $n < \infty$; (see [GJS 1]). Since $\sigma(1+\alpha) \geq 1$ = <u>bare mass</u> in the covariance of the Gaussian expectation $\langle - \rangle^o$,

$\sum_{n=0}^{N} \frac{1}{(2n)!} \langle F_\sigma(\Omega)^{2n}\rangle^o$ <u>diverges</u>, as $N \to \infty$, provided $|\Omega|$ is large enough; (see also [FSS]). Since all terms in (7.35) are positive and by (7.36), the proof of (7.34) is now complete.

<u>Remark</u>. These estimates also hold for $\lambda(\vec{\phi}\cdot\vec{\phi})^2$ interactions in three space-time dimensions and will be used again in Section 8.

To prove (7.33) recall that

$$p^{(\sigma(1+\alpha)/2)}(\lambda) = \lim_{L_1,L_2\to\infty} \frac{1}{|\Lambda|} \log \left\langle e^{-\frac{\sigma(1+\alpha)}{2} \sum_{j\in\Lambda\cap\mathbb{Z}^2} :\phi(j)^2:} e^{-V(\Lambda)} \right\rangle_\Lambda^\circ . \quad (7.37)$$

Since $\sigma < 1$, the factor $e^{\frac{\sigma}{2} : \phi_2^2 : (\Lambda)}$ in $e^{-V(\Lambda)}$ may be absorbed into the Gaussian expectation $\langle - \rangle_\Lambda^\circ$. We then apply the Schwarz inequality to the r.h.s. of (7.37). This yields the inequality

$$p^{(\sigma(1+\alpha)/2)}(\lambda) \le \frac{1}{2} \{p_1 + p_2 + p(2\lambda,(0,\sigma))\} ,$$

where

$$p_1 = \lim_{\Lambda\to\mathbb{R}^2} \frac{1}{|\Lambda|} \log \left\langle e^{-\sigma(1+\alpha)\{\sum_{j\in\Lambda\cap\mathbb{Z}^2} :\phi(j)^2: - :\phi_1^2:(\Lambda)\}} \right\rangle_\Lambda^\circ \quad (7.38)$$

and

$$p_2 = \lim_{\Lambda\to\mathbb{R}^2} \frac{1}{|\Lambda|} \log \langle e^{\frac{\sigma}{2} : \phi_2^2 : (\Lambda)} \rangle_\Lambda^\circ .$$

Since $\sigma < 1$ (= bare mass in the covariance of $\langle - \rangle_\Lambda^\circ$), $p(2\lambda, (0,\sigma))$ is bounded uniformly in λ on any interval $[0,\lambda_\circ]$. The r.h.s. of (7.38) are Gaussian expectations that can be estimated explicitly: Standard formulas show that, for $\sigma < 1$, $\sigma(1+\alpha) \le 2$, p_1 and p_2 are finite.

Q.E.D.

Lemma 7.5 completes the proof of Theorem 7.1(').

<u>Remarks</u>. The following additional results on phase transitions in two dimensional models have been proven (or a proof has been outlined) in [F3]:
Existence of phase transitions for the $P(\phi)_2$-models, where P is an "almost even", positive polynomial; for the <u>pseudo-scalar Yukawa</u> - and the <u>massive sine-Gordon model</u> in two space-time dimensions. The phase transition in the pseudo-scalar Yukawa model

is of interest because it is accompanied by the breaking of a space-time symmetry, namely parity.

7.2 Soliton sectors in the anisotropic $\lambda(\vec{\phi}\cdot\vec{\phi})^2_2$ - theory.

In this subsection we review results showing that the spontaneous breaking of a discrete internal symmetry in two dimensional theories such as the anisotropic $\lambda(\vec{\phi}\cdot\vec{\phi})^2_2$ - model gives rise to soliton sectors. These are Poincare-invariant eigenspaces of the conserved, topological charge

$$\vec{Q} = \int d\vec{x} \left(\frac{\partial}{\partial\vec{x}}\vec{\phi}\right)(\vec{x},t) \tag{7.39}$$

corresponding to a non-zero eigenvalue; (see [F1, F3] for a mathematical theory of soliton sectors, and [Co2] for a general review and references). We derive a lower bound on the mass gap in the energy-momentum spectrum on the soliton sectors in terms of a "surface tension"; (this is a result of [F2']; an estimate on the "surface tension" appears in [BFG]). The analysis of soliton sectors is important for scattering theory.

By Theorem 7.1 we know that, for a proper choice of the bare couplings $(\lambda,\vec{\sigma})$, there are two clustering (pure phase) vacuum states $\omega_+ \neq \omega_-$ which break the $\phi_1 \longrightarrow -\phi_1$ symmetry. Let ρ_π be defined by the equation

$$\rho_\pi(P(\phi_1,\phi_2)) = P(-\phi_1,-\phi_2) , \tag{7.40}$$

where P is an arbitrary polynomial of the identity 1 and the fields $\{\phi_1(f),\phi_2(g) : f,g \text{ in } \mathscr{S}(\mathbb{R}^2)\}$. Then ω_+, ω_- may be constructed such that

$$\omega_+ = \omega_- \circ \rho_\pi , \quad \omega_- = \omega_+ \circ \rho_\pi . \tag{7.41}$$

If $(\lambda,\vec{\sigma})$ is chosen such that the Gaussian expansion about mean field theory developed in [GJS4] converges then

$$\omega_+(\phi_1) = -\omega_-(\phi_1) = \phi_c > 0 . \tag{7.42}$$

(In general (7.13) need however not imply (7.42), although this is expected).

The basic idea behind our construction of soliton sectors is to find states ω_s and $\omega_{\bar{s}}$ that interpolate between ω_+ and ω_- such that in space-time regions very far to the left $\omega_{s(\bar{s})}$ resembles $\omega_{-(+)}$ and in regions far to the right it resembles $\omega_{+(-)}$; $\omega_s(\omega_{\bar{s}})$ is then called a soliton (anti-soliton) state.

The precise definition of such states is somewhat more complicated. The following remarks summarize some standard terminology and some facts used in the construction of $\omega_{s(\bar{s})}$. (They are formulated in a somewhat cavalier way but can easily be made precise). Let $\mathscr{P}$ denote the algebra generated by $\{1, \phi_1(f), \phi_2(g) : f,g \text{ in } \mathscr{S}(\mathbb{R}^2)\}$. Let ρ be some state on $\mathscr{P}$, (i.e. a positive, linear functional on $\mathscr{P}$ with $\rho(1) = 1$) . With the pair $(\rho,\mathscr{P})$ there is associated a Hilbert space $\mathscr{H}_\rho$, and a unit vector $\Omega_\rho \in \mathscr{H}_\rho$ such that

$$\rho(P) = (\Omega_\rho, P\Omega_\rho) \text{ , for all } P \in \mathscr{P}\text{, and}$$

$$\mathscr{H}_\rho = \text{norm closure of } \{P\Omega_\rho : P \in \mathscr{P}\} .$$

Given some test function $\mathbf{f}$ in $\mathscr{S}(\mathbb{R})^2$ and a proper Poincaré transformation $\xi \equiv (\Lambda,\mathbf{a})$, we set $f_\xi(x) \equiv f(\Lambda^{-1}(\mathbf{x}-\mathbf{a}))$. We define a mapping $\tau_\xi : \mathscr{P} \longrightarrow \mathscr{P}$ (a "* automorphism of the *algebra $\mathscr{P}$") by the equations

$$\tau_\xi(1) = 1 \text{ , } \tau_\xi(\phi_i(f)) = \phi_i(f_\xi) \text{ ; } i = 1,2 . \tag{7.43}$$

<u>Definition C</u>:

A state ρ on the algebra $\mathscr{P}$ is said to be <u>Poincaré-covariant</u> iff there exists a continous, unitary representation $\mathcal{U}_\rho$ of the proper Poincaré group on the Hilbert space $\mathscr{H}_\rho$ such that, on some common (Poincaré-invariant) dense domain $\mathscr{D} \subset \mathscr{H}_\rho$

$$\mathcal{U}_\rho(\xi) P \, \mathcal{U}_\rho(\xi)^* = \boldsymbol{\tau}_\xi(P) \quad ,$$

for all $P \in \mathscr{P}$.

We say that ρ satisfies the relativistic spectrum condition iff the joint spectrum of the generators of the space-time translation subgroup $\{\mathcal{U}_\rho(1,a): a \in \mathbb{R}^2\}$ is contained in the forward light cone $\overline{V}_+$.

Let $\mathcal{O}$ be some open region in $\mathbb{R}^2$. We define $\mathcal{P}(\mathcal{O})$ to be the subalgebra of $\mathcal{P}$ generated by

$$\{1, \phi_1(f), \phi_2(g): f,g \text{ in } \mathcal{S}(\mathbb{R}^2);\ \text{supp } f,\ \text{supp } g \subset \mathcal{O}\}\,.$$

A diamond $\overline{\mathcal{O}}$ is defined as the intersection of an open forward and an open backward light cone. Given a diamond $\overline{\mathcal{O}}$ we let $\sim\overline{\mathcal{O}}$ denote its causal complement (all space-time points space-like separated from $\overline{\mathcal{O}}$). In two dimensions $\sim\overline{\mathcal{O}}$ is the union of two wedge regions: $\sim\overline{\mathcal{O}} = \overline{\mathcal{O}}_L \cup \overline{\mathcal{O}}_R$, where $\overline{\mathcal{O}}_L$ opens to the left and $\overline{\mathcal{O}}_R$ to the right.

We are now prepared for a precise definition of soliton states.

Definition S:

A state ω_s on $\mathcal{P}$ is called a soliton state iff

(S1) there is a diamond $\overline{\mathcal{O}}$ such that

$$\omega_s(P) = \omega_-(P)\,, \text{ for all } P \in \mathcal{P}(\overline{\mathcal{O}}_L)$$

$$\omega_s(P) = \omega_+(P)\,, \text{ for all } P \in \mathcal{P}(\overline{\mathcal{O}}_R)\,;$$

(S2) ω_s is Poincaré-covariant and satisfies the relativistic spectrum condition.

The definition of $\omega_{\bar{s}}$ is obtained by interchanging + and − in (S1).

The Hilbert space and the cyclic vector associated with a soliton state ω_s and the algebra $\mathcal{P}$ are denoted by $\mathcal{H}_s$ and

Ω_s, respectively. In the same way $\mathcal{H}_{\bar{s}}$ and $\Omega_{\bar{s}}$ are defined.

From (7.42), (S1) and (S2) we conclude that

$$\vec{Q}\underline{\psi} = \pm\, 2(\phi_c, 0)\, \underline{\psi}\ , \quad \text{for all} \quad \underline{\psi} \in \begin{cases} \mathcal{H}_s \\ \mathcal{H}_{\bar{s}} \end{cases} ,$$

where $\vec{Q}$ is the charge introduced in (7.39). A special case of this is

$$\lim_{\vec{x} \to \pm\infty} \omega_s(\phi_1(\vec{x}, t)) = \pm\, \phi_c\ ,$$

for arbitary t. Thus the function $\omega_s(\phi_1(\vec{x}, t))$ resembles the soliton solution of the classical field equation (explaining the nomenclature).

Starting from the vacuum state ω_+ we now construct explicitly soliton states ω_s and $\omega_{\bar{s}}$ satisfying (S1) and (S2). This is achieved by composing ω_+ with a suitable automorphism of $\mathcal{P}$ (of the type of a Bogoliubov transformation). Let $\phi_i(\vec{x}) \equiv \phi_i(\vec{x},0)$ $\pi_i(\vec{x}) \equiv (\frac{\partial}{\partial t}\, \phi_i)(\vec{x},t=0)$, $i = 1,2$, denote the time 0-fields and their conjugate momenta. It has been shown in [F2] that the following equations uniquely define a mapping $\rho_\theta : \mathcal{P} \longrightarrow \mathcal{P}$; (a *automorphism of $\mathcal{P}$):

$$\begin{aligned} \rho_\theta(\phi_1(\vec{x})) &= \cos\, \theta(\vec{x})\cdot\phi_1(\vec{x}) + \sin\, \theta(\vec{x})\cdot\phi_2(\vec{x}) \\ \rho_\theta(\phi_2(\vec{x})) &= -\sin\, \theta(\vec{x})\cdot\phi_1(\vec{x}) + \cos\, \theta(\vec{x})\cdot\phi_2(\vec{x}) \end{aligned} \tag{7.44}$$

+ identical equations with ϕ_i replaced by π_i, $i = 1,2$. Here θ is some C^∞ function on $\mathbb{R}$ (a space-dependent angle) satisfying

$$\left.\begin{aligned} &\text{supp}(\tfrac{d}{d\vec{x}}\, \theta)(\vec{x}) \quad \text{is compact, and} \\ &\lim_{\vec{x}\to +\infty} \theta(\vec{x}) = 0\ , \quad \lim_{\vec{x}\to -\infty} \theta(\vec{x}) = \pi\ ; \end{aligned}\right\} \tag{7.45}$$

(this is called the "soliton condition").

We set $\overline{\theta}(\vec{x}) \equiv \theta(-\vec{x})$.

We then define ω_s and $\omega_{\overline{s}}$ by

$$\omega_s \equiv \omega_+ \circ \rho_\theta \ , \ \omega_{\overline{s}} = \omega_+ \circ \rho_{\overline{\theta}} \ , \text{ where} \tag{7.46}$$

$$\omega_+ \circ \rho_\theta(P) = \omega_+(\rho_\theta(P)) \ , \text{ for all } P \in \mathcal{P} \ , \text{ etc.}$$

Let # denote s or $\overline{s}$.

Theorem 7.6, [F2, F3].

(1) $\omega_\#$ is an (anti-) soliton state on $\mathcal{P}$, i.e. $\omega_\#$ satisfies properties (S1) and (S2).

(2) The Hilbert space $\mathcal{H}_\#$ and the representation $\mathcal{U}_\#$ of the proper Poincaré group on $\mathcal{H}_\#$ are independent of the choice of $\theta(\overline{\theta})$ as long as θ satisfies the soliton condition (7.45).

(3) The spectrum of the energy-momentum operator $(H_\#, P_\#)$ (defined as the infinitesimal generator of the space-time translations $\{\mathcal{U}_\#(a) : a \in \mathbb{R}^2\}$ is purely continuous, i.e. $\mathcal{H}_\#$ does not contain any vacuum state.

(4) $\vec{Q}\,\underline{\psi} = 2(\phi_c, 0)\underline{\psi}$, for all $\underline{\psi} \in \mathcal{H}_s$,

$\vec{Q}\,\underline{\psi} = -2(\phi_c, 0)\underline{\psi}$, for all $\underline{\psi} \in \mathcal{H}_{\overline{s}}$.

We refer the reader to [F2, F3] for the proof.

Next we propose to analyze the spectrum of $(H_\#, P_\#)$. We intend to show that, under suitable conditions on $(\lambda, \vec{\sigma})$, the mass gap on $\mathcal{H}_\#$ is strictly positive. Since space-reflections intertwine the theories on $\mathcal{H}_s$ and on $\mathcal{H}_{\overline{s}}$, it suffices to do this analysis for (H_s, P_s).

Let $\langle - \rangle_+ \equiv \langle - \rangle_+ (\lambda, \vec{\sigma})$ be the Euclidean expectation corresponding to the vacuum state ω_+ . We assume now that $\langle - \rangle_+$ is constructed by means of the Gaussian expansion about mean field theory developed in [GJS4]. This assumption can be verified for a suitable range of $(\lambda, \vec{\sigma})$. Let $L \times T$ denote (the characteristic function of) the rectangle $[-\frac{L}{2}, \frac{L}{2}] \times [-\frac{T}{2}, \frac{T}{2}]$ and $V(L \times T)$ the Euclidean action of the anisotropic $\lambda(\vec{\phi}\cdot\vec{\phi})_2^2$ - theory introduced in (7.2).

Let $\chi_{+\ell}$ be the characteristic function of the rectangle $[\frac{L}{2}, \frac{\ell}{2}] \times [-\frac{T}{2}, \frac{T}{2}]$, and $\chi_{-\ell}$ the one of $[-\frac{\ell}{2}, -\frac{L}{2}] \times [-\frac{T}{2}, \frac{T}{2}]$, with $0 < L < \ell < \infty$.

Finally, $\langle - \rangle^{o}_{\ell \times T}$ denotes the free field Euclidean (Gaussian) expectation with periodic b.c. at the boundary of $[-\frac{\ell}{2}, \frac{\ell}{2}] \times [-\frac{T}{2}, \frac{T}{2}]$ introduced in 7.1.

Definition:

Choose $\mu > 0$ such that $\phi_1 \equiv \frac{\mu}{4\pi}$ is the value at which the classical Goldstone potential $V_{class.}(\phi_1, \phi_2 = 0)$ takes a minimum. Let

$$\tau(T,L,\ell) \equiv -\frac{1}{T} \log \frac{\langle e^{-V(L\times T)} e^{\mu\phi_1(\chi_{+\ell}-\chi_{-\ell})} \rangle^{o}_{\ell}}{\langle e^{-V(L\times T)} e^{\mu\phi_1(\chi_{+\ell}+\chi_{-\ell})} \rangle^{o}_{\ell\times T}}, \qquad (7.47)$$

$$\tau(T,L) \equiv \lim_{\ell\to\infty} \tau(T,L,\ell), \text{ and}$$
$$\tau(L) \equiv \lim_{T\to\infty} \tau(T,\ell) \qquad (7.48)$$

(and the limits can be shown to exist by standard arguments; [Si, GJS4]).

Obviously $\tau(L)$ is the analogue of the surface tension of a ferromagnetic system in the strip $-\frac{L}{2} \leq \vec{x} \leq \frac{L}{2}$; (+ boundary conditions at $\vec{x} = \frac{L}{2}$, and - boundary conditions at $\vec{x} = -\frac{L}{2}$).

We set $$\bar{\tau} \equiv \overline{\lim_{L\to\infty}}\ \tau(L) \qquad (7.49)$$

$\bar{\tau}$ is the surface tension in the thermodynamic limit; presumably $\bar{\tau} = \lim_{L\to\infty} \tau(L)$ exists. We shall see that $\bar{\tau}$ is always non-negative.

Let m_s denote the <u>mass gap</u> on the soliton sector $\mathcal{H}_s$.

<u>Theorem 7.7</u>, [F2'].

$$m_s \geq \bar{\tau} \tag{7.50}$$

<u>Remarks</u>.

1) Presumably $m_s = \bar{\tau} = \lim_{L\to\infty} \tau(L)$, but we have at present no proof for this equation. The surface tension is given purely in terms of Euclidean path space integrals; (see (7.47)-(7.49)). It is not hard to see that it is a rigorous version of formal expressions for the mass of the quantum soliton that one finds in the physics literature; ([Co 2] and refs. given there). Theorem 7.7 connects our rigorous ("C* algebra"-) construction of soliton sectors with Euclidean techniques; (Euclidean path integrals).

2) Heuristic considerations, based on (7.47)-(7.49) give

$$\bar{\tau} = O(\lambda^{-1}) \text{ , for small } \lambda \text{ ,} \tag{7.51}$$

(with $\vec{\sigma}$ kept fixed, such that [GJS4] applies). This estimate, as well as details of the proof of Theorem 7.7 appear in [BFG].

3) Fairly convincing heuristic arguments indicate that the mass spectrum on $\mathcal{H}_s$ has an <u>upper gap equal to the mass gap m of the vacuum sector</u>, i.e. the spectrum of the mass operator $(H_s^2 - P_s^2)^{1/2}$ on $\mathcal{H}_s$ is as follows:

Fig. 2

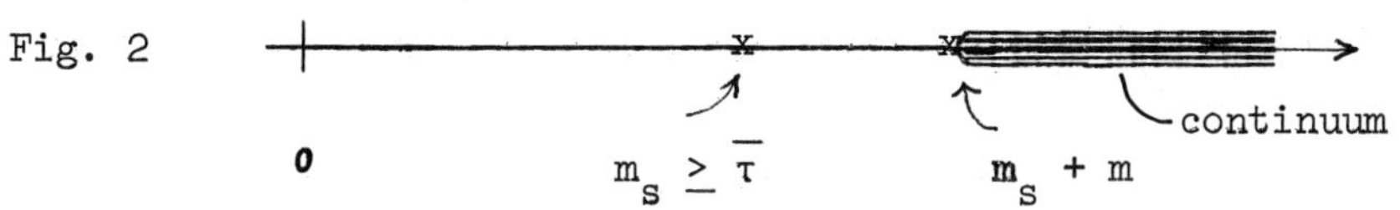

(<u>no</u> spectrum at 0).

If these arguments are correct there are <u>one particle states</u> in the (anti-) soliton sector, the quantum (anti-) solitons. These are identical particles, but have opposite $\vec{Q}$-charge. A space reflection transforms a soliton into an anti-soliton and conversely; [F2]. By (7.51) these particles are very heavy, for small λ .

One can now do Haag-Ruelle scattering theory; (see §4 of [F2]). The total number of quantum solitons and anti-solitons in any scattering state in a vacuum sector is <u>even</u>, and in a scattering state in a soliton sector it is <u>odd</u>; ([F2], §4).

Proof of Theorem 7.7.

We first derive an expression for the mass gap m_s .

Lemma 7.8.

$$m_s = \inf \text{spec } H_s = - \lim_{t\to\infty} \frac{1}{t} \log(\Omega_s, e^{-tH_s}\Omega_s)$$

Note that, by [F2], m_s is independent of the choice of the angle θ in (7.44) as long as θ satisfies the soliton condition (7.45). For the r.h.s. of Lemma 7.8 this is a priori not obvious. Before we prove Lemma 7.8 we state a second result that we shall need.

Definition: Let $\vec{\sigma} = (\sigma(1+\alpha), \sigma)$, $\partial \equiv \frac{\partial}{\partial \vec{x}}$. We define (formally)

$$\begin{aligned}\delta V_{[0,t]} = \int d\vec{x} \int_0^t ds\{ &-(\partial\theta)(\vec{x}) j_1(\vec{x},s) \\ &+ \frac{1}{2}(\partial\theta)^2(\vec{x}) :\vec{\phi}\cdot\vec{\phi}: (\vec{x},s) \\ &+ \frac{\alpha\sigma}{2}\sin^2(\theta(\vec{x}))[:\phi_2^2:(\vec{x},s) - :\phi_1^2:(\vec{x},s)] \\ &+ \frac{\alpha\sigma}{2}\sin(2\theta(\vec{x}))\phi_1(\vec{x},s)\phi_2(\vec{x},s)\},\end{aligned} \tag{7.52}$$

where

$$j_1(x) = \phi_1(x)(\partial\phi_2)(x) - (\partial\phi_1)(x)\phi_2(x). \tag{7.53}$$

Lemma 7.9, [F2'].

$$(\Omega_s, e^{-tH_s}\Omega_s) = \langle e^{-\delta V_{[0,t]}} \rangle_+ ,$$

where $\langle - \rangle_+ \equiv \langle - \rangle_+ (\lambda, \vec{\sigma})$ is the Euclidean expectation corresponding to the vacuum state ω_+ .

Remarks. As they stand the above definition and Lemma 7.9 are somewhat formal, because j_1 is ultraviolet singular. A precise version of Lemma 7.9 involves removing an ultraviolet and a space-time cutoff; (see [BFG]).

A heuristic proof of Lemma 7.9 follows easily from an explicit (though formal) calculation of $\rho_\theta(H)$ - with H the formal Hamiltonian of the $\lambda(\vec{\phi}\cdot\vec{\phi})_2^2$ - theory - and the Feynman-Kac formula.

Proof of Lemma 7.8, given Lemma 7.9:

The main idea is as follows: Let G be some bounded (continuous) functional of the Euclidean field $\vec{\phi}$, G_t the translate of G by the vector $(0,t)$, (i.e. $G_t(\vec{\phi}) = G(\vec{\phi}_t)$, where $\vec{\phi}_t(\vec{x},s) \equiv \vec{\phi}(\vec{x}, s+t))$ and $G_\vartheta(\vec{\phi}) \equiv G(\vec{\phi}_\vartheta)$, where $\vec{\phi}_\vartheta(\vec{x},s) = \vec{\phi}(\vec{x},-s)$. Let

$$||G||_\infty = \sup_{\vec{\phi}} |G(\vec{\phi})| .$$

Clearly

$$\inf \operatorname{spec} H_s \leq - \lim_{t\to\infty} \frac{1}{t} \log(\Omega_s, e^{-tH_s}\Omega_s)$$

$$= - \lim_{t\to\infty} \frac{1}{t} \log\langle e^{-\delta V[0,t]}\rangle_+ ,$$

by the spectral theorem and Lemma 7.9. Using the Osterwalder-Schrader reconstruction theorem [OS] one shows that

$\inf \operatorname{spec} H_s$

$$= \inf_{\{G:||G||_\infty<\infty\}} - \lim_{t\to\infty} \frac{1}{t} \log|\langle \overline{G}_\vartheta\, e^{-\delta V[0,t]} G_t\rangle_+|$$

$$\geq \inf_{\{G:||G||_\infty<\infty\}} - \lim_{t\to\infty} \frac{1}{t} \log \{\langle e^{-\delta V[0,t]}\rangle_+ \quad ||\overline{G}_\vartheta||_\infty ||G_t||_\infty\}$$

$$= - \lim_{t\to\infty} \frac{1}{t} \log \langle e^{-\delta V[0,t]}\rangle_+ ,$$

and the last equation follows from the fact that $||\overline{G}_\vartheta||_\infty = ||G_t||_\infty = ||G||_\infty$; (see [F4] for related results). This completes the proof of Lemma 7.8.

Proof of Theorem 7.7, continued.

Estimate 1: From the spectral theorem and Hölder's inequality we conclude that $\frac{1}{t} \log(\Omega_s, e^{-tH_s}\Omega_s)$ is monotone increasing in t. This combined with Lemmas 7.9 and 7.8 shows that, given $\varepsilon > 0$, there is some $t_o < \infty$ such that

$$m_s + \varepsilon \geq -\frac{1}{t}\log \langle e^{-\delta V_{[0,t]}}\rangle_+ \quad , \tag{7.54}$$

for all $t \geq t_o$.

In order to estimate the r.h.s. of (7.54) we need some more detailed information about the Euclidean expectation $\langle - \rangle_+$. Such information is supplied by the expansion of [GJS4]. Let A be some polynomial in $\{1, \phi_1(f), \phi_2(g) : f, g \text{ in } C_o^\infty\}$. We define

$$\langle A\rangle_+(T,L,\ell) \equiv \frac{\langle A\, e^{-V(L\times T)}\, e^{\mu\phi_1(\chi_{+\ell}+\chi_{-\ell})}\rangle^o_{\ell\times T}}{\langle e^{-V(L\times T)}\, e^{\mu\phi_1(\chi_{+\ell}+\chi_{-\ell})}\rangle^o_{\ell\times T}} \quad , \tag{7.55}$$

where μ is as in (7.47).

The proof of the following three facts are simple; (see e.g. [Si]):

$$\langle A\rangle_+(L,\ell) \equiv \lim_{T\to\infty} \langle A\rangle_+(T,L,\ell) \tag{7.56}$$

$$\langle A\rangle_+(T,L) \equiv \lim_{\ell\to\infty} \langle A\rangle_+(T,L,\ell) \tag{7.57}$$

$$\langle A\rangle_+(L) \equiv \lim_{\ell\to\infty} \langle A\rangle_+(L,\ell) \tag{7.58}$$

$$= \lim_{T\to\infty} \langle A\rangle_+(T,L) \; ,$$

exist, for arbitrary A .

From the strong results of [GJS4] one obtains that, for a suitable choice of λ and $\vec{\sigma}$,

$$\langle A\rangle_+ \equiv \lim_{L\to\infty} \langle A\rangle_+(L) \tag{7.59}$$

exists, for arbitrary A , and the expectation $\langle - \rangle_+$ is Euclidean invariant, satisfies exponential clustering (from which one infers positivity of the mass gap m on the vacuum sector; [GJS4]) and

$$\langle \phi_1\rangle_+ > 0 \tag{7.60}$$

By a limiting argument

$$<e^{-\delta V_{[0,t_o]}}>_+ \;=\; \lim_{L\to\infty} <e^{-\delta V_{[0,t_o]}}>_+(L). \tag{7.61}$$

Estimate 2:

From Estimate 1, (7.54) and (7.61) we conclude that, given $\varepsilon > 0$, there exists $L_o < \infty$ such that

$$\begin{aligned} m_s + 2\varepsilon &\geq -\frac{1}{t_o}\log <e^{-\delta V_{[0,t_o]}}>_+ + \varepsilon \\ &\geq -\frac{1}{t_o}\log <e^{-\delta V_{[0,t_o]}}>_+(L), \end{aligned} \tag{7.62}$$

for all $L \geq L_o$.

Since $<\!-\!>_+(L)$ is invariant under translations in the time direction,

$$<e^{-\delta V_{[0,t_o]}}>_+(L) \;=\; <e^{-\delta V_{[-t_o/2,\,t_o/2]}}>_+(L).$$

Using Osterwalder-Schrader positivity of $<\!-\!>_+(L)$ in the time direction and the Schwarz inequality one shows, (see Section 5, proof of Theorem 5.2),

$$\begin{aligned} &\frac{1}{t_o}\log <e^{-\delta V_{[-t_o/2,\,t_o/2]}}>_+(L) \\ &\leq \lim_{t\to\infty}\frac{1}{2t}\log <e^{-\delta V_{[-t,t]}}>_+(L). \end{aligned}$$

A simple transfer matrix argument, (see [Si]; also [F3]), shows that

$$\begin{aligned} &\lim_{t\to\infty}\frac{1}{2t}\log <e^{-\delta V_{[-t,t]}}>_+(L) \\ &= \lim_{T\to\infty}\frac{1}{T}\log <e^{-\delta V_{[-T/2,\,T/2]}}>_+(T,L). \end{aligned} \tag{7.63}$$

Next

$$<e^{-\delta V[-T/2,T/2]}>_+(T,L)$$

$$= \lim_{\ell\to\infty} \left\langle e^{-\delta V[-T/2,T/2]}\right\rangle_+(T,L,\ell)$$

$$= \lim_{\ell\to\infty} \frac{\left\langle e^{-\delta V[-T/2,T/2]} e^{-V(L\times T)} e^{\mu\phi_1(\chi_{+\ell} + \chi_{-\ell})}\right\rangle^o_{\ell\times T}}{\left\langle e^{-V(L\times T)} e^{\mu\phi_1(\chi_{+\ell} + \chi_{-\ell})}\right\rangle^o_{\ell\times T}} \tag{7.64}$$

We now replace the Euclidean field $\vec{\phi}$ in the numerator on the r.h.s. of (7.64) by a field $\vec{\tilde{\phi}}$, defined by the equations

$$\left.\begin{aligned} \phi_1(\vec{x},t) &= \cos\theta(\vec{x})\,\tilde{\phi}_1(\vec{x},t) - \sin\theta(\vec{x})\,\tilde{\phi}_2(\vec{x},t) \\ \phi_2(\vec{x},t) &= \sin\theta(\vec{x})\,\tilde{\phi}_1(\vec{x},t) + \cos\theta(\vec{x})\,\tilde{\phi}_2(\vec{x},t)\,, \end{aligned}\right\} \tag{7.65}$$

where $-T/2 \leq t \leq T/2$, and θ is the angle introduced in (7.44), (7.45). We then obtain the equation

$$<e^{-\delta V[-T/2,T/2]}>_+(T,L) = \lim_{\ell\to\infty} \frac{\left\langle e^{-V(L\times T)} e^{\mu\phi_1(\chi_{+\ell}-\chi_{-\ell})}\right\rangle^o_{\ell\times T}}{\left\langle e^{-V(L\times T)} e^{\mu\phi_1(\chi_{+\ell}+\chi_{-\ell})}\right\rangle^o_{\ell\times T}} \tag{7.66}$$

To prove this equation, note that the transformation (7.65) is the <u>inverse</u> of (7.44), applied to Euclidean fields, (see (7.52), Lemma 7.9) and that there are <u>no</u> boundary terms at $t = \pm T/2$, because we have introduced <u>periodic boundary conditions</u> at $t = \pm T/2$; (in the proof "free b.c." at $\vec{x} = \pm \ell/2$ are more convenient).

Combining now the definition (7.48) of $\tau(L)$ with Estimate 2, (7.62) and with (7.63), (7.64) and (7.66) we obtain <u>Estimate 3</u>:

$$m_s + 2\varepsilon \geq -\frac{1}{t_o} \log <e^{-\delta V[0,t_o]}>_+(L)$$

$$\geq \lim_{t\to\infty} - \frac{1}{2t} \log \langle e^{-\delta V_{[-t,t]}} \rangle_+ (L)$$

$$= \lim_{T\to\infty} - \frac{1}{T} \log \langle e^{-\delta V_{[-T/2,T/2]}} \rangle_+ (T,L) \tag{7.67}$$

$$= \lim_{T\to\infty} \tau(T,L) = \tau(L) \ , \text{ for all } \ L \geq L_o \ .$$

If we now let $L \to \infty$ in (7.67) and use that $\varepsilon > 0$ may be chosen arbitrarily small we obtain the assertion of Theorem 7.7.

Q.E.D.

Further details concerning this proof and (7.51) and a comparison with more heuristic proposals may be found in [BFG]. (In principle our methods extend to more interesting models, such as scalar QED in three space-time dimensions, once these models are constructible).

8. Symmetry Breaking in $(\vec{\phi}\cdot\vec{\phi})^2_3$ Quantum Field Theories.

In this section we establish symmetry breaking for N-component, $(N = 1,2,3)$, ϕ^4 field theories in three space-time dimensions. We consider the interaction in a region $\Lambda \subset R^3$ defined by

$$V(\Lambda) = \int_\Lambda [\lambda :(\vec{\phi}\cdot\vec{\phi})^2:(x) - \frac{\sigma}{2} :\vec{\phi}\cdot\vec{\phi}:(x) - \mu\phi_1(x)]dx + \text{c.t.} \tag{8.1}$$

It is technically convenient to normal order :-: in (8.1) with respect to zero bare mass. The symbol c.t. denotes an O(N)-symmetric mass counter term <u>which depends only on λ</u> plus a vacuum energy counterterm.

<u>Theorem 8.1</u>. Let $\lambda > 0$ and let Λ be a cube whose sides are of length ≥ 1. There is a constant K such that

$$\langle e^{-V(\Lambda)} \rangle_o \leq e^{K|\Lambda|} \ . \tag{8.2}$$

Moreover if $\mu \neq 0$ and $f_i \in \mathcal{S}(\mathbb{R}^3)$

$$\lim_{\Lambda \uparrow \mathbb{R}^3} \frac{< \prod_{i=1}^{m} \phi_{j_i}(f_i) e^{-V(\Lambda)} >_o}{< e^{-V(\Lambda)} >_o} \equiv < \prod_{i=1}^{m} \phi_{j_i}(f_i) > (\lambda, \sigma, \mu) \tag{8.3}$$

exists and satisfies the Osterwalder-Schrader axioms including clustering. Recall that $<\cdot>^o$ denotes the free field expectation of bare mass 1.

The theorem is based on the phase space cell expansion of Glimm and Jaffe [GJ4] who established (8.2). For small λ and σ Feldman-Osterwalder [FO] and Magnen-Sénéor [MS] showed that the limit (8.3) exists and defines a quantum field model with a mass gap. Feldman-Osterwalder [FO2] have also established this for large μ. If $N = 1, 2$ one can then construct a theory for all $\lambda > 0, \sigma, \mu$ using correlation inequalities. Fröhlich [Fl] has established the theorem for $N = 3$ using Lee-Yang methods, provided $\mu \neq 0$. The limits $\mu \uparrow 0$ and $\mu \downarrow 0$ are denoted by $< \cdot > (\lambda, \sigma, 0\pm)$ respectively. The main result of this section is

Theorem 8.2, [FSS]. Fix $\sigma > 1$ and $N = 1, 2$, or 3. Then

$$\lim_{\lambda \downarrow 0} <\phi_1> (\lambda, \sigma, 0+) = \infty , \tag{8.4}$$

which implies the presence of symmtery breaking. For an even state ($\mu=0$; $N=1,2$) $\lim_{|x| \to \infty} <\vec{\phi}(0) \cdot \vec{\phi}(x)> (\lambda, \sigma, 0) = c > 0$, for all $\lambda \in (0, \lambda_c)$.

Remarks. 1. The restriction on N is needed only to construct Euclidean invariant states.

2. By scaling and renormal ordering we obtain (8.4) in other regions of the coupling constant space, e.g. fixed λ and $\sigma \uparrow \infty$.

3. For $N = 2$, Dunlop and Newman [DN] have shown

$$<\phi_2(0)\phi_2(x)>^2 \leq \text{Const.}\{<\phi_1(0)\phi_1(x)> - <\phi_1(0)>^2\}, |x| \to \infty, \tag{8.5}$$

which means that the transverse two point function decays at least as fast as the square root of the parallel, truncated two point function. Thus ϕ_1 couples the vacuum to two Goldstone Bosons in the regime where $<\phi_1>(\lambda,\sigma,0+) > 0$.

The proof of Theorem 4.2 follows from the Källen-Lehmann representation and the following lemma.

Lemma 8.3. Fix $\sigma > 1$. Then, for small μ ,

$$<:\vec{\phi}\cdot\vec{\phi}:(0)> = \frac{1}{|\Omega|}\int_\Omega <:\vec{\phi}\cdot\vec{\phi}:(x)> dx \geq b(\lambda), \quad (8.6)$$

where $b(\lambda) \to \infty$, as $\lambda \downarrow 0$, uniformly in $\mu \in (-1,1)$.

Proof of Theorem 8.2. The proof is similar to that of 3.1. By a Lee-Yang argument, [S1, DN],

$$\lim_{|x|\to\infty} <\phi_i(0)\phi_i(x)> (\lambda,\sigma,\mu) - <\phi_i(0)>^2(\lambda,\sigma,\mu) \longrightarrow 0 ,$$

for $\mu \neq 0$. The Källen-Lehmann representation shows that

$$<:\phi_i(0)\phi_i(x):> - <\phi_i(0)>^2 = \int \left[\frac{e^{-m|x|}}{|x|} - \frac{1}{|x|}\right] d\sigma_i(m)+c , \quad (8.7)$$

where $\int d\sigma_i(m) = 1$. Note that $c = 0$, for $\mu \neq 0$. As $|x| \to 0$, we see that

$$b(\lambda) \leq \frac{1}{|\Omega|} \int_\Omega <:\vec{\phi}\cdot\vec{\phi}:(x)> dx \leq <\phi_1(0)>^2 , \quad (8.8)$$

for all $\mu \neq 0$. Equation (8.4) now follows by letting μ tend to zero and then letting $\lambda \downarrow 0$. To establish long range order for the symmetric state $(<\vec{\phi}>=0)$, we note that in (8.7) the integrand is always negative. Hence $<:\vec{\phi}\cdot\vec{\phi}:(0)> > 0$ implies $c > 0$. This completes the proof of the theorem.

Proof of Lemma 8.3. Let Ω be a large cube and define

$$:\phi^2:(\Omega) = \int_\Omega :\vec{\phi}\cdot\vec{\phi}:(x)\ d\ x .$$

$$\exp\{p\ (\lambda,\sigma,\mu)\} = \lim_{\Lambda\uparrow R^3} [<e^{-V(\Lambda)}>_0]^{1/|\Lambda|} \quad (8.9)$$

By Theorem 5.2 (generalized by taking the lattice spacing to zero)

$$<e^{-(\sigma/2):\phi^2:(\Omega)}> (\lambda,\sigma,\mu) <e^{(\sigma/2):\phi^2:(\Omega)}> (\lambda,0,\mu)$$

$$\leq \exp[p(\lambda,0,\mu) - p(\lambda,\sigma,\mu) + p(\lambda,\sigma,\mu) - p(\lambda,0,\mu)]|\Omega| = 1 . \tag{8.10}$$

By Jensen's inequality and (8.10),

$$0 \leq \exp[-(\sigma/2) <:\phi^2:(\Omega)> (\lambda,\sigma,\mu)] \leq <e^{-(\sigma/2):\phi^2:(\Omega)}> (\lambda,\sigma,\mu)$$

$$\leq [<e^{+(\sigma/2):\phi^2:(\Omega)}> (\lambda,0,\mu)]^{-1} .$$

By Lemma 7.4 the right hand side tends to 0, as $\lambda \downarrow 0$. To complete the proof, take logarithms.

Next we prove the <u>existence of a critical $(\phi^4)_3$ theory</u>, $(N=1)$, i.e. <u>a field theory with zero mass and no long range order</u>. Let us fix λ small and set $\mu = 0$. We define

$$m(\sigma) = -\lim_{|x|\to\infty} \frac{\log<\phi(0)\cdot\phi(x)>(\lambda,\sigma)}{|x|} ,$$

which is the mass in the single phase region, see [Si]. The critical bare mass is defined by

$$\sigma_c = \sup\{\sigma|m(\sigma) > 0\} .$$

By Theorem 8.2 we know that $\sigma_c < \infty$. We define the critical theory by

$$\lim_{\sigma\uparrow\sigma_c} < \prod_i \phi(x_i) > (\lambda,\sigma) \equiv < \prod_i \phi(x_i) >_c .$$

<u>Theorem 8.4</u> [GJ 3, Ba, McR]. The state $< \cdot >_c$ is a clustering state of zero mass. The mass $m(\sigma)$ is a continuous, monotone decreasing function of σ on the interval $(-\infty, \sigma_c)$.

<u>Sketch of the proof</u>. By Griffiths inequality the mass is a monotone function of σ. The continuity of $m(\sigma)$ relies on the

estimate

$$dm_V(\sigma)^2/d\sigma \leq Z \leq 1 \qquad (8.11)$$

where Z is the residue of the one particle pole, and m_V is the finite volume, periodic mass.

As explained in the lectures of Glimm and Jaffe, the bound follows from correlation inequalities. Since the finite volume masses $m_V(\sigma)$ are easily shown to be smooth in σ, the physical mass $m(\sigma)$, which can be shown to be $= \lim_{V\to\infty} m_V(\sigma)$, is continuous and σ_c is the first point at which it vanishes. To show that there is no long range order we use the Källen-Lehmann representation to note that, for all $\sigma < \sigma_c$,

$$<\phi(0)\phi(x)> (\lambda,\sigma) \leq (4\pi|x|)^{-1} \; ; \qquad (8.12)$$

hence the bound holds in the limit. The bound (8.12) implies uniform upper bounds on all the n-point Euclidean Green's functions, (a consequence of GHS inequalities).

Note that this argument is special to three or more dimensions (because in one or two dimensions there is no uniform upper bound of the form (8.12)).

As discussed in the lectures of Glimm and Jaffe, inequality (8.11) implies that, for $\phi^4_{2,3}$ in the single phase region, there exist stable one particle states for all values of σ for which

$$\frac{dm(\sigma)^2}{d\sigma} > 0 \; .$$

One can use the Lee-Yang theorem to prove that the physical mass of the $\lambda(\vec{\phi}\cdot\vec{\phi})^2 + \mu\phi_1$ theory in two or three space-time dimensions is strictly positive, for $\mu \neq 0$ and $N = 1,2$ or 3 components; see [GRS] (N=1) and [F1] (N=2,3).

Similar results hold for the Ising model in three or more dimensions: The Ising equilibrium state at the critical temperature has no long range order (in fact is clustering), but correlations do not decay exponetially. This follows easily from the results of [McR], the fact that $<\sigma_i\sigma_j>(\beta) \geq 0$ is increasing in β and the FSS uniform upper bound on its Fourier transform (for $\beta < \beta_c$). See also [McR].

References

[B] Baker, G.: Self interacting boson quantum field theory and the thermodynamic limit in d dimensions. J. Math. Phys. 16, 1324, (1973).

[BFG] Bellissard, J., Fröhlich, J., Gidas, V. To appear.

[Co] Coleman, S.: There are no Goldstone bosons in two dimensions. Commun. math. Phys. 31, 259-264, (1973).

[Co2] Coleman, S.: Classical lumps and their quantum descendents; 1975 Erice Lectures.

[D] Dunlop, F.: Correlation inequalities for Multicomponent rotators. Commun. math. Phys. 49, 247-57, (1976).

[DLS] Dyson, F., Lieb, E. and Simon, B. to appear in J. Stat. Phys.

[DN] Dunlop, F., Newman, C.: Multicomponent field theories and classical rotators. Commun. math. Phys. 44, 223-235, (1975).

[DS] Dobrushin, R. L., Shlosman, S. B.: Absence of breakdown of continuous symmetry in two-dimensional models of statistical physics. Commun. math. Phys. 42, 31-40, (1975).

[ES] Ezawa, H., Swieca, A.: Spontaneous breakdown of symmetry and zero mass states. Commun. math. Phys. 5, 330-336, (1967).

[F1] Fröhlich, J.: In "Les Méthodes Mathématiques de la Théorie Quantique des Champs"; Editions du C.N.R.S., Paris 1976.

[F2'] Fröhlich, J.: Unpublished report, ZiF-University of Bielefeld, 1976.

[F3] Fröhlich, J.: In "Current Problems in Elementary Particle & Mathematical Physics", P. Urban, (ed.), Springer-Verlag, Wien-New York, 1976.

[F4] Fröhlich, J.: The Pure phases, the irreducible quantum fields. Ann. of Phys. 97, 1-54, (1976).

[FO] Feldman, J., Osterwalder, K.: The Wightman axioms and the mass gap for weakly coupled $(\phi^4)_3$ quantum field theories. Ann. Phys., 97, (1976).

[FO2] Feldman, J., Osterwalder, K. : To appear.

[FSS] Fröhlich, J., Simon, B. and Spencer, T.: Infrared bounds, phase transitions and continuous symmetry breaking. Commun. math. Phys. 50, 79-95, (1976).

[GJ1] Glimm, J. and Jaffe, A.: On the approach to the critical point. Ann. Inst. H. Poincaré, 22, 109-122, (1975).

[GJ2] Glimm, J. and Jaffe, A.: Three-particle structure of ϕ^4 interactions and the scaling limit. Phys. Rev. D11, 2816-2827, (1975).

[GJ3] Glimm, J. and Jaffe, A.: Critical exponents and elementary particles. Rockefeller preprint, 1976.

[GJ4] Glimm, J. and Jaffe, A.: Positivity of the ϕ^4_3 Hamiltonian. Fortschritte der Physik, 21, 327-376, (1973).

[GJS1] Glimm, J., Jaffe, A. and Spencer T.: In "constructive quantum field theory", G. Velo and A. Wightman, (eds.). Springer-Verlag, Berlin-Heidelberg-New York, (1973).

[GJS2] Glimm, J., Jaffe, A. and Spencer, T.: Ann.of Math. 100, 585-632, (1974).

[GJS3] Glimm, J., Jaffe. A. and Spencer, T.: Phase transitions for ϕ^4_2 quantum fields. Commun. math. Phys. 45, 203-216, (1975).

[GJS4] Glimm,J., Jaffe, A. and Spencer, T.: An expansion about mean field theory. To appear Ann. of Phys.

[G] Goldstone, J.: Nuovo Cimento 19, 154, (1961).

[GRS1] Guerra, F., Rosen, L., Simon, B.: The $P(\phi)_2$ euclidean quantum field theory as classical statistical mechanics. Ann. Math. 101, 111-259,(1975).

[GRS2] Guerra, F., Rosen, L., Simon, B.: Nelson's symmetry and the infinite volume behavior of the vacuum in $P(\phi)_2$. Commun. math. Phys. 27, 10-22 (1972).

[Gr] Griffiths, R.: Phase transitions. In "Statistical mechanics and quantum field theory",C.De Witt and R.Stora, (eds.), Gordon and Breach, New York-London, 1970.

[Gu] Guerra, F.: In Proceedings of the Bielefeld Symposium, L. Streit, (ed.), Bielefeld, 1975.

[K] Kac, M.: On applying mathematics: Reflections and examples. Quart. Appl. Math. 30, 17-29, (1972).

[KL] Klein, A., Landau, L.: In "Les Méthodes Mathématiques..."; see ref. [F1].

[KPV] Kunz, H., Pfister, Ch.-Ed., Vuillermot, P.A.: Inequalities for some classical spin vector models. Phys. Lett 54A, 428, (1975).

[Me] Mermin, N.D.: Absence of ordering in certain classical systems. J. Math. Phys. 8, 1061-1064,(1967).

[MR] McBryan, O. and Rosen, J. To appear.

[McS] McBryan, O. and Spencer, T. On the decay of correlations for O(N) symmetric rotators. To appear.

[MS] Magnen, J., Sénéor, R.: The infinite volume limit of the ϕ_3^4 model: Ann. Inst. Henri Poincaré, (1976).

[OS] Osterwalder,K. and Schrader, R.: Axioms for Euclidean Green's functions. Commun math. Phys. 42, 281-305, (1975).

[Pa] Park, Y. M.: In "Current Problems...." (see ref. [F3] and refs. given there.

[Pe] Peierls, R.: Proc. Cambridge Philos. Soc. 32, 477, (1936).

[PS1] Pirogov, S. A., Sinai, Ya.G.: Phase transitions of the first kind for small perturbations of the Ising model. Funct. Anal. Appl. 8, 21-25, (1974).

[PS2] Pirogov, S. A., Sinai, Ya.G.: Phase diagrams of classical lattice systems. Th. and Math. Phys. 26, 39,(1976).

[S1] Simon, B. and Griffiths, R.: The ϕ_2^4 field theory as a classical Ising model. Commun. math. Phys. 33, 145-164, (1973).

[Si] Simon B.: The $P(\phi)_2$ Euclidean (Quantum) Field Theory, Princeton University Press, Princeton, 1973. Moreover: Correlation inequalities and the mass gap in $P(\phi)_2$, II. Uniqueness of the vacuum in a class of strongly coupled theories. Ann. Math. 101, 260-267, (1975).

[Sp1] Spencer, T.: Commun. math. Phys. 39, 63-76, (1974).

[Sp2] Spencer, T.: Commun. math. Phys. 44, 143-164, (1975).

[St] Stanley, H.: Phys. Rev. 176, 718, (1968).

[Sy] Symanzik, K.: Commun. math. Phys. 6, 288, (1967).

CONTINUOUS SYMMETRY BREAKDOWN IN 2+ε DIMENSION : THE NON-LINEAR σ MODEL

J. Zinn-Justin

Centre d'Etudes Nucléaires
Saclay, France

I. INTRODUCTION

In these lectures we shall briefly discuss the existence and structure of spontaneous symmetry breaking and phase transitions at low dimensions. Around four dimensions and above, these questions can be understood, as we have learned from K.G. Wilson[1] through the use of renormalization group techniques, applied to an expansion around mean field theory. The solutions to the renormalization group equations have been generated under the form of the Wilson-Fisher[2] $\varepsilon = 4-d$ expansion[3]

At lower dimensions the situation was until recently not so clear. In one euclidean dimension (quantum mechanics with a finite number of degrees of freedom) it was known that spontaneous symmetry breaking (S.S.B) was impossible . It should be remarked that nevertheless the situation of discrete symmetries and continuous symmetries is somewhat different. For continous symmetries, S.S.B. is forbidden by the would be Goldstone bosons which express that even classically when the ground state is degenerated, it costs no energy to go from one ground state to another. Therefore perturbation theory in the broken phase, breaks down at first order in $\hbar$. In the case of discrete symmetries the situation is more subtle. In the case of degenerate classical minima, perturbation theory around one classical minimum exists. It is the tunnelling between the different vacua, which is an effect of order $\exp -a/\hbar$, which is responsible of the absence of phase transition. A way of characterizing this tunneling is to calculate the quantity $\exp-\ell H$, where H is the hamiltonian of the system. For ℓ large, $\exp-\ell H$ is a projector on the ground state. In the configuration representation, matrix elements of $\exp-\ell H$

can be written under the form of a path integral :

$$\langle x|\exp-\ell H|x'\rangle = \int_{q(o)=x}^{q(\ell)=x'} [dq(\tau)] \exp -\mathcal{A}[q(\tau)] \qquad (1)$$

where $\mathcal{A}$ is the euclidean action :

$$\mathcal{A}[q(\tau)] = \int_0^\ell [\frac{1}{2}\dot{q}^2(\tau)+ V(q(\tau))]\, d\tau \qquad (2)$$

If we calculate the path integral through the steepest descent method, the integral is dominated for ℓ large, by classical trajectories $q_{cl}(\tau)$ going from one classical minimum to another .

Taking the standard example :

$$V(q) = -\frac{1}{2}q^2 + \frac{\lambda^2}{4}q^4 \quad . \qquad (3)$$

we see that the classical solutions are

$$q_{cl}(\tau) = \pm\frac{1}{\lambda} \qquad (4)$$

$$q_{cl}(\tau) = \pm\frac{1}{\lambda}\, \text{th}\, \frac{(\tau-\tau_o)}{\sqrt{2}} \quad . \qquad (5)$$

The fact that the asymptotic values of $q_{cl}(\tau)$ are only $\pm\frac{1}{\lambda}$, corresponds to the fact that, for ℓ large, $\exp-\ell H$ projects onto two states $|+\rangle$ and $|-\rangle$ for which the wave functions are concentrated around $x = \pm\frac{1}{\lambda}$. The solution $q_{cl}(\tau) = \pm\frac{1}{\lambda}$ gives the matrix elements $\langle +|\exp-\ell H|+\rangle$ and $\langle -|\exp-\ell H|-\rangle$ which are equal. The solution $q_{cl}(\tau) = \pm\frac{1}{\lambda}\,\text{th}\,\frac{(\tau-\tau_o)}{\sqrt{2}}$ contributes to the transition matrix element $\langle +|\exp-\ell H|-\rangle$. At leading order one has :

$$\langle +|\exp-\ell H|-\rangle = \langle +|\exp-\ell H|+\rangle \exp -\frac{2\sqrt{2}}{3\lambda^2} \qquad (6)$$

The fact that, in the large ℓ limit, the ratio :

$$\frac{\langle +|\exp-\ell H|-\rangle}{\langle +|\exp-\ell H|+\rangle} \sim \exp\left\{ [\mathcal{A}[q_{cl}(\tau)= \pm\frac{1}{\lambda}] - \mathcal{A}[q_{cl}(\tau)= \pm\frac{1}{\lambda}\,\text{th}\,\frac{\tau}{\sqrt{2}}]\right\} \qquad (7)$$

has a finite limit, shows $|+\rangle$ and $|-\rangle$ are not eigenstates of H, and that the 2×2 matrix $\langle\pm|H|\pm\rangle$ has to be diagonalized, so that the symmetry is not broken. The solution $q_{cl}(\tau)=\pm\frac{1}{\lambda}\,\text{th}\,\frac{\tau}{\sqrt{2}}$ which has a finite euclidean action, is the simplest example of an instanton [4].

Now in higher dimensions, the problem of S.S.B. has been first solved approximatively by A.M. Polyakov[5] using approximate renormalization group equations, and A.A. Migdal [6] using approximate recursion formulae on a lattice. We are not going to repeat their arguments here, but rather look for generalized instantons, which will give us a qualitative idea of what is happening, and then study a field theory with a continuous O(n) the symmetry; the non linear σ-model[7]

The use of low temperature series expansion, renormalization group equations, and of a $\varepsilon = d-2$ expansion, will allow us [8] to describe quantitatively the structure of phase transitions for continuous symmetries, around dimension two.

II. SPONTANEOUS SYMMETRY BREAKING AT LOW DIMENSIONS: QUALITATIVE ARGUMENTS

We shall now give some arguments which will indicate when, in low dimensions, in field theoretical models for example, spontaneous symmetry breaking is possible. We shall study an arbitrary dimension d, d being the total number of dimensions, space + time.

A - Discrete Symmetries

We shall discuss the problem on the path integral representation for exp-ℓH. The matrix elements are expressed in terms of classical fields $\varphi(\underset{\sim}{x})$, where $\underset{\sim}{x}$ is a vector in (d-1) dimensions

$$<\varphi(\underset{\sim}{x})|\exp-\ell H|\varphi'(\underset{\sim}{x})> = \int_{\varphi(\underset{\sim}{x},0)=\varphi(\underset{\sim}{x})}^{\varphi(\underset{\sim}{x},\ell)=\varphi'(\underset{\sim}{x})} [d\varphi(\underset{\sim}{x},t)]\exp-\mathcal{A}[\varphi] \tag{8}$$

$$A[\varphi(\underset{\sim}{x},\tau)] = \int_0^\ell d\tau\, d\underset{\sim}{x} \left\{ \frac{1}{2}[\partial\mu\, \varphi(\underset{\sim}{x},\tau)]^2 - \frac{1}{2}\varphi^2(x,t) + \frac{\lambda^2}{4}\varphi^4(\underset{\sim}{x},\tau)\right\} \tag{9}$$

Because, in the large ℓ limit, exp-ℓH becomes a projector on a 1 or 2 dimensional space (depending on the fact that the symmetry is spontaneously broken or not) it is not necessary to calculate all matrix elements of exp-ℓH. It is sufficient to calculate the matrix elements of this operator in a 2-dimensional subspace invariant under the symmetry φ in $-\varphi$.

We shall take $\varphi(\underset{\sim}{x})$ and $\varphi'(\underset{\sim}{x})$ constant and equal to $\pm 1/\lambda$.

We shall also calculate in a finite volume of linear size ℓ. The leading saddle points are :

$$\varphi(\underset{\sim}{x},\tau) = \pm \frac{1}{\lambda} \qquad\qquad \varphi(\underset{\sim}{x},\tau) = \pm \frac{1}{\lambda} \,\mathrm{th}\, \frac{(\tau-\tau_o)}{\sqrt{2}} \tag{10}$$

The calculation is almost identical to the calculation of the 1-dimensional case. The only difference comes from the integration on x, which gives an additional factor ℓ^{d-1}.

$$<+|\exp-\ell H|+> = <-|\exp-\ell H|-> \sim \exp-\ell^{d} E(\lambda^2) \tag{11}$$

$$<+|\exp-\ell H|-> \sim <+|\exp-\ell H|+> \quad \exp - \frac{2\sqrt{2}}{3\lambda^2}\,\ell^{d-1} \tag{12}$$

Therefore, for $d > 1$, in the large ℓ limit, the 2×2 matrix becomes diagonal, and the ground state is degenerate.

It is easy to verify that this is a general feature of models with discrete symmetries. The symmetry can be spontaneously broken in any dimension larger than one, i.e. for systems with an infinite number of degrees of freedom [9]

B - Continuous Symmetries

We have already seen in one dimension that it appeared more difficult to break the continuous symmetries than the discrete one. This will be confirmed in higher dimensions. For simplicity we shall take the example of the O(n) symmetry.

$$<\varphi_i(\underset{\sim}{x})|\exp-\ell H|\varphi'_i(\underset{\sim}{x})> = \int_{\underset{\sim}{\varphi}(x,o)=\underset{\sim}{\varphi}(x)}^{\underset{\sim}{\varphi}(x,\ell)=\underset{\sim}{\varphi}'(x)} [d\underset{\sim}{\varphi}(x,\tau)]\exp-A$$

$$A[\underset{\sim}{\varphi}(x,\tau)] = \int d\tau\, d\underset{\sim}{x} \left\{ \frac{1}{2}\sum_i [(\partial\mu\varphi_i(x,\tau))^2 - \varphi_i^2(x,\tau)] + \frac{\lambda^2}{4}[\sum_i \varphi_i^2(x,\tau)]^2 \right\} \tag{13}$$

We shall take here the boundary conditions :

$$\begin{aligned} \varphi_i(\underset{\sim}{x}) &= S_i \\ \varphi'_i(\underset{\sim}{x}) &= S'_i \end{aligned} \tag{14}$$

with

$$\sum_i S_i^2 = \sum_i S_i'^2 = \frac{1}{\lambda^2} \tag{15}$$

We shall call θ the angle between $\underset{\sim}{S}$ and $\underset{\sim}{S}'$.

$$(\underset{\sim}{S}, \underset{\sim}{S}') = \theta \tag{16}$$

The solution which minimizes the action, corresponds to a vector $\underset{\sim}{\varphi}(x,\tau)$ belonging to the plane $(\underset{\sim}{S}, \underset{\sim}{S}')$ of length $\frac{1}{\lambda}$. If α is the angle between $\underset{\sim}{S}$ and $\underset{\sim}{\varphi}$:

$$\alpha = (\underset{\sim}{S}, \underset{\sim}{\varphi}) . \tag{17}$$

Then we have :

$$\alpha = \Theta \frac{\tau}{\ell} \tag{18}$$

Because the symmetry is continuous, the solution which minimizes the action, has a derivative of order $1/\ell$. So $(\partial_\mu \underset{\sim}{\varphi})^2$ is of order θ^2/ℓ^2. As a consequence :

$$\langle \underset{\sim}{S} | \exp - \ell H | \underset{\sim}{S}' \rangle \sim \exp - \frac{1}{\lambda^2} \theta^2 \ \ell^{d-2}. \tag{19}$$

A spontaneous symmetry breaking is only possible for a dimension d larger then 2 [10]. In dimension 2, the solution (18) is an instanton which corresponds to the tunnelling effect between wave functions concentrated around vectors pointing towards different directions. As mentioned in the introduction, these features about phase transitions have first pointed out by A.M. Polyakov [5] , using approximate renormalization group equations, and A.A. Migdal [6] using approximate recursion formulae on a lattice.

Through its recursion formulae, Migdal has also pointed out an intringuing similarity between global symmetries in d dimension, and gauge symmetries in 2d dimension.

III. A FIELD THEORETICAL MODEL : THE NON-LINEAR σ-MODEL

In order to study in a more detailed way the structure of S.S.B. in the case of continuous symmetries, we shall examine a field theoretical model, the non-linear σ-model. It is easy to show [8] that this model is also the continuous limit, at low temperature, of a lattice model of Statistical Mechanics, a spin model with short range positive interaction and continuous O(n) symmetry whose partition function is given by :

$$Z = \int \Pi\, \delta(\underset{\sim}{S}_i^2 - 1)\, d\underset{\sim}{S}_i \exp - \frac{1}{T}\mathcal{H}(S)$$

$$\mathcal{H}(S) = \sum_{ij} V_{ij}\, \underset{\sim}{S}_i \cdot \underset{\sim}{S}_j \tag{20}$$

The potential V_{ij} is short range (for instance nearest neighbor coupling) and positive. The indices i, j label lattice sites.

The spin correlation functions are defined by :

$$G^{(N)}_{\alpha_1 \ldots \alpha_N}(i_1, \ldots i_N) = < S_{i_1}^{\alpha_1} S_{i_2}^{\alpha_2} \ldots S_{i_N}^{\alpha_N} > \tag{21}$$

where α_k are the component indices of the vectors S_{i_k}.

A - The non linear σ-model

If one is only interested in long range correlations, i.e, all the lattice sites $i_1, \ldots i_N$ are largely separated, and at low temperature, it is equivalent to calculate the euclidean (with imaginary time) Green functions of the continuous model whose euclidean action reads :

$$A(S) = \frac{1}{2}[\partial\mu\, \underset{\sim}{S}(x)]^2 \text{ with } \sum_\alpha [S^\alpha(x)]^2 = 1 . \tag{22}$$

The generating function of the euclidean Green's functions is given by the path integral :

$$Z(K) = \int [d\underset{\sim}{S}(x)\, \delta[\underset{\sim}{S}^2(x) - 1]] \exp - \frac{1}{t} A(S) + \frac{1}{t}\int \underset{\sim}{K}(x)\underset{\sim}{S}(x)dx , \tag{23}$$

We shall be interested by the euclidean dimension 2 and the neighborhood of dimension 2. We shall calculate integral (23) again by the steepest descent method as a power series in t. A set of saddle points is given by $\underset{\sim}{S}(x)$ constant and pointing towards a fixed direction. But we have already seen that we cannot expand around one of these saddle points only, but we have to consider all of them. Furthermore we have to put the system in a large but finite box. Indeed it is easy to see that the naive perturbation expansion around one of these saddle points lead to I.R. divergences of the form :

$$\int \frac{d^d p}{p^2} = \infty \quad \text{for } d \leqslant 2 \quad .$$

Infortunately the correct calculation in a box is not very simple. So we shall do something somewhat perverse and break explicitly but in a soft way, the symmetry by adding a source $\int \underset{\sim}{h}\underset{\sim}{S}(x)dx$ to the lagrangian, $\underset{\sim}{h}$ being some constant vector. In the magnetic language, h is a uniform magnetic field. In presence of this source, $\underset{\sim}{S}(x)$ has a vacuum expectation value in the direction of $\underset{\sim}{h}$, and the Goldstone modes disappear.

We shall call $\sigma(x)$ the component of $\underset{\sim}{S}(x)$ along $\underset{\sim}{h}$ and $\underset{\sim}{\pi}(x)$ the component of $\underset{\sim}{S}(x)$ orthogonal to $\underset{\sim}{h}$. In the same way we shall call $\underset{\sim}{J}(x)$ the component of $\underset{\sim}{K}(x)$ orthogonal to $\underset{\sim}{h}$.

$$Z(J) = \int [d\underset{\sim}{\pi}(x)(1-\underset{\sim}{\pi}^2(x))^{-1/2}]\exp-\frac{1}{t}\left\{[A(\underset{\sim}{\pi},\sigma)-h\int\sigma(x)dx] - \int \underset{\sim}{J}(x)\underset{\sim}{\pi}(x)dx\right\} \quad . \tag{24}$$

We have used the constraint $\underset{\sim}{S}^2(x)=1$. So that $\bar{\sigma}(x)$ just means $\sqrt{1-\underset{\sim}{\pi}^2(x)}$. The action A thus reads :

$$A(\underset{\sim}{\pi},\sigma) = \frac{1}{2}\int\left\{[\partial\mu\underset{\sim}{\pi}(x)]^2+[\partial\mu\sigma(x)]^2\right\}dx$$

$$= \frac{1}{2}\int dx\left\{[\partial\mu\underset{\sim}{\pi}(x)]^2+\frac{[\pi(x).\partial\mu\underset{\sim}{\pi}(x)]^2}{1-\underset{\sim}{\pi}^2(x)}\right\} \quad . \tag{25}$$

We shall now study the zero h limit. If this limit exists in perturbation theory, in the small t power series expansion, then the symmetry is spontaneously broken at low t. At the value t_c (which depends on d) at which the perturbation series breaks down, the symmetry is restored. To explore this question we shall use a set of renormalization group equations (R.G.E.) that we shall derive now.

B - Renormalization and Renormalization group Equations

It is easy to see that the action (25) is renormalizable in two dimension. The π-field has canonical dimension zero and therefore the interaction has canonical dimension 2 as the free part. In dimension d higher than 2, the π field has dimension $\frac{d-2}{2}$ and the theory is no longer renormalizable. In 2+ε dimension we shall therefore use a trick which we cannot justify here [8]: We shall perform a double series expansion in t and ε, and then the theory becomes again renormalizable. Of course we shall have then, using

R.G.E., to argue about the resummation in ε.

Because the theory is renormalizable, we can in the usual way derive R.G.E. We shall first render the theory finite by an O(n) invariant regularization : we can use for example lattice or dimensional regularization. It is easy to show that the theory can be renormalized by two renormalization constants [11], a wave function renormalization Z for the field $\underset{\sim}{\pi}$ and a coupling constant renormalization Z_t.

The renormalized theory is defined by renormalization conditions or by a minimal substraction procedure. The scale parameter of the renormalized theory will be called μ, and the dimensionless renormalized coupling constant τ.

Therefore :

$$t = \mu^{\varepsilon} Z_{t}(\tau)\tau \tag{26a}$$

$$\underset{\sim}{\pi} = Z(\tau)\underset{\sim}{\pi}_{Ren.} \tag{26b}$$

$$\frac{h}{t} = h_R \quad \mu^{\varepsilon}/\tau\sqrt{Z} \tag{26c}$$

Equation (26c) is a consequence of Ward-Takahashi identities which express the O(n) symmetry of the model.

The renormalized Green's functions $G_R^{(N)}$ are related to the bare one by the equation :

$$G^{(N)}_{Ren}(\underset{\sim}{p}_i,\mu,\tau,h_R) = Z^{-N/2}(\tau)\ G^{(N)}_{bare}\ (p_i,t,h)\ . \tag{27}$$

We now differentiate equation (28) with respect to μ at bare parameters t, h (an the cut-off if the bare theory has been defined in this way) fixed. We obtain the R.G.E. :

$$\left\{\mu\frac{\partial}{\partial\mu} + \beta(\tau)\frac{\partial}{\partial\tau} + \frac{N}{2}\gamma(\tau) + \left[\frac{1}{2}\gamma(\tau) + \frac{\beta(\tau)}{\tau} - \varepsilon\right] h_R \frac{\partial}{\partial h_R}\right\} G_R^{(N)}(\underset{\sim}{p}_i,\mu,\tau,h_R) = 0 \tag{28}$$

where we have defined :

$$\beta(\tau) = \mu\frac{\partial}{\partial\mu}\tau\Big|_{t,h} = \varepsilon\left[\frac{\partial \ell n(\tau Z_t(\tau))}{\partial\tau}\right]^{-1} \tag{29}$$

$$\gamma(\tau) = \mu\frac{\partial}{\partial\mu}\ell n\ Z\Big|_{t,h} = \beta(\tau)\frac{\partial \ell n Z(\tau)}{\partial\tau} \tag{30}$$

In order to explore the consequences of these equations, we

shall calculate at one-loop order the renormalization group functions $\beta(\tau)$ and $\gamma(\tau)$.

C - One Loop Calculation and Consequences [8]

At one loop order the two functions $\beta(\tau)$ and $\gamma(\tau)$ are :

$$\beta(\tau) = \varepsilon\,\tau - (n-2)\,\frac{\tau^2}{2\pi} + O\,(\tau^3) \tag{31}$$

$$\gamma(\tau) = (n-1)\,\tau + O(\tau^2) \tag{32}$$

1) The case of dimension two . Let us first consider the case of the dimension 2, $\varepsilon=0$. Then for n larger than 2, the coefficient of τ^2 is negative and the theory is asymptotically free. On the other hand the origin is a I.R. repulsive fixed point for the coupling constant and therefore the spectrum of the theory is not given by perturbation theory for h=0. The special case n = 2 is the abelian case for which it is easy to show that the non linear σ-model is equivalent to a free field theory, and $\beta(\tau)$ vanishes identically in two dimensions.

In the non-abelian case the field theoretical analysis confirms therefore that the symmetry is not spontaneously broken for any value of the coupling constant. Further confirmation of this fact can be found in the large n expansion for this model. In perturbation theory, the spectrum of the theory is given by the (n-1) massless π particles. The resummed theory is O(n) invariant, and the particles of the theory are (n-1) massive π and a massive σ degenerated in mass with the π [12]

The integration of equation (28) for h = 0 leads to :

$$G^{(N)}(\underset{\sim}{p_i},\mu,\tau) = \sigma^N(\tau)\,[m(\tau)]^{d(1-N)}\,\phi^{(N)}\left(\frac{p_i}{m(\tau)}\right) \tag{33}$$

where $(\sigma(\tau)$ and $m(\tau)$ are given by :

$$\sigma(\tau) = \exp - \frac{1}{2}\int^{\tau} \frac{\gamma(\tau')}{\beta(\tau')}\,d\tau' \tag{34}$$

$$m(\tau) = \mu\,\exp - \int^{\tau} \frac{d\tau'}{\beta(\tau')} \tag{35}$$

From equation (33) we see that $m(\tau)$ is proportional to the mass of π and σ. It behaves for small τ as :

$$m(\tau) \sim \mu\,\exp - \frac{2\pi}{(n-2)\tau}\ . \tag{36}$$

It has an exponential singularity in the coupling constant .

2) The dimension 2+ε. For ε small positive the function $\beta(\tau)$ has an U.V. stable zero τ_c :

$$\tau_c = \frac{2\pi\varepsilon}{n-2} + O(\varepsilon^2) \, . \tag{37}$$

Therefore for τ smaller than τ_c the field σ has a vacuum expectation value given by :

$$< \sqrt{1-\underset{\sim}{\pi}^2}(x) > = \sigma(\tau) \tag{38}$$

and the theory is spontaneously broken. At τ_c, $\sigma(\tau)$ vanishes, and the symmetry is restored :

$$\sigma(\tau) \sim (\tau_c - \tau) \text{ for } \tau \lesssim \tau_c \tag{39}$$

$$b = \frac{1}{2} \frac{\gamma(\tau_c)}{\beta'(\tau_c)} = \frac{n-1}{2(n-2)} + O(\varepsilon) \, . \tag{40}$$

The quantity b corresponds to the critical exponent β of statistical mechanics.

On the other hand, because τ_c is an U.V. fixed point, the large momentum behavior of Green's functions is not given by perturbation theory, but by this fixed point.

$$G^{(N)}(\lambda \underset{\sim}{p}_i, \tau, \mu) \sim \lambda^{d - \frac{N}{2}(d+2-\eta)} \tag{41}$$

$$\lambda \to +\infty$$

where η is given by:

$$\eta = d-2 + \gamma(\tau_c) = \frac{\varepsilon}{n-2} + O(\varepsilon^2) \tag{42}$$

Therefore the non linear σ model which, in perturbation theory, is not renormalizable for d > 2, can be defined in higher dimensions The large n expansion and various arguments seem to indicate that the model can actually be defined for dimension d between two and four.

Before presenting the results for the various critical exponents at two loop order a last comment is in order. This property of asymptotic freedom is shared by non abelian color gauge theories in dimension four. The results explained here for the non linear σ model confirm the analogies found by A.A. Migdal[6] between this model and gauge theories, and makes even more plausible that the spectrum of color gauge theories is qualitatively given by strong coupling expansions on a lattice which lead to confinement [13]

D - Results for Critical Exponents

For η larger than 2, various critical exponents have been calculated up to two loop order, and therefore at order ε^2. For completeness we shall give here some results :

The exponent η defined in the previous section is given by[8]

$$\eta = \frac{(d-2)}{n-2} - \frac{(n-1)}{(n-2)^2}(d-2)^2 + O(d-2)^3 \quad .$$

If we define the exponent ν by the behaviour of $m(\tau)$ near τ_c:

$$m(\tau) \sim (\tau_c - \tau)^\nu \ .$$

Then ν is given by[8]:

$$\frac{1}{\nu} = (d-2) + \frac{(d-2)^2}{n-2} + O(\ (d-2)^3) \quad .$$

Other critical exponents can be obtained from ν and η by scaling relations.

Some exponents which govern corrections to scaling laws have also been calculated [13] like ω which is defined near four dimension as $\omega = \beta'(g^*)$ where g is ϕ^4 coupling constant

$$\omega = 4 - d - 2\,\frac{(d-2)}{(n-2)} + O(d-2)^2 \ .$$

Effect of longe range forces have also been explored [14]

REFERENCES

1. K.G. Wilson, Phys. Rev. B4, 3174, 3184 (1971)
2. K.G. Wilson and M.E. Fisher, Phys. Rev. Lett. 28, 240 (1972).
3. For a review on the role of the renormalization group in critical phenomena see : K.G. Wilson and J.B. Kogut; Phys. Rep. 12c 75 (1974). E. Brezin, J.C. Le Guillou and J. Zinn-Justin in "Phase transitions and critical phenomena" Vol. VI edited by C. Domb and M.S. Green (to be published - Academic Press).
4. The role of instantons has been emphazised in : A.M. Polyakov Phys. Lett. 59B, 82, 85 (1975).
5. A.M. Polyakov; Phys. Lett. 59B, 79 (1975).
6. A.A. Migdal; Zh. E.T.F. 69, 1457 (1975).
7. M. Gell-Mann and M. Levy ; Nuovo Cimento 16, 705 (1960)
8. E. Brézin and J. Zinn-Justin ; Phys. Rev. Lett. 36, 691 (1976) and Phys. Rev. B14, 3110 (1976).
9. Strictly speaking, only possible to eliminate completely the possibility of finding instantons for d>2. Because the case 1 d 2 has no physical relevance we are not going to discuss it further here.
10. The absence of S.S.B. with ordering, for continuous symmetries, in two dimensions have been shown by N.D. Mermin and H. Wagner, Phys. Rev. Lett. 17, 1133 (1960) P.C. Hohenberg Phys. Rev. 158 383(1973), S. Coleman Commun. Math. Phys. 31, 259(1973).
11. E. Brézin, J.C. Le Guillou and J. Zinn-Justin; Phys. Rev. D to be published.
12. See ref. [8], [10] and W. Bardeen, B.W. Lee and Schrock; Phys. Rev. D to be published.
13. K.G. Wilson ; Phys. Rev. D, 10, 2445 (1974). R. Balian, J.M. Drouffe and C. Itzykson ; Phys. Rev. D10, 3336 (1974)
14. E. Brézin, J.C. Le Guillou and J. Zinn-Justin ; Phys. Rev. B to be published
15. E. Brézin, J.C. Le Guillou and J. Zinn-Justin, Journal of Physics A 9, L119 (1976).

QUANTUM CHROMODYNAMICS ON A LATTICE

Kenneth G. Wilson

Lab. of Nuclear Studies, Cornell University, Ithaca,

New York 14850: and Fairchild Scholar, Calif. Inst. of Tech.

INTRODUCTION

The phenomenological description of hadrons in terms of quarks continues to be successful; the most recent advance was the description of the new particles as built from charmed quarks. Meanwhile theoretical advances have led to the formulation of a specific field theory of quarks, namely "quantum chromodynamics[1]." In quantum chromodynamics, the standard quarks of the quark model are each xeroxed twice to make three different "colors" of quarks, say red, yellow and blue quarks. The three colors form the basis for an SU(3) group (this group is a second SU(3) group, in addition to the Gell-Mann-Ne'eman SU(3).) The quarks interact with an octet of colored vector mesons called gluons. The theory is renormalizable and in some ways is similar to quantum electrodynamics. There is one very crucial difference however: in quantum chromodynamics the gluons interact with themselves. A consequence of this interaction is "asymptotic freedom[2]." Asymptotic freedom arises as follows. The fundamental interactions of quarks and gluons are modified by "radiative" corrections of higher order in the quark-gluon coupling constant. These radiative corrections depend on the quark and gluon momenta. A careful analysis shows that the cumulative effect of radiative corrections to all orders can be characterized by a <u>momentum-dependent</u> effective coupling constant. The effective coupling is found to vanish in the limit of large momenta (to be precise, large momentum transfers between the quarks and gluons). This is called asymptotic freedom. As a result of asymptotic freedom the quarks can behave as nearly free particles at short distances; this is required to explain the high energy electron scattering experiments[3]. Meanwhile the interactions of quarks at long distances can be strong enough to bind the quarks into the observed bound states; protons, mesons, etc.

Unfortunately, it is not known yet whether the quarks in quantum chromodynamics actually form the required bound states. To establish whether these bound states exist one must solve a strong coupling problem and present methods for solving field theories don't work for strong coupling. The only way known at present to solve quantum chromodynamics is to use renormalized field theory, and at low momenta one gets infrared divergent logarithms to each order in perturbation theory; the sum to all orders of these logarithms is not known[4].

In order to make available a wider range of methods for solving the quark theory, it has been formulated on a discrete space-time lattice[5]. The ultimate aim is to let the lattice spacing go to zero. The lattice is to be understood as an aid to solving the theory much as a discrete mesh is used when a partial differential equation is solved numerically. The continuum limit of the lattice theory should give back the continuum asymptotically free theory.

One of the methods which is available to solve lattice theories but not continuum theories is the block spin method borrowed from statistical mechanics[6]. I am currently trying to carry out a block spin calculation for the lattice version of quantum chromodynamics: I have no results yet.

The lattice gauge theory has already been discussed in the Erice Lecture notes[7]. In addition, various versions of the lattice theory are being investigated by Kogut and Susskind[8] et al. Bardeen et al.[9], and others.

In these lectures three specific topics will be discussed. First, the detailed definition of the space-time lattice theory will be given including the arguments showing that the theory has a Hermitian Hamiltonian in a Hilbert space with positive metric[10]. Secondly, a qualitative discussion of quark confinement will be given. The emphasis will be on how quark confinement might arise due to specific properties of the gauge theory, including asymptotic freedom. Finally, the block spin method will be formulated and the reasons for pursuing this method explained.

An important feature of the lattice theory is that gauge invariance is an exact symmetry of the lattice action. This means the lattice action is invariant to separate SU(3) color groups at each space-time lattice point. The role of this symmetry in quark confinement is explained in the Erice lecture notes[7] and will not be reviewed here.

I. EUCLIDEAN LATTICE THEORY

In this lecture the space-time version of the lattice theory will be considered. The time lattice is a lattice in imaginary time. It is not obvious that normal quantum mechanics can be developed on an imaginary time lattice, so in this lecture it will be shown that a normal quantum mechanics can be defined from the Euclidean space-time lattice theory. However, there is a restriction on the nearest neighbor quark coupling constant in order to have positive energies and a positive metric. This restriction, fortunately, is satisfied by the asymptotically free continuum limit of the gauge theory.

The action of the space-time lattice gauge theory is[7]

$$A = -\sum_n \bar{\psi}_n \psi_n + K \sum_n \sum_\mu \bar{\psi}_n (1-\gamma_\mu) U_{n\mu} \psi_{n+\hat{\mu}} + K \sum_n \sum_\mu \bar{\psi}_{n+\hat{\mu}} (1+\gamma_\mu) U^+_{n\mu} \psi_n$$

$$+ \frac{1}{g_0^2} \sum_n \sum_{\substack{\mu\nu \\ \nu\neq\mu}} \mathrm{Tr}\, U_{n\mu} U_{n+\hat{\mu},\nu} U^+_{n+\hat{\nu},\mu} U^+_{n\nu} \quad , \tag{1}$$

where g_0 is the gauge field coupling constant. (The normalization of the last term is different from Ref. 7; it is now consistent with the definition of g_0 in Ref. 2.

ψ_n is the quark field;

$U_{n\mu}$ is the exponential of the gauge field; $U_{n\mu}$ is a unitary 3 x 3 color SU(3) matrix.

K is the nearest neighbor quark coupling; to be precise there are three separate couplings for the nonstrange, strange, and charmed quarks but in this lecture only a single quark coupling will be used.

For a positive definite metric and real energies (see below) one must have $0 \leq K \leq 1/6$. Vacuum expectation values in the lattice theory involve integrals of e^A over all the quark and gauge field variables. For example the π meson propagator (ignoring isotopic spin) is

$$D_\pi(m) = \frac{1}{Z} \Pi_n \int_{(\psi_n,\bar{\psi}_n)} \Pi_{n\mu} \int_{U_{n\mu}} \bar{\psi}_m \gamma_5 \psi_m \bar{\psi}_0 \gamma_5 \psi_0 \; e^A \tag{2}$$

where

$$Z = \Pi_n \int_{(\psi_n,\bar{\psi}_n)} \Pi_{n\mu} \int_{U_{n\mu}} e^A \quad . \tag{3}$$

$\int_{U_{n\mu}}$ means invariant group integration over all SU(3) matrices

$U_{n\mu}$ and $\int_{(\psi_n,\bar{\psi}_n)}$ is fermion integration[11] over pure anticommuting variables ψ_n and $\bar{\psi}_n$ (to be precise, one integrates over all spin, flavor, and color components of ψ_n and $\bar{\psi}_n$).

The quantum mechanical definition of $D_\pi(m)$ is

$$D_\pi(m) = \langle\Omega|T\,\bar{\psi}_m\gamma_5\psi_m\bar{\psi}_0\gamma_5\psi_0|\Omega\rangle \quad , \tag{4}$$

where $|\Omega\rangle$ is the vacuum state and $\bar{\psi}_m$, etc., are quark operators. Due to the lattice being in imaginary time the Heisenberg field $\bar{\psi}_m$ is defined with H replacing iH:

$$\bar{\psi}_m = e^{Hm_0a}\,\bar{\psi}_{\vec{m}}\,e^{-Hm_0a} \quad , \tag{5}$$

where a is the lattice spacing (for both the space and time axes), H is the Hamiltonian of the theory and $\psi_{\vec{m}}$ is the Schrödinger quark field operator. Thus, for $m_0 > 0$ the time ordering gives

$$D_\pi(m) = \langle\Omega|e^{Hm_0a}\,\bar{\psi}_{\vec{m}}\gamma_5\psi_{\vec{m}}\,e^{-Hm_0a}\,\bar{\psi}_0\gamma_5\psi_0|\Omega\rangle \tag{6}$$

Assume (as usual) that H is defined so that the vacuum has energy zero; then (still for $m_0 > 0$)

$$D_\pi(m) = \langle\Omega|\bar{\psi}_{\vec{m}}\gamma_5\psi_{\vec{m}}\,e^{-Hm_0a}\,\bar{\psi}_0\gamma_5\psi_0|\Omega\rangle \quad . \tag{7}$$

The problem now is to extract from the path integral formula for $D_\pi(m)$ a definition of the Hamiltonian H, a definition of the Hilbert space on which H acts, and a definition of the operators $\psi_{\vec{m}}$ and ψ_0, such that the path integral form for $D_\pi(m)$ is equivalent to the quantum mechanical form (4). To show how this problem is solved we shall first discuss a very much simpler version of the action A. Throw out the gauge field, the spatial lattice and all but one component of the quark field. Let γ_0 be +1 for the remaining component. Then the simplified action reads

$$A = -\sum_n \bar{\psi}_n\psi_n + 2K\sum_n \bar{\psi}_{n+1}\psi_n \quad , \tag{8}$$

where n is a single integer measuring location on the imaginary time axis and ψ_n is a single component anticommuting field.

It is convenient to introduce now the Hilbert space that will be used in the simplified model. A state $|f\rangle$ is defined by a wave function $f(\bar{\psi})$. $\bar{\psi}$ is a single anticommuting variable; this means that $\bar{\psi}^2$ is 0. The most general such function $f(\bar{\psi})$ is

$$f(\bar{\psi}) = b + c\bar{\psi} \quad , \tag{9}$$

where b and c are numerical coefficients (commuting, not anti-commuting). The conjugate state $\langle f|$ is defined by the conjugate wave function

$$f^*(\psi) = b^* + c^*\psi \quad , \tag{10}$$

(since $\gamma_0 = +1$, $(\bar{\psi})^*$ is the same as ψ).

The scalar product $\langle f|f\rangle$ is defined to be $|b|^2 + |c|^2$. This formula can be obtained from a fermion integral:

$$\langle f|f\rangle = \int_{(\psi,\bar{\psi})} f^*(\psi) e^{-\bar{\psi}\psi} f(\bar{\psi}) \quad . \tag{11}$$

The fermion integral has been defined elsewhere[11]. We review it briefly. It is defined by the rules that

$$\int_{(\psi,\bar{\psi})} \psi\bar{\psi} = 1 = -\int_{(\psi,\bar{\psi})} \bar{\psi}\psi \quad , \tag{12}$$

$$\int_{(\psi,\bar{\psi})} (\alpha + \beta\psi + \gamma\bar{\psi}) = 0 \quad , \tag{13}$$

where α, β, and γ are arbitrary constants. The conditions that define this integral are (1) it must be translationally invariant, namely,

$$\int_{(\psi,\bar{\psi})} F(\psi+\eta, \bar{\psi}+\bar{\eta}) = \int_{(\psi,\bar{\psi})} F(\psi,\bar{\psi}) \quad ,$$

where η and $\bar{\eta}$ are other anticommuting variables, and (2) a normalization condition which is arbitrary and is contained in eq.(12). The normalization chosen is the convenient one for the present purposes. Given the formulae (12) and (13), the integral (11) becomes

$$\int_{(\psi,\bar{\psi})} (a^* + b^*\psi)(1 - \bar{\psi}\psi)(a + b\bar{\psi}) \quad , \tag{14}$$

where we have expanded $\exp(-\bar{\psi}\psi)$ obtaining only $1 - \bar{\psi}\psi$ because $(\bar{\psi}\psi)(\bar{\psi}\psi) = -\bar{\psi}^2\psi^2 = 0$.

The integral (14) simplifies to

$$\int_{(\psi,\bar{\psi})} (a^*a - a^*a\,\bar{\psi}\psi + a^*b\bar{\psi} + b^*a\psi + b^*b\,\psi\bar{\psi}) = |a|^2 + |b|^2 \; . \tag{15}$$

Hence it reproduces the metric.

It will be useful later on to consider a more general form for the metric, namely the metric

$$\langle f|f\rangle = \int_{(\psi,\bar{\psi})} f^{*}(\psi)\ e^{-r\bar{\psi}\psi}\ f(\bar{\psi}) \quad . \tag{16}$$

This integral gives

$$\langle f|f\rangle = r|a|^{2} + |b|^{2} \quad . \tag{17}$$

This metric is positive definite if and only if r is positive. This simple constraint on r will be the origin of the restriction $K \leq 1/6$ in the full action in order to have a positive metric.

Consider the quark propagator (for simplicity we study the quark propagator rather than the meson propagator) in the simplified model:

$$S(m) = \langle\Omega|T\ \psi_m\bar{\psi}_0|\Omega\rangle \quad . \tag{18}$$

To be specific consider S(1)

$$S(1) = \langle\Omega|\psi\ \ e^{-Ha}\ \bar{\psi}\ |\Omega\rangle \quad . \tag{19}$$

The path integral formula for S(1), written out, is

$$S(1) = \frac{1}{Z}\prod_n \int_{(\psi_n,\bar{\psi}_n)} \cdots\ e^{-\bar{\psi}_2\psi_2}\ e^{2K\bar{\psi}_2\psi_1}\ \psi_1\ e^{-\bar{\psi}_1\psi_1}\ e^{2K\bar{\psi}_1\psi_0}$$
$$e^{-\bar{\psi}_0\psi_0}\ \bar{\psi}_0\ e^{2K\bar{\psi}_0\psi_{-1}}\ e^{-\bar{\psi}_{-1}\psi_{-1}}\ e^{2K\bar{\psi}_{-1}\psi_{-2}}\ \cdots \tag{20}$$

where

$$Z = \Pi_n \int_{(\psi_n,\bar{\psi}_n)} e^{A} \tag{21}$$

It is not necessary to form the complete product of exponentials in eq. (20) before doing any integrals. One can perform the integral over $(\psi_{-1}, \bar{\psi}_{-1})$, $(\psi_{-2}, \bar{\psi}_{-2})$, etc., keeping only the terms to the righ of $\bar{\psi}_0$ in eq. (20). The result of these integrals (over ψ_{-1}, $\bar{\psi}_{-1}$, etc is a function $f(\bar{\psi}_0)$ which (as will be shown) is the vacuum state wave function. Likewise the integrals over $(\psi_2, \bar{\psi}_2)$, $(\psi_3, \bar{\psi}_3)$, etc., can be performed keeping only the exponentials to the left of ψ_1, the result being a function $g(\psi_1)$ which will be the wave function for $\langle\Omega|$. To be specific

$$f(\bar{\psi}_0) = \Pi_{n<0} \int_{(\psi_n,\ \bar{\psi}_n)} e^{2K\bar{\psi}_0\psi_{-1}}\ e^{-\bar{\psi}_{-1}\psi_{-1}}\ e^{2K\bar{\psi}_{-1}\psi_{-2}} \ldots \tag{22}$$

$$g(\psi_1) = \Pi_{n>1} \int_{(\psi_n,\ \bar{\psi}_n)} \ldots\ e^{-\bar{\psi}_2\psi_2}\ e^{2K\bar{\psi}_2\psi_1} \quad . \tag{23}$$

Now

$$S(1) = \frac{1}{Z} \int_{(\psi_1,\bar{\psi}_1)} g(\psi_1)\ e^{-\bar{\psi}_1\psi_1} \{\int_{(\psi_0,\bar{\psi}_0)} e^{2K\bar{\psi}_1\psi_0}\ e^{-\bar{\psi}_0\psi_0} \bar{\psi}_0 f(\bar{\psi}_0)\}. \tag{24}$$

The multiplication of $f(\bar{\psi}_0)$ by $\bar{\psi}_0$ defines the operator $\bar{\psi}$: it takes the wavefunction $f(\bar{\psi}_0)$ into the new wavefunction $\bar{\psi}_0 f(\bar{\psi}_0)$. The integration over ψ_0 and $\bar{\psi}_0$ results in a function of $\bar{\psi}_1$; this operation defines the operator exp(-Ha). To be specific, suppose

$$|h_1\rangle = e^{-Ha}\ |h_0\rangle \quad ,$$

where $h_0(\bar{\psi})$ and $h_1(\bar{\psi})$ are two wavefunctions. Then the definition of exp(-Ha) gives

$$h_1(\bar{\psi}_1) = \int_{(\psi_0,\bar{\psi}_0)} e^{2K\bar{\psi}_1\psi_0}\ e^{-\bar{\psi}_0\psi_0}\ h_0(\bar{\psi}_0) \quad . \tag{25}$$

Finally, in eq. (24) the exponential $\exp(-\ \bar{\psi}_1\psi_1)$ and the integral over $(\psi_1,\ \bar{\psi}_1)$ define the scalar product of $\langle g|\psi$ and the state $\exp(-Ha)\bar{\psi}|f\rangle$. So in operator notation eq. (24) becomes

$$S(1) = \langle g\,|\psi e^{-Ha}\ \bar{\psi}|f\rangle/Z \quad . \tag{26}$$

In the simplified model being discussed, the operator exp(-Ha) can be diagonalized explicitly. To start with, the action of exp(-Ha) on the general wavefunction $h_0(\bar{\psi}) = b + c\bar{\psi}$ will be computed. The exponentials in eq. (25) can be expanded (dropping terms which are due to $\psi^2 = 0$, etc.) giving

$$h_1(\bar{\psi}_1) = \int_{(\psi_0,\bar{\psi}_0)} \{1 + 2K\ \bar{\psi}_1\psi_0\}\ \{1 - \bar{\psi}_0\psi_0\}\ \{b + c\bar{\psi}_0\}$$

$$= \int_{(\psi_0,\bar{\psi}_0)} \{b + 2K\ \bar{\psi}_1\psi_0\ b - b\bar{\psi}_0\psi_0 + c\bar{\psi}_0 + 2Kc\ \bar{\psi}_1\psi_0\bar{\psi}_0\} \ . \tag{27}$$

Using the rules for the fermion integral, this gives

$$h_1(\bar{\psi}_1) = b + 2Kc\bar{\psi}_1 \tag{28}$$

From this it is clear that exp(-Ha) has two eigenfunctions and eigenvalues:

1) eigenfunction 1, eigenvalue 1
2) eigenfunction $\bar{\psi}$, eigenvalue $2K$.

The eigenfunctions of exp(-Ha) are also the eigenfunctions of H itself. The eigenvalues of H are obtained from

$$H = -a^{-1} \ln(e^{-Ha}) \quad , \tag{29}$$

so the two energy eigenvalues are 0 and $[\ln(2K)]/a$ respectively. In order that the second eigenvalue be real, K must be positive (this is the reason for the restriction $K \geq 0$ in the full lattice theory action). If, more precisely, $0 < 2K < 1$, then the ground state of H is the state with eigenfunction 1. Only this case will be discussed here.

Consider the state $|f\rangle$ in eq. (22). The exact definition of $|f\rangle$ depends on the boundary condition one imposes on the path integral at large negative time. For simplicity, suppose one stops the path integral at the lattice site -N, where N is large. Then the definition of $|f\rangle$ is

$$f(\bar{\psi}_0) = \int_{(\psi_{-1},\bar{\psi}_{-1})} \cdots \int_{(\psi_{-N},\bar{\psi}_{-N})} e^{2K\bar{\psi}_0\psi_{-1}} \, e^{-\bar{\psi}_{-1}\psi_{-1}} \cdots$$

$$e^{2K\bar{\psi}_{-N+1}\psi_{-N}} \, e^{-\bar{\psi}_{-N}\psi_{-N}} \quad . \tag{30}$$

This formula for the wavefunction $f(\bar{\psi}_0)$ involves the operator exp(-Ha) multiplied N times and acting on the wavefunction 1:

$$|f\rangle = e^{-NHa}|1\rangle \quad . \tag{31}$$

Since the state $|1\rangle$ is the ground state $|\Omega\rangle$ of H, we obtain

$$|f\rangle = |\Omega\rangle \quad . \tag{32}$$

Likewise one finds $\langle g| = \langle\Omega|$.

Finally, the "partition function" Z can be calculated: it is the same as the numerator in eq. (26) except that ψ and $\bar{\psi}$ are removed:

$$Z = \langle g|e^{-Ha}|f\rangle = \langle\Omega|e^{-Ha}|\Omega\rangle = 1 \quad . \tag{33}$$

The results $|f\rangle = |\Omega\rangle$ and $Z = 1$ are special to the simplified model. In the full lattice theory eq. (31) still holds but $|1\rangle$ is no longer an eigenstate of H. In the limit of large N, $\exp\{-NHa\}$ acting on $|1\rangle$ projects out the vacuum component of $|1\rangle$:

$$|f\rangle = c\, e^{-E_0 Na} |\Omega\rangle \tag{34}$$

where

$$c = \langle\Omega|1\rangle \quad , \tag{35}$$

and E_0 is the ground state energy of H. Likewise $|g\rangle$ is proportional to $|\Omega\rangle$. The first equality of eq. (33) for Z also holds for the full theory. The division by Z in eq. (26) ensures that the normalization of the states $|f\rangle$ and $|g\rangle$ is cancelled out, leaving

$$S(1) = \langle\Omega|\psi e^{-(H-E_0)a}\, \bar{\psi}|\Omega\rangle \quad . \tag{36}$$

Thus one is left with a vacuum expectation value with the ground state energy of H automatically subtracted away.

So far only the upper components of ψ and $\bar{\psi}$ have been discussed. There is a clear analogy between the wavefunction $\bar{\psi}$ and the one quark state created by the operator $\bar{\psi}$. Thus the energy $-[\ln(2K)]/a$ of the state $\bar{\psi}$ is the quark mass in the model discussed above. In the full theory one also has antiquark states. In ordinary Dirac theory the antiquark states are created by ψ, not $\bar{\psi}$. In the lattice theory antiquark states are created using the lower components of ψ. To illustrate this consider the simplified theory but with ψ and $\bar{\psi}$ being one of the lower components for which $\gamma_0 = -1$. In this case the action is

$$A = -\sum_n \bar{\psi}_n \psi_n + 2K \sum_n \bar{\psi}_n \psi_{n+1} \quad . \tag{37}$$

In particular the K term involves ψ_{n+1} rather than $\bar{\psi}_{n+1}$. This means the wavefunction f in eq. (22) will now depend on ψ_0 instead of $\bar{\psi}_0$. A general wavefunction is $b + c\,\psi_0$. The conjugate wavefunction is $b^* + c^*\,\psi_0^*$ which (because $\gamma_0 = -1$) is $b^* - c^*\,\bar{\psi}_0$. The metric factor is still $\exp(-\bar{\psi}_0\psi_0)$; one finds that

$$\int_{(\psi_0,\bar{\psi}_0)} (b^* - c^*\,\bar{\psi}_0)\, e^{-\bar{\psi}_0\psi_0}\, (b + c\,\psi_0) = |b|^2 + |c|^2 \, . \tag{38}$$

The operator $\exp(-Ha)$ can be constructed as before.

To illustrate the role of the gauge field and local gauge symmetry, a somewhat more complicated model will not be discussed. The new model includes one upper component $\bar{\psi}_u$ and one lower component $\bar{\psi}_\ell$ (also ψ_u and ψ_ℓ) and the time component U_{n0} of the gauge field. The gauge field will be simplified to be a phase factor $\exp(iA_{n0})$, not a color matrix. The action is

$$A = -\sum_n \bar{\psi}_n\psi_n + 2K\sum_n \{\bar{\psi}_{n+1,u}\, e^{-iA_{n0}}\, \psi_{nu} + \bar{\psi}_{n\ell} e^{iA_{n0}} \psi_{n+1,\ell}\} \quad . \tag{39}$$

The Hilbert space consists of the functions $f(\bar{\psi}_u,\psi_\ell)$. The transfer matrix $\exp(-Ha)$ is defined by $|h_1\rangle = \exp(-Ha)|h_0\rangle$ with

$$h_1(\bar{\psi}_{1u},\psi_{1\ell}) = \frac{1}{2\pi}\int_0^{2\pi} dA_{00} \int_{(\psi_0,\bar{\psi}_0)} \exp\{2K[\bar{\psi}_{1u} e^{-iA_{00}}\psi_{0u} + \bar{\psi}_{0\ell}\, e^{iA_{00}}\psi_{1\ell}]\} \times \exp\{-(\bar{\psi}_{0u}\psi_{0u}+\bar{\psi}_{0\ell}\psi_{0\ell})\} h_0(\bar{\psi}_{0u},\psi_{0\ell}) \tag{40}$$

The variable A_{00} is not included in the definition of the Hilbert space; instead the integration over A_{00} is included in the definition of $\exp(-Ha)$. This is possible because there are no coupling terms coupling A_{00} to A_{10} or A_{-10}. (If there were $A_{00}A_{10}$ and $A_{00}A_{-10}$ terms present the A_{00} integration would result in a function of A_{10} and A_{-10}: one could not avoid an A dependence in the wave functions h_1 and h_0.)

The most general wavefunction is

$$h_0(\bar{\psi}_{0u},\psi_{0\ell}) = b + c\,\bar{\psi}_{0u} + d\,\psi_{0\ell} + e\,\bar{\psi}_{0u}\psi_{0\ell} \quad . \tag{41}$$

When $\exp(-Ha)$ is applied to this state and all the fermion integrals are performed (in this case

$$\int_{(\psi_0,\bar{\psi}_0)} \psi_{0u}\bar{\psi}_{0u}\psi_{0\ell}\bar{\psi}_{0\ell} = 1$$

and all integrals of lower order polynomials vanish) one obtains

$$h_1(\bar{\psi}_{1u},\psi_{1\ell}) = \frac{1}{2\pi}\int_0^{2\pi} dA_{00}\{b + 2K\, e^{-iA_{00}}\, \bar{\psi}_{1u}\, c + 2K\, e^{iA_{00}}\psi_{1\ell} d$$

$$+ (2K)^2\, \bar{\psi}_{1u}\psi_{1\ell}\, e\} = b + (2K)^2\, \bar{\psi}_{1u}\psi_{1\ell}\, e\ . \qquad (42)$$

The eigenfunctions and eigenvalues of H are summarized in the table below:

Eigenfunction	Eigenvalue	Type of State
1	0	vacuum
$\bar{\psi}_{1u}$	∞	one quark
$\psi_{1\ell}$	∞	one antiquark
$\bar{\psi}_{1u}\psi_{1\ell}$	$2\{-\frac{1}{a}\ln 2K\}$	quark + antiquark

The presence of the gauge field in this simplified example has a remarkable effect. The single quark states have infinite energy (exp(-Ha) has eigenvalue 0 for the single quark states, hence H has eigenvalue ∞). The operator exp(-Ha) has eigenvalue 0 due to the A_{00} integral giving 0. However, the energy of the quark-antiquark state is finite, and moreover is the sum of the energies the single quark and the single antiquark would have had in the absence of the gauge field.

The reason exp(-Ha) gives 0 for the single quark states is local gauge invariance. A quark cannot propagate in time because that would violate the conservation of SU(3) separately at each time (see the Erice Lecture notes[7]). Hence exp(-Ha) applied to the single quark state gives 0. (To be precise a quark cannot propagate in imaginary time; for real time the question of propagation is more ambiguous.)

This completes the discussion of simplified models.

For the full action, eq. (1), we can only define the metric and the Hamiltonian. The eigenvectors and eigenvalues are not known. The definitions are as follows. A wave function is a function $h(\bar{\psi}_{\vec{n}u}, \psi_{\vec{n}\ell}, U_{\vec{n}i}, U^{+}_{\vec{n}i})$ where $U_{\vec{n}i}$ $(1 \leq i \leq 3)$ are the spacelike components of the gauge field. h depends on these variables for all $\vec{n}$ and i. There are of course two upper components for $\bar{\psi}_{nu}$ and two lower components for $\psi_{\vec{n}\ell}$; in addition these variables have ordinary SU(3) and color SU(3) indices. The metric is defined to be

$$\langle h|h\rangle = \Pi_{\vec n} \int_{\{\psi_n,\bar\psi_n\}} \Pi_{\vec n,i} \int_{U_{\vec n i}} h^*(\psi_{\vec n u}, -\bar\psi_{\vec n \ell}, U^+_{\vec n i}, U_{\vec n i}$$

$$\exp -\{\sum_{\vec n} \bar\psi_{\vec n}\psi_{\vec n} + K \sum_{\vec n} \sum_{\mu=1}^{3} [\bar\psi_{\vec n} U_{\vec n\mu} \psi_{\vec n+\hat\mu} + \bar\psi_{\vec n+\hat\mu} U^+_{\vec n\mu} \psi_{\vec n}]\}$$

$$h(\bar\psi_{\vec n u}, \psi_{\vec n\ell}, U_{\vec n i}, U^+_{\vec n i}) \quad . \tag{43}$$

The exponential contains a subset of those terms from the action which involve fermion fields at a single time, e.g., $\bar\psi_{\vec n}\, \psi_{\vec n+\hat\mu}$ terms are included for spacelike but not timelike μ. It is somewhat strange to find that the metric depends on the dynamics (through the parameter K), but this is necessary in order to have a Hermitian Hamiltonian (see below) and there is no physical reason to forbid a metric with this dependence. The metric is positive definite if and only if $|K| < 1/6$; one proves this by diagonalizing the quadratic form in the fermion variables holding the $U_{\vec n i}$ and $U^+_{\vec n i}$ fixed. Positivity requires the eigenvalues all be negative (see eqs. (16) and (17)). Since the U_{ni} are unitary matrices, they are bounded by 1 and simple upper bounds show that the eigenvalues are all negative for $|K| < 1/6$ (the factor 6 comes from the 6 nearest neighbor terms). If $|K| > 1/6$ one can construct explicitly a state with negative metric using a state which is non-zero only for $U_{\vec n i} \simeq 1$ for all n and i. Asymptotic freedom gives $K = 1/8$ which is OK.

The operator $\exp(-Ha)$ is defined so that $|h_1\rangle = \exp(-Ha)|h_0\rangle$ if (note: $\gamma_\mu(\mu \neq 0)$ connects u to ℓ components and vice versa only)

$$h_1(\bar\psi'_{\vec n u}, \psi'_{\vec n\ell}, U'_{ni}, U^{+\prime}_{ni}) =$$

$$\Pi_{\vec n} \int_{\{\psi_{\vec n},\bar\psi_{\vec n}\}} \Pi_{\vec n,\mu} \int_{U_{\vec n\mu}} \exp \Big\{ K \sum_{\vec n} [\bar\psi_{\vec n} (1-\gamma_0) U_{\vec n 0} \psi'_{\vec n} + \bar\psi'_{\vec n} (1+\gamma_0) U^+_{\vec n 0} \psi_{\vec n}$$

$$+ K \sum_{\vec n} \sum_{\mu=1}^{3} [-\bar\psi'_{\vec n u} \gamma_\mu U'_{\vec n\mu} \psi'_{\vec n+\hat\mu\ell} - \bar\psi_{\vec n\ell} \gamma_\mu U_{\vec n\mu} \psi_{\vec n+\hat\mu u} + \bar\psi'_{\vec n+\hat\mu u} \gamma_\mu U'^{+}_{n\mu} \psi'_{\vec n\ell}$$

$$+ \bar\psi_{\vec n+\hat\mu\ell} \gamma_\mu U^+_{\vec n\mu} \psi_{\vec n u}] + \frac{1}{g_0^2} \sum_{\vec n} \sum_{\mu=1}^{3} \mathrm{Tr}[U_{\vec n\mu} U_{\vec n+\hat\mu,0} U'^{+}_{\vec n\mu} U^+_{\vec n 0}$$

$$+ U_{\vec n 0} U'_{\vec n\mu} U^+_{\vec n+\hat\mu,0} U^+_{n\mu}] + \frac{1}{2g_0^2} \sum_{\vec n} \sum_{\mu=1}^{3} \sum_{\nu=1}^{3} \mathrm{Tr}[U_{\vec n\mu} U_{\vec n+\hat\mu,\nu} U^+_{\vec n+\nu,\mu} U^+_{n\nu}$$

$$+ U'_{n\mu} U'_{n+\hat{\mu},\nu} U'^{+}_{n+\hat{\nu},\mu} U'^{+}_{n\nu}] \Bigg\}$$

$$\times \exp \left\{ - \sum_{\vec{n}} \bar{\psi}_n \psi_n + K \sum_{\vec{n}} \sum_{\mu=1}^{3} [\bar{\psi}_n U_{\vec{n}\mu} \psi_{n+\hat{\mu}} + \bar{\psi}_{n+\hat{\mu}} U^{+}_{\vec{n}\mu} \psi_n] \right\}$$

$$h_0 \, (\bar{\psi}_{\vec{n}u}, \psi_{\vec{n}\ell}, U_{ni}, U^{+}_{ni}) \quad . \tag{44}$$

This definition was arrived at using the following requirements.

(1) The path integral (3) must be interpretable as an infinite product of exp(-Ha) operators, i.e., the terms in the action (1) must all be reproduced when one forms exp(-NHa) for large N by iterating eq. (44).

(2) The operator exp(-Ha) must be Hermitian, so that H is Hermitian. This requirement is verified by considering matrix elements $\langle h_2 | \exp(-Ha) | h_0 \rangle = \langle h_2 | h_1 \rangle$ and making sure that

$$\langle h_2 | e^{-Ha} | h_0 \rangle = \langle h_0 | e^{-Ha} | h_2 \rangle^* \quad . \tag{45}$$

Now

$$\langle h_2 | h_1 \rangle = \prod_{\vec{n}} \int_{\{\bar{\psi}'_{\vec{n}}, \psi'_{\vec{n}}\}} \Pi_{\vec{n},i} \int_{U'_{\vec{n}i}} h_2^* (\psi'_{\vec{n}u}, \bar{\psi}'_{\vec{n}\ell}, U'^{+}_{\vec{n}i}, U'_{\vec{n}i})$$

$$\exp \left\{ - \sum_{\vec{n}} \bar{\psi}'_{\vec{n}} \psi'_{\vec{n}} + K \sum_{\vec{n}} \sum_{\mu=1}^{3} [\bar{\psi}'_{\vec{n}} U'_{n\mu} \psi'_{\vec{n}+\hat{\mu}} + \bar{\psi}'_{\vec{n}+\hat{\mu}} U'^{+}_{\vec{n}\mu} \psi'_n] \right\}$$

$$h_1 (\bar{\psi}'_{\vec{n}u}, \psi'_{\vec{n}\ell}, U'_{\vec{n}i}, U^{+\prime}_{\vec{n}i}) \quad . \tag{46}$$

When this formula is combined with eq. (44) for h_1, one can verify explicitly the requirement (45). The metric plays a crucial role in this verification: it ensures that the expression for $\langle h_2 | \exp(-Ha) | h_0 \rangle$ is symmetric when the primed and unprimed variables are interchanged (apart from the functions h_2^* and h_0 themselves), in particular the exponential in $\bar{\psi}'$ and ψ' coming from the metric (43) balances the last exponential in eq. (46).

The final requirement on exp(-Ha) is that its eigenvalues should be positive or 0. The proof of this is complicated and will not be given (see Ref. 10); the result is that exp(-Ha) has positive or 0 eigenvalues only if $K \geq 0$, exactly as in the first two simplified examples.

II. QUALITATIVE PICTURE OF QUARK CONFINEMENT

The central paradox in the phenomenological quark picture is that for short times and short distances the quarks must behave like free particles, while they must be completely trapped on longer time and length scales. The trapping is necessary because no quarks have been seen experimentally.

We know, from continuum perturbation theory, that the coupling strength of quarks to gluons is large for low energy quarks and gluons. This however, does not guarantee confinement. In this lecture I shall present a qualitative discussion of how quark confinement might come about in the asymptotically free gauge theory. The argument shows that confinement is a reasonable possibility but not certain; whether it occurs can only be settled by a detailed calculation. The argument was developed in conversations with V. Gribov and A. Casher and is based on earlier work of Kogut and Susskind[12]. This lecture is concerned with the continuum gauge theory but will provide a physical motivation for the "block spin" approach discussed in Lecture III.

Kogut and Susskind have offered a simple picture of quark confinement making use of a "lines of force" picture. They propose that the gauge field generated by two quarks (to be precise, one quark and one antiquark) can be characterized by lines of force similar to the lines of force of classical electrodynamics. For weakly coupled quarks, e.g., quarks sufficiently close to each other, the lines of force have the same distribution as for the Coulomb field of two point charges (see Fig. 1).

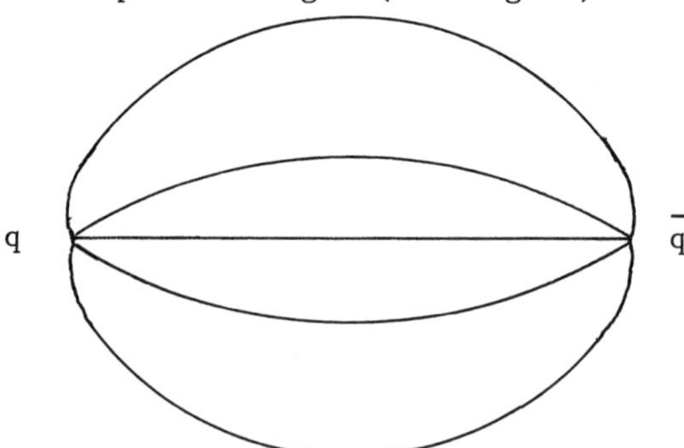

Fig. 1. Lines of force between a quark and an antiquark: weak coupling case.

In this case the potential energy of two quarks has the usual Coulomb form g^2/r where g is the coupling constant and r is the quark-antiquark separation. However, when the quarks are far apart, Kogut and Susskind suggest that the lines of force are trapped into a sausage-like region connecting the two quarks (see Fig. 2). The energy density of the gluon field (in the central region where the lines of force are parallel) is constant along each line of force, by translational invariance. Hence the total field energy is linearly proportional to distance. This means the potential energy of the two quarks is proportional to r, and becomes infinite for infinite separation r. (There is also a constant term in the energy for each end of the sausage.)

What does one mean by lines of force? Why are the lines of force confined to a tube? The purpose of this lecture is to propose answers to these questions.

The fundamental question is: what is the minimum energy configuration of two quarks in the limit of large separation? If the minimum energy is finite and bounded, then quarks are free. If the minimum energy goes to infinity as the separation goes to infinity, then the quarks are confined. This suggests that one use a variational approach. A set of two quark states will be constructed corresponding to various arrangements of the lines of force between the quarks. The minimum energy configuration will then be found. The variational calculation given below will be very qualitative and incomplete but hopefully can distinguish infinity from a finite bound.

We should include in the variational calculation quark states which would naively be expected to have a finite energy for large separation. The simplest such state is simply $\bar{\psi}(x)\psi(0)|\Omega$. The field $\bar{\psi}(x)$ (with $x_0 = 0$) creates a quark at the point x, while $\psi(0)$ creates an antiquark at the origin.

Fig. 2. Lines of force between a quark and antiquark: strong coupling case.

The analysis given below depends crucially on manifest gauge invariance. The state $\bar{\psi}(x)\psi(0)|\Omega\rangle$ is certainly not manifestly gauge invariant. The typical procedure for making the product $\bar{\psi}(x)\psi(0)$ gauge invariant is to insert an exponential of the gauge field between $\bar{\psi}(x)$ and $\psi(0)$, namely to consider the product

$$\bar{\psi}(x) P \exp \{ig \int_{LF} A_{\mu}^{a}(z)\, T^{a} \cdot dz^{\mu}\}\psi(0) \quad , \tag{47}$$

where $A_{\mu}^{a}(x)$ is the color gauge field (a is the color octet index and is summed over). LF is a path from x to 0. LF will be called a line of force. T^a is the 3 x 3 Hermitian SU(3) matrix generator for octet index a. P is a path-ordering symbol(necessary because the matrices T^a do not commute). The exponential of A is a quantum operator, since A is an operator. It will be called a string operator, LF being the location of the string.

To set up the variational calculation the single term (47) will be replaced by an arbitrary sum over many lines of force connecting x to 0. Using the pictures (1) and (2) as a guide, imagine a central plane bisecting the line from $\vec{x}$ to $\vec{0}$. Then any line of force can be labelled by the point $y_{\perp}$ where it intersects the central plane: $y_{\perp}$ is a two-dimensional vector in the central plane. To completely specify a line of force one must specify the complete line $LF(y_{\perp})$. Each line of force has a weight factor $\rho(y_{\perp})$. The resulting state is

$$\bar{\psi}(x) \int d^2 y_{\perp} \rho(y_{\perp})\, P \exp \{ig \int_{LF(y_{\perp})} A_{\mu}^{a}(z) T^{a} \cdot dz^{\mu}\}\psi(0)|\Omega\rangle. \tag{48}$$

The variational procedure is to vary the weight function $\rho(y_{\perp})$ and the locations of the lines $LF(y_{\perp})$ (subject to the constraints that the line $LF(y_{\perp})$ begins at x, passes through the point $y_{\perp}$ on the central plane, and ends at 0).

How do the string operators affect the energy of a state? A string operator creates particles (since any operator in a quantum field theory creates particles). Because a single string operator is infinitely thin, the particles it creates will have infinite momentum in the directions perpendicular to the string, which means infinite energy. If a sum of strings of finite width can act coherently, they can create particles with finite momentum and hence finite energy. Thus the lines of force want to be spread out in order to minimize the energy of the particles they create. In the weak coupling case the particles created are zero mass gluons. For two quarks of separation L the lines of force can easily be spread out over a region of transverse size L, so that the energy of the gluons in the central region between the two quarks is of order 1/L. When the quarks are pulled away further, the lines of force spread out further, thus reducing the energy created in the central region. The total energy of the two quarks clearly would be finite out to infinite separation, if this picture is correct.

An obvious way to spoil this picture is to eliminate the coherence between strings. In particular, suppose there is a finite coherence length ξ beyond which the string operators are uncorrelated.

If the strings are spread out over a region of size L, but the coherence length ξ is only L/2, then we can write the sum of strings as two operators O_1 and O_2, each representing the sum of strings in a region of size L/2. The lack of coherence means the energy of the sum is the average of the energy of each term: if each term has energy 2/L (corresponding to the size L/2) then so does the sum. (Mathematically, the energy of a state created by $O_1 + O_2$ is

$$E = \frac{\langle\Omega|(O_1 + O_2)^+ H(O_1 + O_2)|\Omega\rangle}{\langle\Omega|(O_1 + O_2)^+ (O_1 + O_2)|\Omega\rangle} \tag{49}$$

If O_1 and O_2 are operators in different regions of space separated by more than a coherence length, then one can simplify eq. (40) using the standard cluster decomposition rules. The result is that all terms involving both O_1 and O_2 vanish, giving

$$E \simeq \frac{\langle\Omega|O_1^+ H O_1|\Omega\rangle + \langle\Omega|O_2^+ H O_2|\Omega\rangle}{\langle\Omega|O_1^+ O_1|\Omega\rangle + \langle\Omega|O_2^+ H O_2|\Omega\rangle} \tag{50}$$

which is an average of the energies of O_1 and O_2 separately. It is assumed here that the vacuum expectation values of O_1 and O_2 have been subtracted from O_1 and O_2.) Thus if there is a finite coherence length there is no energy gained if the lines of force spread out indefinitely: the Coulomb-type distribution does not have a lower energy than the sausage type.

We must determine the coherence (or correlation) length of the gauge field. To discuss this question consider a set of lines of force forming closed loops (see Fig. 3).

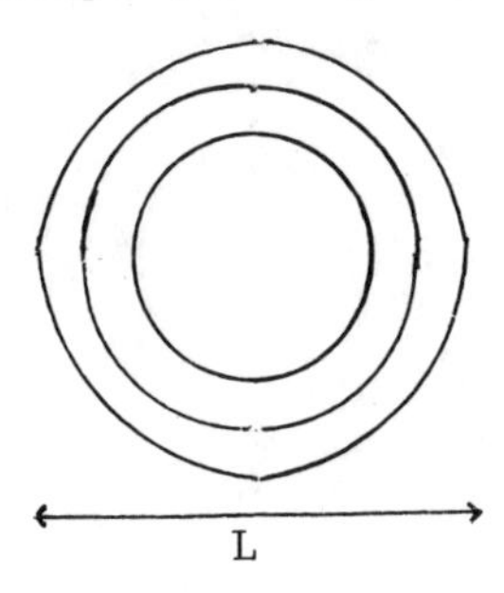

Fig. 3. Closed loops of lines of force

A closed line of force on a line LF corresponds to a trace of a string operator:

$$\mathrm{Tr}\ \exp\left\{ig \int_{LF} A_\mu^a(x) \cdot T^a\ dx^\mu\right\} \quad , \tag{51}$$

where the trace is over the color SU(3) indices. A sum of lines of force is considered in order to produce an operator with a single size scale (L). This means its qualitative behavior is determined by the effective coupling constant for the size scale L.

To discuss coherence we shall think of the sum of closed loop strings as a product of two halves, the halves being operators A and B (see Fig. 4). Strictly speaking one must have all lines of A connected to all lines of B at the joints between the two in order that one have a strict product; this problem will be ignored here.

The question is, are there correlations between the two operators A and B. To determine this, the fluctuations in A, B and the gauge invariant product Tr AB will be considered. For weak effective coupling the string operators have the value (Tr 1) + $0(g_{eff})$, as can be seen by expanding the exponential. That is, the expectation value of Tr AB is of order 1 plus a small correction and the fluctuations of Tr AB about its expectation value are also small (of size g_{eff}). At first sight this would also be true of A and B separately. However, A and B are not gauge invariant. A gauge transformation can change A or B by a large phase factor. Physics must be invariant to gauge transformations; therefore the fluctuations in A or B are of order 1, not of order g_{eff}. (The fluctuations in A and B can be artificially reduced by using gauge fixing, but by reducing the strength of the gauge fixing term one can increase the fluctuations in A and B so that they are larger than g_{eff} in size.) Since the fluctuations in the product Tr AB are much smaller than for the factors A and B separately, there must be strong correlations between the two. Hence if the effective coupling constant for size L is weak, any finite coherence length must be larger than L.

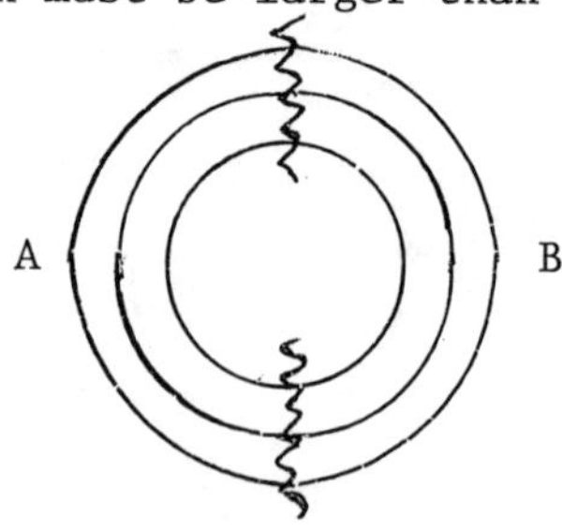

Fig. 4. Closed loops as a product of two operators

If the effective coupling constant for size scale L is large, the situation is different. The fluctuations in the product Tr AB are as large as the gauge-generated fluctuations of A and B separately. Hence it is possible that A and B are uncorrelated operators. It is not certain that A and B are uncorrelated. It is perfectly possible for the fluctuations of Tr AB to be large and yet not of the precise form they would have if A and B are independent.

Thus the existence of a strong effective coupling for large distances in the color gauge theory makes it possible that the correlation length for the gauge field is finite, but this is not guaranteed to happen. If the correlation length is finite then there is no advantage to spreading out the lines of force to a width larger than the correlation length. In this case there is a finite energy per unit length in the strings connecting the two quarks. In fact the strings connecting the two quarks (see Fig. 2) can be thought of roughly as a product of unit lengths of string, the unit of length being the correlation length. Each unit length of string defines an independent operator uncorrelated with other lengths of string; each unit length of string creates particles with an average energy, μ, say. The total energy for the whole length of strings connecting the quarks is μ times the separation of the quarks measured in units of the correlation length.

This analysis is qualitative and incomplete. A more quantitative analysis depends on completion of the "block spin" calculations described in Section III. However, the analysis of this Section should provide a pictorial framework for understanding the block spin approach and its relation to confinement.

III. BLOCK SPIN FORMALISM

One would like to be able to solve quantum chromodynamics. One would like to compute particle masses (pion, K, proton, etc.) (in terms of the fundamental parameters of the theory). One would like to find out whether quarks are confined in quantum chromodynamics. The block spin formalism provides a basis which may allow all these questions to be answered. However, very lengthy numerical calculations are required and these have not been performed as yet. In this Section the basic ideas of the block spin approach will be described.

The continuum gauge theory will be assumed to be the zero lattice spacing limit of the lattice gauge theory; the block spin method will be used to solve the lattice theory with an arbitrarily small lattice spacing.

The block spin approach originated with L. Kadanoff's paper on scaling in statistical mechanics[13]. The first quantitative block spin formalism was developed by Niemeijer and Van Leeuwen[14], for the Ising model. The first block spin calculations for the gauge theory were developed by Migdal[15] and extended by Kadanoff[16]. Unfortunately the Migdal-Kadanoff procedures involve very rough approximations. The approach defined here is an exact formalism which may be calculable with more controllable approximations.

The Feynman path integral of Section I is the starting point for the block spin calculation. To be precise we shall discuss the calculation of the generating functional for vacuum expectation values of a current, say the π meson field $\bar{\psi}\gamma_5\vec{\tau}\psi$. The path integral is

$$Z(h) = \Pi_n \Pi_\mu \int_{U_{n\mu}} \int_{\{\psi_n,\bar{\psi}_n\}} \exp\left\{A + \sum_n h_i(n)\bar{\psi}_n\gamma_5\tau_i\psi_n\right\} \tag{52}$$

where $h_i(n)$ is an arbitrary external source and i is an isotopic spin index. Vacuum expectation values are generated by differentiating Z(h) with respect to h.

It is not possible to compute Z(h) directly from eq. (52) because there are too many integrations to do. The block spin method is designed to reduce the number of integrations to a manageable size. Even after this reduction there will be thousands of integrations to perform but this may be feasible using Monte Carlo integration methods.

In the block spin approach one computes Z(h) in a sequence of steps. Each step consists in replacing the lattice of integration variables by a new lattice of integration variables. The new lattice has twice the spacing of the old lattice. In consequence, in a given volume, say 1 fermi on a side, there are many fewer sites of the new lattice (and hence fewer integration variables) than for the old lattice. The procedure for doubling the lattice spacing is to introduce a kernel into the formula for Z(h). The kernel K depends on both the new lattice variables and the old lattice variables. Let $U_{m\mu}$ and ψ_m be the new lattice variables, with m labelling the new lattice sites. Let $V_{n\mu}$ and ψ'_n be the old lattice variables. One then defines an action functional for the new lattice by

$$\exp\left\{A_1(\psi,U,h)\right\} = \Pi_{n\mu} \int_{V_{n\mu}} \Pi_n \int_{\{\psi'_n,\bar{\psi}'_n\}} K(U,\psi;V,\psi')$$

$$\times \exp\left\{A(\psi',V) + \sum_n h_i(n)\bar{\psi}'_n\gamma_5\tau_i\psi'_n\right\} , \tag{53}$$

The new action A_1 depends on the variables $U_{m\mu}$ and ψ_m of the new lattice. A_1 also depends on the source function $h_i(n)$ for the old lattice.

The fundamental requirement on the kernel (discovered by Kadanoff) is that the integral of the kernel over the new variables be 1:

$$\Pi_m \Pi_\mu \int_{U_{m\mu}} \Pi_m \int_{\{\psi_m,\bar{\psi}_m\}} K(U,\psi;V,\psi') = 1 \quad , \tag{54}$$

(this must be true for all values of the old variables $V_{n\mu}$ and ψ_n'). As long as Kadanoff's condition is satisfied, one has

$$Z(h) = \Pi_m \Pi_\mu \int_{U_{m\mu}} \Pi_m \int_{\{\psi_m,\bar{\psi}_m\}} \exp\left\{A_1(U,\psi,h)\right\} \quad . \tag{55}$$

This means that A_1 is an effective action on the new lattice: it is the integrand for a path integral just as A is the integrand for the original path integral. This means that the same method can be used to define subsequent steps in the calculation: the second step is obtained by using eq. (53) over again except that $A_1(\psi',V,h)$ is substituted for $A(\psi',V,h)$ and on the left-hand side a new action $A_2(\psi,U,h)$ replaces A_1.

The new lattice sites m are related to 2x2x2x2 blocks of lattice sites of the old lattice. This relation is made explicit through the form of the kernel K. The first step in defining the kernel is to define "block fields" on the old lattice. For example, a simple block quark field is

$$\chi_m = \eta \sum_{n\varepsilon m} \psi_n' \quad , \tag{56}$$

where "$n\varepsilon m$" means all old sites n which are in the block m. It is convenient to define the block $m = (m_1, m_2, m_3, m_4)$ to consist of the sites n for which $n_\mu = 2m_\mu$ or $2m_\mu - 1$. The constant η can be adjusted; see below. The quark part of the kernel K can now be a simple function of the new quark field ψ_m and the block field χ_m; for example

$$K_{quark}(\psi,\psi') = \exp\{-\zeta \sum_m (\bar{\psi}_m - \bar{\chi}_m)(\psi_m - \chi_m)\} \quad . \tag{57}$$

(The constant ζ is adjustable.)
However, for practical calculations a generalization of this form is used: see below. Because the kernel depends only on the difference $\psi_m-\chi_m$, the integral of $K_{quark}(\psi,\psi')$ over the new variables ψ gives

a constant independent of the old variables ψ' ; one then divides the exponential by this constant to give a kernel satisfying the Kadanoff condition.

The gauge field part of the kernel is more complicated. First of all, it is desirable to maintain gauge invariance. A gauge transformation among the new variables has the form

$$\psi_m \rightarrow \Phi_m \psi_m \tag{58}$$

$$\bar{\psi}_m \rightarrow \bar{\psi}_m \Phi_m^+ \tag{59}$$

$$U_{m\mu} \rightarrow \Phi_m U_{m\mu} \Phi_{m+\hat{\mu}}^+ \tag{60}$$

where Φ_m is an arbitrary SU(3) matrix for each site m. To ensure that A_1^m is invariant to these gauge transformations it is necessary to know how the old variables transform. The logical rule is that Φ_m is the gauge transformation for every site n in the block m. Thus

$$\psi_n' \rightarrow \Phi_m \psi_n' \qquad \text{(for } n \in \text{block } m) \tag{61}$$

$$V_{n\mu} \rightarrow \Phi_m V_{n\mu} \Phi_{m+\hat{\mu}}^+ \qquad \text{(if } n \in \text{block } m \text{ and } n+\hat{\mu} \in \text{block } m+\hat{\mu}) \tag{62}$$

However if $n_\mu = 2m_\mu - 1$ then n and $n+\hat{\mu}$ are both in the block m and in this case

$$V_{n\mu} \rightarrow \Phi_m V_{n\mu} \Phi_m^+ \quad . \tag{63}$$

This is called a block gauge transformation. We need a block gauge field defined to transform the same way as $U_{m\mu}$. The simplest (but not only) choice is to define the block field $W_{m\mu}$ to be

$$W_{m\mu} = \sum_{n\epsilon m,\ n+\hat{\mu}\,\epsilon\, m+\hat{\mu}} V_{n\mu} \quad . \tag{64}$$

That is, the block field $W_{m\mu}$ is a sum over the links $(n, n+\hat{\mu})$ which connect the block m to the adjacent block $m+\hat{\mu}$. A relatively simple kernel is

$$K_{gauge}(U,V) = \exp\left\{\beta \sum_{m\mu} \mathrm{Tr}(U_{m\mu}^+ W_{m\mu} + W_{m\mu}^+ U_{m\mu})\right\} \quad . \tag{65}$$

To ensure that this kernel satisfies the Kadanoff criterion, one must divide it by the integral

$$\Pi_{m\mu} \int_{U'_{m\mu}} \exp\{\beta \sum_{m\mu} \mathrm{Tr}(U'^{+}_{m\mu} W_{m\mu} + W^{+}_{m\mu} U'_{m\mu})\} \quad . \tag{66}$$

The constant β is adjustable. The complete kernel is the product of the gauge kernel times the quark kernel.

The effective actions A_1, A_2, etc., contain, in principle, every possible interaction that is gauge invariant and invariant to the symmetries of the four-dimensional lattice. One cannot calculate the coefficient of all possible interactions in practice. Hence it is essential to the calculation of A_1, etc., that only a finite number of interactions be important.

There are several hypotheses one can use to limit the number of interactions that are important. The first is locality. The locality hypothesis is that the most important interactions are those which involve the fewest and closest lattice sites. For example the term $\bar{\psi}_m \psi_{m+\hat{\mu}}$ should be more important than the term $\bar{\psi}_m \psi_{m+2\hat{\mu}}$, although both will be present[17]. The justification for the locality hypothesis is that, for example, a second nearest neighbor coupling on the new lattice is roughly analogous to a fourth nearest neighbor coupling ($\bar{\psi}'_n \psi'_{n+4\hat{\mu}}$) on the original lattice, due to the factor 2 change in the lattice spacing from the old to the new lattice. Thus unless one has appreciable fourth nearest neighbor coupling on the original lattice, one will not get much second nearest neighbor coupling on the new lattice. Unfortunately, there is another way to generate second nearest neighbor coupling on the new lattice: the product of four separate nearest neighbor couplings on the old lattice can simulate a single fourth nearest neighbor coupling on the old lattice. To partially prevent this one tries to adjust the kernel to reduce the strength of the nearest neighbor coupling. For example, in the original action (eq. 1) the coupling term $\bar{\psi}'_n \psi'_{n+\hat{\mu}}$ (without γ_μ) has a positive coefficient K. The kernel $K_{quark}(\psi,\psi')$ also has nearest neighbor terms $-\zeta\eta^2 \bar{\psi}'_n \psi'_{n+\hat{\mu}}$, provided that n and $n+\hat{\mu}$ are in the same block m. Since these terms are of opposite sign, (negative) to the original nearest neighbor term, they weaken the effects due to products of several nearest neighbor terms and hence reduce the strength of nonlocal terms in the effective action A_1. This argument suggests that one should also like to have $\bar{\psi}'_n \gamma_\mu \psi'_{n+\hat{\mu}}$ terms in the kernel, again of opposite sign from those in the original action. This can be achieved by redefining χ_m and $\bar{\chi}_m$ in eq. (56) to be

$$\chi_m = \eta \sum_{n\varepsilon m} [1 - (-1)^{n_\mu} \eta' \gamma_\mu]\psi'_n \quad , \tag{67}$$

$$\bar{\chi}_m = \eta \sum_{n\varepsilon m} \bar{\psi}'_n [1 + (-1)^{n_\mu} \eta' \gamma_\mu] \quad , \tag{68}$$

where η' is another adjustable constant. (The sign change for the η' term between χ_m and $\bar{\chi}_m$ is necessary and contradicts no fundamental requirements, even though it means that

$$\bar{\psi}_n' \neq (\psi_n')^+ \gamma_0 \quad . \tag{69}$$

In the Euclidean metric $\bar{\psi}_n'$ and ψ_n' are independent integration variables and the Hermitian conjugate $\psi_n'^+$ has no meaning.)

In practice one selects a few nonlocal interactions to compute such as the coefficients of

$$\bar{\psi}_m U_{m\mu} U_{m+\hat{\mu},\mu} \psi_{m+2\hat{\mu}} \text{ or } \mathrm{Tr}\, U_{m\mu} U_{m+\hat{\mu},\mu} U_{m+2\hat{\mu},\nu} U^+_{m+\hat{\mu}+\hat{\nu},\mu} U^+_{m+\hat{\nu},\mu} U^+_{m\nu}$$

and then the adjustable parameters (ζ,η,η',β) in the kernel are adjusted to minimize these coefficients. If the minimization is successful enough the nonlocal interactions are discarded.

The next simplification has to do with complicated functions of variables from a few (or even only one) lattice
one can have products of four or more quark fields at one site or two nearest neighbor sites; these cannot be ruled out by locality. However, since phenomenologically the systems of two or three quarks seem to be the important ones, it is hoped that higher powers of the quark fields will also not be important. There is no theoretical evidence either for or against this hypothesis at present. Secondly, for the gauge field one can have an interaction which is an arbitrary function of the simplest trace of U's. Here the aim will be to construct a polynomial fit to this function using only a few parameters; this should be feasible since the trace of unitary matrices is bounded. Thus the argument of the function has a bounded range.

The above hypotheses, if correct, allow the construction of effective actions with a reasonable set of parameters, say 50 - 100.

To determine the parameters of a new action (A_1, A_2, etc.) one must compute the $V_{n\mu}$, ψ_n', and $\bar{\psi}_n'$ integrations in eq. (53). These computations are carried out for specific sets of values of the new variables $\{U_{m\mu}, \psi_m, \bar{\psi}_m\}$. One must pick (essentially at random), say, 200 sets of these values, which means 200 separate calculations of the integrations over $V_{n\mu}$, ψ_n' etc. The result will be 200 values of A_1, which are to be fitted by 50 - 100 parameters. The redundancy can be used to check whether these parameters are sufficient.

It is not possible to do the integrations in eq. (53) exactly. One has to truncate the number of lattice sites involved to a manageable level and then approximate the integrals that are left. For example, to compute the term in A_1 involving the trace of four U's around an elementary square on the new lattice involves at a minimum the four blocks (on the old lattice) corresponding to the four corners of the square. Since each block contains sixteen old lattice sites, one has a minimum total of sixty-four lattice sites. One may also want to include other nearby blocks in the calculation, but hopefully these will be less important, especially if the kernels have been successfully chosen to minimize the coherent effects of many nearest neighbor interactions. Between sixty-four lattice sites there are 198 links - giving 198 simultaneous gauge field integrations. These integrals must be computed either by Monte Carlo methods or by weak or strong coupling methods. At the present time it appears that the Monte Carlo calculations may be feasible, which would avoid the necessity of expansions.

There are several attractive features of the block spin approach. One is that each effective action in the sequence A_1, A_2, etc. is associated with a definite length scale, namely the associated lattice spacing. If the initial lattice spacing is a then the scale for A_ℓ is 2^ℓ times a. For each such scale there is a different effective coupling g_{eff}, which can be determined from the coefficient of the trace of the elementary gauge loop square in A_ℓ. (A_ℓ will have a term identical to the last term in eq.(1) except that g_{eff} replaces g_0.) This means the behavior of the coupling constant vs. distance can be determined even in the infrared region where perturbation theory breaks down, assuming the Monte Carlo calculations are feasible.

Secondly, the gauge field variables of A_ℓ are very similar in spirit to the sums of string operations discussed in Sec. II, more precisely to the sum over strings in a box of size $2^\ell a$. Thus the physical question of whether the string variables are correlated over distances $\sim 2^\ell a$ becomes the question of whether the $U_{m\mu}$ are correlated in A_ℓ. The simplest way to avoid such correlations at large distances is for A_ℓ to become independent of the $U_{m\mu}$ for large ℓ, which means in particular that the effective coupling g_{eff} must go to ∞ for $\ell \to \infty$. Thus the confinement problem translates into a precise statement about the limit of A_ℓ for large ℓ. Migdal[15] and Baaquie[18] have already verified the existence of confinement in the SU(2) pure gauge theory of color using Migdal's approximate recursion formula, but one awaits confirmation of this result in more accurate approximations.

The effective action A_ℓ is a function of the original external field h_n on the original lattice. It complicates matters if one has to compute the dependence of A_ℓ on the external field. Fortunately the S matrix can be computed without reference to any external fields. The reason for this is the following: To compute the S matrix what is required is the vacuum expectation values of fields for large separations. This means that the generating functional $Z(h)$ is required to be known only to linear order in the field h_n from any particular region of space of nuclear size. Terms of order h^2 and higher are needed only for products of well separated external fields Now if one considers only the actions A_ℓ for effective lattice spacing of order the nuclear size or smaller, these actions, due to locality, will not contain any products of well separated h fields. Hence one needs to compute A_ℓ only to linear order in h. To linear order A_ℓ has to have the form

$$A_\ell(U,\psi,h) = \sum_n h_n\, O_n\,(U,\psi) + A_\ell(U,\psi,\ h{=}0) \quad , \tag{70}$$

for some function O_n depending on the fields U, ψ. The fields U, ψ are labelled by lattice sites with separation $2^\ell a$ while n labels lattice sites of the original lattice; by locality, therefore, O_n will depend primarily on $U_{m\mu}$ and ψ_m for $m \sim 2^{-\ell}n$. Now suppose $Z(h)$ were computed directly from the action A_ℓ, using eq. (55). Then $Z(h)$ would simply be the generator of vacuum expectation values of the composite field O_n, at least for well separated products of the O_n's. But since the S-matrix is independent of the interpolating fields used to generate it, one does not have to use the complicated composite field O_n; it is equally valid to use simple fields like $\bar\psi_m\gamma_5\tau\psi_m$ instead. In other words one can calculate the S matrix for the effective action A_ℓ with $h=0$ just as if it were itself the fundamental action. The idea that an action on a lattice with spacing of order the nuclear size can have the same S matrix as a continuum theory seems strange at first glance but involves no fundamental contradictions: see Ref. 19.

In order to compute the S matrix or $Z(h)$ one must stop iterating the equation for A_ℓ at some finite value of ℓ. Fortunately if confinement occurs one expects the effective coupling to become very large for sufficiently large ℓ, in which case the strong coupling methods discussed in Ref. 7 can be used to solve A_ℓ directly and one can stop iterating. Hence the block spin calculations would be repeated only until the effective coupling is large enough to use the strong coupling expansions.

There is one further complication to be discussed, namely gauge fixing. No gauge fixing is required in principle. But in practice the block spin calculations cannot be carried out when the effective coupling is weak except by using gauge fixing. When the effective coupling is weak, the action A_ℓ is numerically very large and rapidly varying. In consequence, only a small range of values of a particular integration variable $V_{n\mu}$ are important in the block spin integration. However, without gauge fixing, entire surfaces generated by gauge transformations are important. It is very difficult to integrate over these entire surfaces when the integrand varies rapidly in directions perpendicular to the surface. Hence one uses gauge fixing to shrink the region where A_ℓ is a maximum down to a manageable size.

One cannot use complete gauge fixing in the block spin integral because this would spoil the gauge invariance of the new effective action. To preserve the gauge invariance of the block spin action one must preserve block gauge invariance on the original lattice.

Only one of many possible forms of gauge fixing will be mentioned here. A procedure which is very simple but unfortunately destroys the lattice symmetry is to set some of the integration variables $V_{n\mu}$ equal to the unit matrix. In each block there are sixteen sites n, hence sixteen gauge matrices to be fixed. However, one of these gauge matrices must be identified with a block transformation; thus only fifteen remain to be fixed. This can be done by setting fifteen separate $V_{n\mu}$ matrices equal to one in each block. One must insist that each $V_{n\mu}$ be within a single block, that is, n and $n+\hat{\mu}$ must be in the same block m. This ensures that setting $V_{n\mu}$ equal to one is invariant to block gauge transformations (63). There is no way to choose fifteen links within a block symmetrically distributed in direction along four coordinate axes. Hence one will break manifest hypercubic symmetry. However one can still show that gauge invariant vacuum expectation values are unchanged by the presence of the gauge fixing so no symmetries of these vacuum expectation values are destroyed. For each action A_ℓ one first introduces gauge fixing and secondly carries out the block spin calculation using the gauge fixed action; neither of these changes Z(h) in any way if h_n is the external source for a gauge invariant field ($\bar{\psi}_n\gamma_5\tau_i\psi_n$ summed over colors to make a color singlet is a gauge invariant field).

One problem caused by gauge fixing is that the S matrix of the effective actions A_ℓ may have extra ghost particles which are not gauge invariant states of the original theory. One will have to be able to separate ghost states from physical states. This problem arises only when one replaces the correct composite fields O_n by simple fields on the effective lattice.

Using computers it is practical to iterate the block spin calculation hundreds of times; this means the initial lattice spacing can be as small as 10^{-40} cm or smaller, and one can iterate until the effective lattice spacing is of order 1 fermi or larger, where strong coupling expansions should be applicable. Thus one can easily determine whether one is solving a continuum theory: one simply asks whether the physical results (masses, etc.) are independent of the initial lattice spacing. If the same results are obtained for, say, initial lattice spacings of 10^{-20}, 10^{-30}, and 10^{-40} cm there is no doubt that one is solving a continuum theory. In Ref. 6, Sec. IV there is a more thorough discussion of the continuum limit of lattice theories and its relation to block spin calculations, using the concept of the "triangle of renormalization."

One aim of the block spin approach is to lay the basis for understanding whether quantum chromodynamics has confined quarks, and if so whether the bound state spectrum is recognizable as the observed spectrum of mesons and baryons. The author fully recognizes that a table of masses produced by hours of computer Monte Carlo calculations does not represent understanding. To obtain understanding one has to go through the calculation step by step, finding out what the crucial terms in each action A_ℓ are and why. One has to extract from each Monte Carlo integration a few important configurations which can then be checked by hand and try to understand from this the qualitative results of the integral.

REFERENCES

1. See, e.g., H. Fritzsch, M. Gell-Mann, and H. Leutwyler, Phys. Lett. 47B, 365 (1973).

2. H. D. Politzer, Phys. Rev. Lett. 30, 1346 (1973);
D. J. Gross and F. Wilczek, ibid 30, 1343 (1973);
G. 't Hooft, unpublished.

3. See, e.g., F. Gilman, Phys. Repts. 4C, 96 (1972).

4. See, e.g., T. Kinoshita and A. Ukawa, Cornell Univ.-Newman Lab. preprint CLNS-349.

5. F. Wegner, J. Math. Phys. 12, 2259 (1971);
K. Wilson, Phys. Rev. D10, 2445 (1974);
A. M. Polyakov, Phys. Lett. 59B, 82 (1975);
R. Balian, J. M. Drouffe, and C. Itzykson, Phys. Rev. D10, 3376 (1974); ibid D11, 2098, 2104 (1975).

6. See, e.g., K. Wilson, Revs. Mod. Phys. 47, 773 (1975), Sec. VI.

7. K. Wilson, Erice Lecture notes (1975).

8. J. Kogut and L. Susskind, Phys. Rev. D11, 395 (1975).

9. J. Bardeen, Proceedings of the Tbilisi conference (1976). See also S. D. Drell, M. Weinstein, and S. Yankielowitz, SLAC-PUB-1752 (1976).

10. M. Lüscher, Hamburg Univ. preprint (1976).

11. F. A. Berezin, The Method of Second Quantization, (Academic, New York, 1966) pp. 52 ff.

12. L. J. Tassie, Phys. Lett. 46B, 397 (1973);
H. Nielsen and P. Olesen, Nucl. Phys. B57, 367 (1973); ibid B61, 45 (1973);
J. Kogut and L. Susskind, Phys. Rev. D9, 3501 (1974).

13. L. Kadanoff, Physics 2, 263 (1966).

14. Th. Niemeijer and J. M. J. Van Leeuwen, Physica (Utr) 71,17 (1974).

15. A. A. Migdal, Zh. E.T.F. 69, 810, 1457 (1975).

16. L. Kadanoff, unpublished lecture notes.

17. For simplicity in this discussion all gauge field variables (U and V) have been set equal to 1. This is a good approximation when the coupling is weak. The example being discussed is not applicable in the strong coupling domain.

18. B. Baaquie, Ph.D. thesis, Cornell University (1976), unpublished

19. K. Wilson in New Pathways in High Energy Physics, Vol. II, A. Perlmutter, ed. (Plenum, New York, 1976).

GAUGE THEORIES ON THE LATTICE[†]

Konrad Osterwalder [□*]

Jefferson Laboratory, Harvard University

Cambridge, Massachusetts 02138, U.S.A.

I. INTRODUCTION

In 1954 Yang and Mills first introduced a non-Abelian generalization of the gauge symmetry of electrodynamics [1]. They studied a Lagrangian that was invariant under arbitrary space-time dependent rotations in isospin space. In the twenty years that followed, much of the progress in the understanding of elementary particle physics was connected in one way or another with the rich structure of quantum field theories involving non-Abelian gauge fields. The connection between broken symmetries, Goldstone particles and gauge fields [2], the unification of weak, electromagnetic and strong interactions [3], the renormalization of Yang Mills fields [4], asymptotic freedom [5] and the recent progress in the understanding of quark trapping [6] are some of the milestones in the developments of that period.

---

† Work supported in part by the National Science Foundation under Grant PHY 75-21212.

* Alfred P. Sloan Foundation Fellow.

□ Everything that is original in this lecture describes work that was carried out in collaboration with Erhard Seiler.

On a formal level, gauge fields seem to be very natural objects, because they force a definite interaction upon us and leave not much freedom except for the choice of a coupling constant. It is also this feature that makes gauge theories so difficult to deal with. While we feel that we understand ordinary boson and fermi fields quite well - see e.g. the contributions of Fröhlich, Glimm, Jaffe and Spencer to these proceedings - our knowledge of Yang-Mills fields is still on a rather rudimentary level. The purpose of this talk is to discuss some ideas that might eventually lead to a rigorous definition and discussion of quantum field theories involving non-Abelian gauge fields.

A rigorous approach usually involves a regularization of the theory through the introduction of cutoffs and a subsequent discussion of the limits as the cutoffs are removed. The cutoffs always destroy some of the symmetries of the models which then have to be recovered in the limit. Also, some cutoffs destroy the positivity of the metric of the Hilbert space of physical states. This should be avoided because it seems to be rather complicated to exhibit positivity as a property that holds in the no cutoff limit only. We conclude that a good cutoff should preserve positivity and most of the symmetries. In particular it should preserve gauge invariance. The lattice approximation first introduced by Wilson [6] and by Polyakov has these properties.

In Section II we introduce lattice approximations for Euclidean Yang-Mills fields [6] and for fermion fields [7] and show that the physical Hilbert space has a positive metric. We follow closely the general methods set up in [8] and in [9]. In Section III we discuss a simplified version of a pure Yang-Mills theory on the lattice with arbitrary coupling and discuss the mass as a function of the coupling strength. In Section IV we construct Yang-Mills models for strong coupling; to prove the existence of an infinite volume limit we use a cluster expansion. This expansion also allows us to get a lower bound for the mass and to verify Wilson's "confinement bound" [6]. In Section V finally we discuss the duality transformation as a possible approach to the weak coupling region, see [10].

II. EUCLIDEAN FIELD THEORIES ON THE LATTICE

Euclidean Green's functions of an arbitrary fermion-boson model uniquely define a relativistic quantum field theory if they satisfy the conditions of physical positivity, (anti)symmetry, Euclidean covariance and a regularity condition [9].

In this section we introduce a lattice approximation for Euclidean quantum field theories and show that it satisfies the physical positivity condition (i.e. positive metric for the Hilbert space of physical states). This condition will continue to hold in the continuum limit, i.e. as the lattice spacing tends to zero.

We will always assume that we have a hypercubical lattice Λ in $\mathbb{R}^d$ with lattice spacing a. For technical reasons (see II.3) we choose Λ to be symmetric with respect to the coordinate axes with no points on these axes. A general point in Λ is thus of the form

$$x = (x^1, \cdots x^d), \quad x^i = a(\tfrac{1}{2} + n), \quad n \text{ integer}, \quad |x^i| < L_i .$$

The values of L_i determine the size of Λ. Furthermore we let e_μ be the vector of length a parallel to the μ'th coordinate axis. If $x, y \in \Lambda$ and $|x - y| = a$, we write xy for the <u>directed</u> lattice bond from x to y. For simplicity we assume $a = 1$ unless specified otherwise.

II.1 Lattice Yang-Mills Fields [6]

Let $\mathcal{G}$ be a compact gauge group, χ a character on $\mathcal{G}$, and $g_{\cdot}$ a map from directed lattice bonds into $\mathcal{G}$ such that

$$g_{xy} = g_{yx}^{-1} .$$

An elementary square of the lattice is called a plaquette and we associate with every plaquette $P = {}^{x_4}_{x_1}\square^{x_3}_{x_2}$ the expression

$$A_P = \tfrac{1}{2}(\chi(g_{x_1x_2} g_{x_2x_3} g_{x_3x_4} g_{x_4x_1}) + \text{complex conj.})$$

Notice that A_P is independent of the orientation and of the starting point of the path around the plaquette P. More generally with a closed path $C = (x_1, x_2, \cdots, x_n = x_1)$ on the lattice, $|x_i - x_{i+1}| = a$, we associate the expression $\chi(C) = \chi(g_{x_1 x_2} \cdots g_{x_{n-1} x_n})$, where χ is again some character on $\mathcal{G}$.

The Yang-Mills action on the lattice Λ is now defined by

$$A_{\mathcal{P}} = \frac{1}{2g_0^2} \sum_{\mathcal{P}} A_P ,$$

where g_0 is the YM coupling constant and the sum $\sum_{\mathcal{P}}$ runs over all plaquettes P in $\mathcal{P}$ (= set of pl. in Λ). For $\mathfrak{F}$ an arbitrary function of the bond variables g_{xy} we define the average

$$\langle \mathfrak{F} \rangle_{\mathcal{P}} = \frac{1}{Z_{\mathcal{P}}} \int \mathfrak{F}\, e^{A_{\mathcal{P}}} \prod_{\Lambda} dg_{xy}$$

$$Z_{\mathcal{P}} = \int e^{A_{\mathcal{P}}} \prod_{\Lambda} dg_{xy} .$$

Here dg_{xy} denotes Haar measure on $\mathcal{G}$. The product $\prod_{\Lambda}$ runs over all bonds xy in Λ. Let h. be a map from lattice points into the group $\mathcal{G}$. Then a <u>gauge transformation</u> associated with h is a change of variables

$$g_{xy} \to g'_{xy} = h_x g_{xy} h_y^{-1} . \tag{2.1}$$

Both the action $A_{\mathcal{P}}$ and the measure $\prod dg_{xy}$ are invariant under such transformations because for any closed path $C = (x_1, x_2, \cdots x_n = x_1)$

$$\chi(g_{x_1 x_2} \cdots g_{x_{n-1} x_n}) = \chi(g'_{x_1 x_2} \cdots g'_{x_{n-1} x_n})$$

and because Haar measure is invariant under left and right translations. As in the continuum case we may use local gauge transformations to eliminate redundant degrees of freedom. E.g. we can eliminate the variable g_{rs} from the action. Postponing integration over g_{rs} we choose a local gauge transformation

$$h_x = \begin{cases} \mathbb{1} & \text{for } x \neq s \\ g_{rs}^{-1} & \text{for } x = s \end{cases} .$$

The substitution (2.1) eliminates g_{rs} from the action (and from $\mathfrak{F}$ if it is a gauge invariant function) and integration over g_{rs} can now be carried out easily. In effect this procedure amounts to <u>fixing the bond variable g_{rs} to $\mathbb{1}$</u> (for gauge invariant $\mathfrak{F}$). Obviously we cannot fix <u>all</u> the bond variables. It is straightforward to show that by appropriate gauge transformations

> you can fix to $\mathbb{1}$ the variables of as many bonds as you can draw <u>without ever closing a loop</u>.

<u>Example 1:</u>

<u>Radiation gauge:</u> Pick a "time axis" parallel to one of the lattice generators. Fix to $\mathbb{1}$ all the variables of bonds parallel to this direction.

<u>Example 2:</u>

<u>Two-dimensional gauge theories:</u> Here the "plaquette variables" p can be introduced as independent variables and the measure

$$e^{\frac{1}{2g_0^2}\sum_P A_P} \prod dg_{xy} \rightarrow e^{\frac{1}{2g_0^2}\sum_P \frac{1}{2}(\chi(p)+c.c.)} \prod dp$$

factorizes. Here, given a plaquette $P = \begin{smallmatrix} x_3 & x_2 \\ x_4 & x_1 \end{smallmatrix}$ (square), the plaquette variable p is defined to be

$$p = g_{x_1x_2}\, g_{x_2x_3}\, g_{x_3x_4}\, g_{x_4x_1} ,$$

see Figure 1. This means that the two dimensional model is exactly solvable, the infinite volume limit is trivial, there is zero correlation, i.e. infinite mass, and there is not very much that can be learned from the study of the case d = 2.

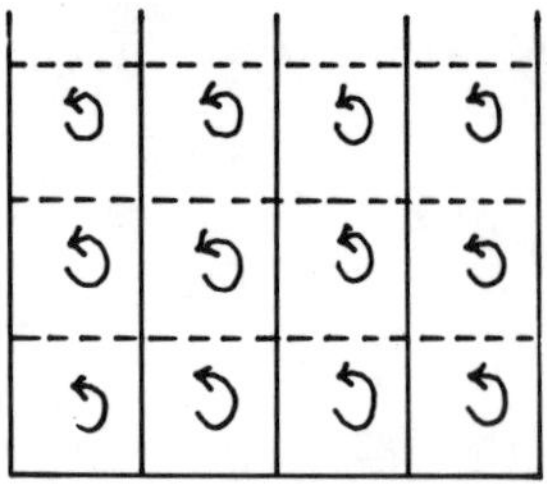

Figure 1. The variables belonging to ——— bonds have been fixed to 1. The variables belonging to ---- bonds are the remaining independent variables, which in turn can be replaced by the plaquette variables ↺ .

II.2 Euclidean Fermion Fields on the Lattice

Free Euclidean fermion fields Ψ^{k}_{μ} were introduced in [8]. They were defined such that

$$\langle \Psi^{1}_{\alpha}(x)\Psi^{2}_{\beta}(y)\rangle = \langle \overline{T}\, \psi_{\alpha}(ix_0, \vec{x})\psi^{\dagger}_{\beta}(iy_0, \vec{y})\rangle$$

$$= \frac{1}{(2\pi)^4}\int \left(\frac{m - i\gamma^{E}_{\mu}p_{\mu}}{p^2 + m^2}\right)_{\alpha\beta} e^{ip(x-y)}d^4p \qquad (2.2)$$

where $\{\gamma^{E}_{\mu}, \gamma^{E}_{\nu}\} = 2\delta_{\mu\nu}$, $\gamma^{E}_{\mu}p_{\mu} \equiv \sum\gamma^{E}_{\mu}p_{\mu}$, etc. E.g.

$$\gamma^{E}_{0} = \gamma_0 = \begin{pmatrix} \sigma_0 & 0 \\ 0 & -\sigma_0 \end{pmatrix}, \quad \gamma^{E}_{k} = i\gamma_k = i\begin{pmatrix} 0 & \sigma_k \\ -\sigma_k & 0 \end{pmatrix}, \; k = 1, 2, 3 .$$

ψ and $\psi^{\dagger} = \gamma_0\psi^{*}$ are free relativistic fermion fields.

To define a lattice approximation of this theory we interpret $\Psi^{k}_{\alpha}(x)$, $k = 1, 2$; $\alpha = 0, \ldots, 3$, $x \in \Lambda$, as generators of a Grassmann algebra $\mathcal{G}_{\Lambda}$, see [11]. On $\mathcal{G}_{\Lambda}$ we define a linear functional, written as an "integral", by first setting

$$⨍_{\Lambda} \bigwedge_{\alpha, x}(\Psi^{1}_{\alpha}(x) \wedge \Psi^{2}_{\alpha}(x)) = 1$$

$$\oint_{\Lambda} \begin{pmatrix} \text{monomial in } \Psi^k_a(x), \ \underline{\text{not}} \text{ containing} \\ \text{all the variables } \overline{\Psi^k_a(x)} \end{pmatrix} = 0 .$$

Then we extend $\oint_{\Lambda}$ to all of $\mathcal{G}_{\Lambda}$ by linearity. The symbol $\wedge$ means exterior multiplication and will be dropped in the following.

Now we define the Gibbs density μ_{Λ} by

$$\mu_{\Lambda} = e^{\sum_{x\in\Lambda}[m\Psi^1(x)\Psi^2(x)-\frac{1}{2}\Psi^1(x)\gamma^E_{\mu}(\Psi^2(x-e_{\mu})-\Psi^2(x+e_{\mu}))]} . \quad (2.3)$$

This is a lattice version of the formal expression $e^{\int\Psi^1(x)(m-\gamma^E_{\mu}\partial_{\mu})\Psi^2(x)d^4x}$ which formally gives (2.2). Terms with $x\pm e_{\mu} \notin \Lambda$ are to be left out in the sum in the exponent of (2.3). Notice that m can be any real number, including zero. In a theory coupling fermions to gauge fields, we have to pick a finite dimensional unitary representation $U(\cdot)$ of the gauge group $\mathcal{G}$. For $g \in \mathcal{G}$, $U(g)$ is then a unitary operator on some finite dimensional Hilbert space. The fermion variables $\Psi^k_a(x)$ are now vectors in that space. The fermion action is

$$A_F = \sum_{x\in\Lambda}[m\Psi^1(x)\Psi^2(x) - \tfrac{1}{2}\Psi^1(x)\gamma^E_{\mu}\{U(g_{x,\,x-e_{\mu}})\Psi^2(x-e_{\mu}) -$$

$$- U(g_{x,\,x+e_{\mu}})\Psi^2(x+e_{\mu})\}] . \quad (2.4)$$

Then, for $\mathfrak{F}$ an element in the Grassmann algebra $\mathcal{G}_{\Lambda}$ whose coefficients may be functions of the bond variables g_{xy}, we define the expectation by

$$\langle\mathfrak{F}\rangle_{\rho} = \frac{1}{Z_{\rho}}\int\prod_{\Lambda} dg_{xy}\, e^{A_{\rho}}\oint_{\Lambda}\mathfrak{F}\, e^{A_F}$$

$$Z_{\rho} = \int\prod_{\Lambda} dg_{xy} \oint_{\Lambda} e^{A_{\rho}+A_F} . \quad (2.5)$$

A local gauge transformation now consists in (2.1) plus a substitution

$$\Psi^1_\alpha(x) \to \Psi^{1'}_\alpha(x) = \overline{U}(h_x)\Psi^1_\alpha(x)$$

$$\Psi^2_\alpha(x) \to \Psi^{2'}_\alpha(x) = U(h_x)\Psi^2_\alpha(x) \quad . \tag{2.5}$$

The action $A = A_\rho + A_F$ is invariant under (2.1/5). We leave it as an exercise to verify that the functional $\oint_\Lambda$, too, is invariant under (2.5), see [11].

II.3 Physical Positivity

In this section we show how to use the expectation $\langle \cdot \rangle$ to define a (positive metric) Hilbert space $\mathcal{H}$, the Hilbert space of physical states. On $\mathcal{H}$ we will then construct the "transfer matrix" e^{-H}, where H is the Hamiltonian. For simplicity we assume that we have already constructed the infinite volume limit $\Lambda \to \mathbb{Z}^d$.

We call the e_d-direction the "time direction". The set $\mathfrak{a}^+$ of functions at positive times is defined to be the Grassmann algebra generated by $\{\Psi^k(x), x^d > 0\}$ with coefficients which are functions $f(\ldots g_{xy} \ldots)$ of the bond variables g_{xy}, $x^d > 0$, $y^d > 0$. Correspondingly we define $\mathfrak{a}^-$. An antilinear map $\Theta: \mathfrak{a}^+ \to \mathfrak{a}^-$ is defined as follows:

$$\Theta f(\ldots g_{x,y} \ldots)\Psi^{k_1}(x_1) \cdots \Psi^{k_n}(x_n)$$

$$= \overline{f}(\ldots g_{\vartheta x, \vartheta y} \ldots)(\gamma_0^E \Psi^{3-k_n})(\vartheta x_n) \cdots (\gamma_0^E \Psi^{3-k_1})(\vartheta x_1)$$

where $\vartheta x = (x^1, x^2, \ldots, -x^d)$ and $k_i = 1$ or 2.

The main result of this section is contained in the following theorem [7].

<u>Theorem:</u> With $\langle \cdot \rangle$ defined as in (2.5) and $\mathfrak{F}$ a gauge invariant function in $\mathfrak{a}^+$,

$$\langle \Theta\mathfrak{F} \cdot \mathfrak{F} \rangle \geq 0 \;.$$

Postponing the proof of this theorem we now define $\mathcal{H}$ to be the Hilbert space completion of $\mathfrak{A}^+_{g.i.}$, the gauge invariant part of $\mathfrak{A}^+$, equipped with the scalar product

$$(\mathfrak{F}, \mathfrak{F}') \equiv \langle \Theta\mathfrak{F} \cdot \mathfrak{F}' \rangle \; .$$

Define T to be the operator on $\mathfrak{A}^+_{g.i.}$ that shifts everything by two lattice units in time direction. This determines an operator $T_{\mathcal{H}}$ on $\mathcal{H}$ by

$$(\mathfrak{F}, T_{\mathcal{H}}\mathfrak{F}') = \langle \Theta\mathfrak{F} \cdot (T\mathfrak{F}') \rangle .$$

Assuming that $(\mathfrak{F}, T^n_{\mathcal{H}}\mathfrak{F}')$ grows at most like a power of n (this is usually very easy to show) one readily proves that $T_{\mathcal{H}}$ is a positive self-adjoint operator of norm less than 1. Hence we can write

$$T_{\mathcal{H}} = e^{-2H}$$

where H is the Hamiltonian and $H = H^* \geq 0$. e^{-H} is the transfer matrix, see [9].

Remark: We define T as shift by 2 lattice units, because that makes it easy to prove that $T_{\mathcal{H}}$ is positive. The operator "shift by one lattice unit" is of course a square root of T but it isn't necessarily positive. This is a peculiarity of the lattice approximation.

The proof of the theorem goes in four easy steps:

1) Pass to the radiation gauge where all the variables belonging to bonds parallel to the time direction are fixed to $\mathbb{1}$.
2) Show that for $\mathfrak{F}_1, \cdots, \mathfrak{F}_n$ in $\mathfrak{A}^+$ (but not necessarily gauge invariant)

$$\Pi[\Theta\mathfrak{F}_i \cdot \mathfrak{F}_i] = \Theta(\Pi\mathfrak{F}_i) \cdot \Pi\mathfrak{F}_i$$

(exercise)

3) Using 2) show that $e^{A_\rho + A_F}$ is of the form $\sum a_k \Theta\mathfrak{F}_k \cdot \mathfrak{F}_k$ with $\mathfrak{F}_k \in \mathfrak{A}^+$ and $a_k > 0$.

This is evident except for the part of A_ρ that comes from plaquettes which intersect the $x^d = 0$ plane. It suffices to consider one factor of the form

$$\rho = e^{\frac{1}{2g_0^2}(\chi(g_{\vartheta y \vartheta x} \cdot g_{xy}) + \text{compl. conj.})}$$

with $x^d = y^d = \frac{1}{2}$.

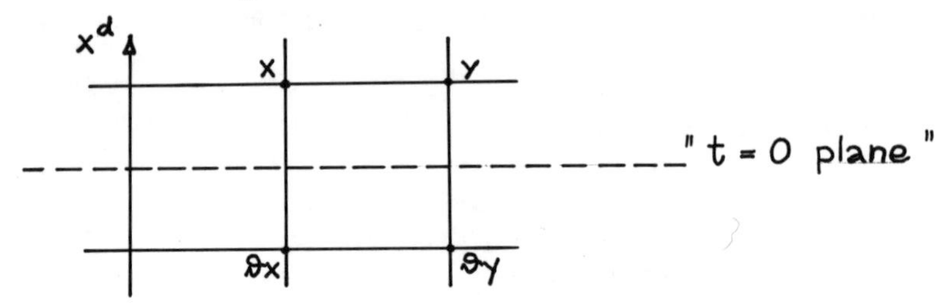

(Remember that we are in the radiation gauge, hence $g_{x,\vartheta x} = g_{y,\vartheta y} = 1$.) By the theorem of Peter and Weyl [12] we can (Fourier) expand

$$\rho = \sum_\sigma \lambda_\sigma(g_0)\chi_\sigma \tag{2.6}$$

where $\{\chi_\sigma\}$ are the characters of a complete system of inequivalent representations $D^{(\sigma)}$ of the group $\mathcal{G}$. The coefficients $\lambda_\sigma(g_0)$ are all nonnegative because ρ is a function of positive type, see [13, p.253]. Finally we observe that

$$\chi_\sigma(g_{\vartheta y \vartheta x} g_{xy}) = \sum_{ij} D_{ij}^{(\sigma)}(g_{\vartheta y \vartheta x}) D_{ji}^{(\sigma)}(g_{xy})$$

$$= \sum_{ij} \overline{D_{ji}^{(\sigma)}}(g_{\vartheta x \vartheta y}) D_{ji}^{(\sigma)}(g_{xy})$$

$$= \sum_{ij} \Theta D_{ji}^{(\sigma)}(g_{xy}) \cdot D_{ji}^{\sigma}(g_{xy}) \quad . \tag{2.7}$$

Combining (2.6) and (2.7) we find the desired result.

4) To prove the theorem it now suffices to show that for $\mathfrak{F} \in \mathfrak{A}^+$, in the radiation gauge,

$$\int \Pi \, dg_{xy} \oint (\Theta \mathfrak{F} \cdot \mathfrak{F})$$

is nonnegative. But this expression equals

$$\left| \int^{+} \Pi \, dg_{xy} \oint^{+} \mathfrak{F} \right|^2$$

where the + indicates that we integrate only over "positive time" variables.
This concludes the proof of the theorem.

Remarks: 1. Sections II.2/3 were not part of the original talk. They are motivated by Wilson's remarks on lattice fermions, see his contribution to these proceedings.
2. The construction proposed here is a straightforward discretization of a construction in [8]. It is different from Wilson's in the following respects: our lattice has no t = 0 plane where one easily runs into troubles, because Euclidean fermion operators always anticommute, while the relativistic $\psi^{+}(0, \vec{x})$ and $\psi(0, \vec{x})$ don't; we have no restrictions on the allowed values of the parameters in the action (except for some signs). As long as A is of the form $\Theta A^{+} A^{+}$ our construction works. Of course we could also replace γ_{μ}^{E} by $\gamma_{\mu}^{E} \pm 1$ (a method to project out half the number of degrees of freedom).
3. Recently M. Lüscher has informed us that he has constructed lattice fermions along the lines proposed by Wilson.

III. SIMPLIFIED YANG-MILLS MODEL

Our idea is to study the masses m_i of physical objects in the lattice theory as functions of the coupling constants g_i and of the lattice spacing a. Then we would like to pass to the continuum limit while keeping the physical masses fixed. This should determine the values of the (bare) coupling constants as functions of the lattice spacing. Scaling arguments show that

$$m_i = \frac{f_i(g)}{a}$$

hence we can fix a = 1 and study the functions f_i. For example, for a non-abelian Yang-Mills theory, one expects a

picture of the following type

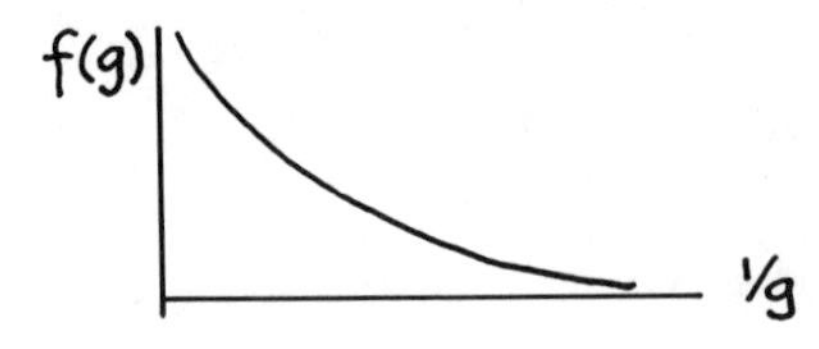

with $\lim_{g\to\infty} f = \infty$, $\lim_{g\to 0} f = 0$, while the abelian case should look different, e.g.

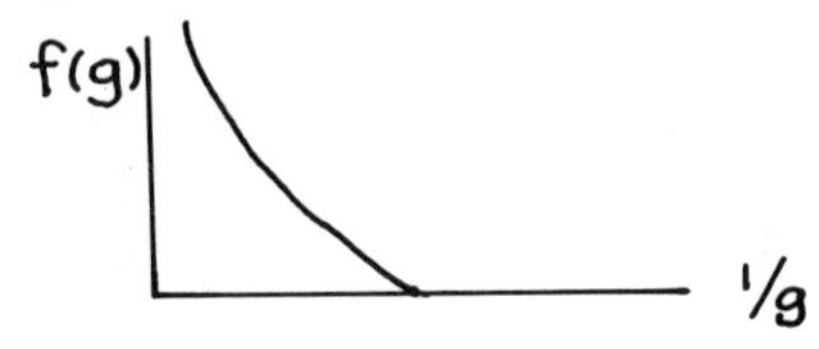

Little is known about the true behaviour of f(g), except for g sufficiently small; that case will be discussed in the next section. If we fix m and solve $m = \frac{f(g_0)}{a}$ we find that $g_0 = g_0(a)$ has to go to zero in the continuum limit (from asymptotic freedom one expects $g_0 \sim -1/\ln a$ as $a \searrow 0$). To get a better idea of what might happen, we first consider a simplified Yang-Mills model which is obtained from the full model by elimination from the action of all the plaquettes which are perpendicular to the time axis. Furthermore, we want to work in the radiation gauge, where all the bonds in time direction have their variables fixed to $\mathbb{1}$. What remains in the action is of the form

$$A_{\text{simplified}} = \frac{1}{2g_0^2} \sum_{P \| x^d} \tfrac{1}{2}(\chi(g_P h_P^{-1}) + \text{c.c.}) \tag{3.1}$$

The transfer matrix T corresponding to (3.1) is the tensor product of copies of the operator t, given by the kernel

$$\hat{t}(h, g) = e^{\frac{1}{4g_0^2}(\chi(hg^{-1}) + \text{c.c.}) + \alpha}$$

where α is a normalization factor, chosen such that the largest eigenvalue of t is 1. As in (2.6) we use the theorem of Peter and Weyl [12] to expand

$$\hat{t}(h) = \sum_{\sigma} \lambda_{\sigma}(g_0)\chi_{\sigma}(h)$$

where

$$\lambda_{\sigma}(g_0) = \int \hat{t}(h)\overline{\chi_{\sigma}}(h)dh$$

and $\{\chi_{\sigma}\}$ are the characters of a complete system of inequivalent, irreducible, unitary representations D_{σ} of the group $\mathcal{G}$. Let d_{σ} be the dimension of the representation D_{σ}, then

$$\hat{\lambda}_{\sigma} = \frac{\lambda_{\sigma}}{d_{\sigma}} \text{ are the } \underline{\text{eigenvalues}} \text{ of t}$$

and all <u>eigenvectors</u> of t with eigenvalue $\hat{\lambda}_{\sigma}$ are linear combinations of the matrix elements of D_{σ}. Because $\hat{t}$ is a function of positive type [13], its Fourier coefficients λ_{σ} are all nonnegative. We order the parameters σ such that

$$1 = \hat{\lambda}_{\sigma_1} \geq \hat{\lambda}_{\sigma_2} \geq \hat{\lambda}_{\sigma_3} \cdots \quad .$$

Interpreting T as e^{-H} we find that H has eigenvalues of the form $-\sum_{\sigma} n_{\sigma} \ln \hat{\lambda}_{\sigma}$, n_{σ} nonnegative integers. The smallest nonzero eigenvalue of H is therefore

$$m = -\ln \hat{\lambda}_{\sigma_2} = \ln \lambda_{\sigma_1} - \ln \lambda_{\sigma_2} .$$

If we restrict the action of the Hamiltonian to gauge invariant states, then it can be shown that

$$m_{g.i.} = -4 \ln \hat{\lambda}_{\sigma_2} . \tag{3.2}$$

For example, let $\mathcal{G} = U(1)$ and choose $\chi(\varphi) = e^{i\varphi}$, then for $\sigma = 0, \pm 1, \pm 2, \ldots$

$$\lambda_{\sigma}(g_0) = \frac{1}{\sqrt{2\pi}} \int e^{\frac{1}{2g_0^2}\cos\varphi} e^{i\sigma\varphi} d\varphi = \sqrt{2\pi} \sum_{k=0}^{\infty} \left(\frac{1}{4g_0^2}\right)^{\sigma+2k} \frac{1}{k!\,(\sigma+k)!} = \tag{3.3}$$

$$= \sqrt{2\pi}(-i)^{\sigma} J_{\sigma}\left(\frac{i}{4g_0^2}\right), \text{ see [14, p. 15]} .$$

This gives

$$m \sim \begin{cases} -\ln \frac{1}{4g_0^2} & \text{for } g_0 >> 1 \\ g_0^2 & \text{for } g_0 << 1 \end{cases}$$

Replacing U(1) by SU(2) we get essentially the same answers. As we will see in the next section, the large g_0-result of the simplified model comes close to what we find for an exact model; the expression we found for small g_0 would be exact in two dimensions, in four dimensions the predictions from asymptotic freedom are that $m \sim \exp(-\text{const.}/g_0)$. This discrepancy is not particularly surprising, because our simplified model is essentially two-dimensional.

IV. PURE YANG-MILLS THEORY WITH LARGE COUPLING: The Infinite Volume Limit and the Mass Gap [7]

In this section we construct the infinite volume limit of a pure Yang-Mills theory with arbitrary (compact) gauge group $\mathcal{G}$ and large coupling g_0 and prove a lower bound on the mass gap. We use a cluster expansion which is a well-known technique both in statistical mechanics [15] and in constructive quantum field theory [16]. We follow closely the methods of [16]. The basic idea of the expansion is that for large g_0

$$e^{\frac{1}{2g_0^2} A_P} \sim 1$$

because $|A_P| \leq D$, where D is the dimension of the representation of which χ is the character (see Section II.1). Hence we write

$$e^{\frac{1}{2g_0^2} \sum_{P \in \mathcal{P}} (A_P + D)} = \prod_{P \in \mathcal{P}} \left[1 + \left(e^{\frac{1}{2g_0^2}(A_P+D)} - 1 \right) \right]$$

$$= \sum_{Q} \prod_{P \in Q} \rho(P) \qquad (4.1)$$

Here

$$\rho(P) \equiv e^{\frac{1}{2g_0^2}(A_P+D)} - 1$$

is small,

$$0 \le \rho\,(P) \le \text{const.}\,\frac{1}{g_0^2} \tag{4.2}$$

and the sum $\sum_Q$ runs over all subsets Q of $\mathcal{P}$. (Remember that $\mathcal{P}$ stands for the set of all plaquettes in the lattice Λ).

Now let $\mathfrak{F}$ be a function of finitely many bond variables g_{xy} and let $\mathcal{P}_0$ be the set of plaquettes which have bond variables occuring in $\mathfrak{F}$. Then the expectation of $\mathfrak{F}$ is given by

$$\langle \mathfrak{F} \rangle_{\mathcal{P}} = \frac{1}{Z_{\mathcal{P}}} \int \mathfrak{F}\, e^{\frac{1}{2g_0^2} \sum_{P\in\mathcal{P}} (A_P + D)}$$

$$= \sum_Q \frac{1}{Z_{\mathcal{P}}} \int \mathfrak{F} \prod_{P\in Q} \rho\,(P) \tag{4.3}$$

where $Z_{\mathcal{P}} = \int e^{\frac{1}{2g_0^2} \sum_{P\in\mathcal{P}} (A_P + D)}$ is the partition function. Let us consider a single term in the sum of (4.3). Q is a collection of plaquettes in $\mathcal{P}$ and can therefore be written as the union of connected components:

$$Q = Q_1^c \cup Q_2^c \cup \cdots\cdot \cup Q_n^c\,.$$

Some of these components, $Q_1^c, \cdots, Q_k^c$, say, have bonds in common with $\mathcal{P}_0$ while the others, $Q_{k+1}^c, \cdots, Q_n^c$ don't. We now fix $Q' \equiv Q_1^c \cup \cdots \cup Q_k^c$ and resum over all terms in (4.3) with $Q_{k+1}^c, \cdots, Q_n^c$ having no bond in common with $\mathcal{P}_0$. If we carry out the integration over variables belonging to bonds not in $Q' \cup \mathcal{P}_0$ then we just get the partition function $Z_{\mathcal{P}\setminus Q'\cup\mathcal{P}_0}$, where $\mathcal{P}\setminus Q'\cup\mathcal{P}_0$ is the set of plaquettes in $\mathcal{P}$ which have no bond in common with $Q' \cup \mathcal{P}_0$. We thus obtain from (4.3)

$$\langle \mathfrak{F} \rangle_{\mathcal{P}} = \sum_{Q'} \int \mathfrak{F} \prod_{P\in Q'} \rho\,(P) \cdot \frac{Z_{\mathcal{P}\setminus Q'\cup\mathcal{P}_0}}{Z_{\mathcal{P}}}\,. \tag{4.4}$$

This is called the cluster expansion. We will show below that (4.4) converges absolutely, the rate of convergence being governed by the size of $\mathcal{P}_0$ and of g_0.

Lemma 1: For g_0 sufficiently large, there are constants α, β depending on $\mathfrak{F}$, d, $\mathcal{G}$ and χ only such that

$$\sum_{\substack{Q' \\ |Q'|>K}} \left| \int^{\mathfrak{F}} \prod_{P \in Q'} \rho(P) \cdot \frac{Z_{\mathcal{P} \setminus Q' \cup \mathcal{P}_0}}{Z_{\mathcal{P}}} \right| \leq \alpha \left(\frac{\beta}{g_0^2} \right)^K . \quad (4.5)$$

Here $|Q'|$ is the number of plaquettes in Q'.

For a proof of this lemma we need three estimates:

(a) $$0 < \frac{Z_{\mathcal{P} \setminus Q' \cup \mathcal{P}_0}}{Z_{\mathcal{P}}} \leq 1 \quad \text{for all } \mathcal{P},\ Q' \cup \mathcal{P}_0 .$$

This follows immediately if we write both numerator and denominator as integrals over all the bond variables and observe that

$$0 \leq (\text{integrand})_{\text{numerator}} \leq (\text{integrand})_{\text{denominator}} .$$

It was because of this inequality that we added D to A_P in (4.1), thus making (4.1) a function which is always larger or equal to 1.

(b) Let k be the number of plaquettes in Q'. Then

$$0 \leq \prod_{P \in Q'} \rho(P) \leq \left(\frac{c_1}{g_0^2} \right)^k$$

where c_1 is a constant that depends on the group and the choice of the character χ only. This follows from (4.2).

(c) Denote the number of possible choices of Q' with $|Q'| = k$, all components of Q' connected with $\mathcal{P}_0$, by N(k). Then

$$N(k) \leq c_2 \cdot c_3^k .$$

Here c_2 depends on the size of $\mathcal{P}_0$ while c_3 is a function of the dimension of space time only. This bound is purely combinatorial and will not be proven here, see however [16] and [7].

Combining (a), (b), and (c) we find that the left hand side of (4.5) is bounded by

$$\sum_{k>K} N(k)\left(\frac{c_1}{g_0^2}\right)^k \int |\mathfrak{F}| \leq c_2 \int |\mathfrak{F}| \sum_{k>K}\left(\frac{c_1 c_3}{g_0^2}\right)^k$$

$$= c_2 \int |\mathfrak{F}| \left(\frac{c_1 c_3}{g_0^2}\right)^K \left(1 - \frac{c_1 c_3}{g_0^2}\right)^{-1} .$$

This proves the lemma. It is crucial that the bound in (4.5) holds uniformly in the size of the lattice Λ. The next step is to prove clustering. Let $\mathfrak{F}$ be as before and v a lattice vector of length $|v|$. (We are assuming here that the lattice constant is one.) Then we define the translate function $\mathfrak{F}^v$ by

$$\mathfrak{F}^v(\cdots\cdot g_{xy} \cdots\cdot) = \mathfrak{F}(\cdots\cdot g_{x+v\, y+v} \cdots\cdot) .$$

Lemma 2: Let $\mathfrak{F}_1$ and $\mathfrak{F}_2$ be functions of finitely many lattice bond variables. Suppose g_0 is sufficiently large. Then there is a constant $c = c(\mathfrak{F}_1, \mathfrak{F}_2)$ such that

$$|\langle \mathfrak{F}_1 \mathfrak{F}_2^v \rangle - \langle \mathfrak{F}_1 \rangle \langle \mathfrak{F}_2 \rangle| \leq c\, e^{-m|v|} \tag{4.6}$$

uniformly in the size of Λ as $\Lambda \to \mathbb{Z}^d$. m is a constant satisfying

$$m \geq 4(\ell n\, g_0^2 - \ell n \beta) \tag{4.7}$$

with β as in Lemma 1.

To prove this lemma we follow Ginibre [17], see also [16], and introduce an artificial symmetry. Let g^*_{xy} denote a second variable corresponding to the bond xy, let $\mathfrak{F}^*_i$ be the function $\mathfrak{F}_i$ but depending on the g^*_{xy} variables rather than on g_{xy} and associate with every plaquette P the function $\tilde{A}_P = A_P + A^*_P$. Define a new expectation by

$$\langle \cdot \rangle_{\sim} = \frac{1}{\tilde{Z}_{\mathcal{P}}} \int \prod_{\Lambda} dg_{xy} dg^*_{xy} \, e^{\sum_{P \in \mathcal{P}} \tilde{A}_P} (\cdot) .$$

Notice that $\tilde{Z}_{\mathcal{P}} = (Z_{\mathcal{P}})^2$. The theory is now symmetric under $g_{xy} \leftrightarrow g^*_{xy}$. We can rewrite the left hand side of (4.6) as

$$\langle \mathfrak{F}_1 \mathfrak{F}_2^v \rangle - \langle \mathfrak{F}_1 \rangle \langle \mathfrak{F}_2 \rangle = \langle (\mathfrak{F}_1 - \mathfrak{F}_1^*)(\mathfrak{F}_2 - \mathfrak{F}_2^*)^v \rangle_{\sim} \tag{4.8}$$

Next we apply the cluster expansion (4.4) to (4.8). For v sufficiently large, $\mathcal{P}_0$ now consists in two widely separated components; the supports of $\mathfrak{F}_1$ and $\mathfrak{F}_2^v$, denoted by $\mathcal{P}_{01}$ and $\mathcal{P}_{02}^v$ resp. Consider a term in (4.4). If Q' consists in connected components Q_i^c all of which have bonds in common with only one of the components of $\mathcal{P}_0$, then we can apply the symmetry $g_{xy} \leftrightarrow g^*_{xy}$ in $\mathcal{P}_{01}$ and in all the Q_i^c's connected with it (but not in the rest). Because $\mathfrak{F}_1 - \mathfrak{F}_1^*$ is odd under this symmetry, but all the $\tilde{A}_P$ are even, terms with such Q' must vanish. The nonvanishing terms in (4.4) thus have a Q' that connects $\mathcal{P}_{01}$ and $\mathcal{P}_{02}^v$. Such a Q' contains at least $K = (|v| - r)$ plaquettes, where r is a fixed number depending on the size of $\mathcal{P}_{01}$ and $\mathcal{P}_{02}$. Applying Lemma 1 we find

$$|\langle (\mathfrak{F}_1 - \mathfrak{F}_1^*)(\mathfrak{F}_2 - \mathfrak{F}_2^*) \rangle| \le a\left(\frac{\beta}{g_0^2}\right)^{|v| - r}$$

$$\le c(\mathfrak{F}_1, \mathfrak{F}_2) \cdot e^{-|v|(\ell n \, g_0^2 - \ell n \, \beta)}$$

This proves Lemma 2 except for the factor 4 in (4.7). We leave it as an exercise to show that Q' must actually contain at least $4(|v| - r)$ plaquettes in order to give a nonvanishing contribution to (4.4), see also [7].

Inequality (4.7) gives a lower bound for the mass gap of the model, which is independent of the gauge group and in agreement with the exact result of the simplified model, see (3.2). We now show that the infinite volume limit $(\Lambda \to \mathbb{Z}^d)$ exists for expectations $\langle \mathfrak{F} \rangle$. Let $\mathfrak{F}$ be a function of finitely many plaquette variables and define $\mathcal{P}_0$ as before. We order all the plaquettes P in $\mathbb{Z}^d \sim \mathcal{P}_0$ such that the distance of P_i from $\mathcal{P}_0$ is a nondecreasing function of i. Finally set $\mathcal{P}_k = \mathcal{P}_0 \cup \{P_1, \cdots, P_k\}$ and study

$$\Delta_k = \langle \mathfrak{F} \rangle_{P_{k+1}} - \langle \mathfrak{F} \rangle_{P_k} .$$

Set

$$e^{\frac{1}{2g_0^2}(A_{P_{k+1}}+D)} = \mathfrak{F}_{k+1} ,$$

then

$$\Delta_k = \frac{\langle \mathfrak{F} \cdot \mathfrak{F}_{k+1} \rangle_{P_k}}{\langle \mathfrak{F}_{k+1} \rangle_{P_k}} - \langle \mathfrak{F} \rangle_{P_k}$$

$$= \langle \mathfrak{F}_{k+1} \rangle_{P_k}^{-1} [\langle \mathfrak{F} \cdot \mathfrak{F}_{k+1} \rangle_{P_k} - \langle \mathfrak{F} \rangle_{P_k} \langle \mathfrak{F}_{k+1} \rangle_{P_k}] .$$

The factor $|\langle \mathfrak{F}_{k+1} \rangle_{P_k}^{-1}|$ is always smaller than 1 and the expression in the square brackets decreases exponentially with the distance between P_0 and P_{k+1}. It follows that

$$\langle \mathfrak{F} \rangle_{\Lambda=Z^d} = \langle \mathfrak{F} \rangle_{P_0} + \sum_{k=0}^{\infty} \Delta_k \tag{4.9}$$

is well defined; the sum over k converges absolutely.

We summarize our results in a theorem.

Theorem: For g_0 sufficiently large, the infinite volume limit of the Yang-Mills lattice gauge theories exists, has a mass gap given by (4.7), satisfies physical positivity, and is invariant under translations by lattice vectors.

Remark: Translational invariance follows from uniqueness of the infinite volume limit (4.9).

The cluster expansion and Lemma 1 can be used to give a rigorous proof of Wilson's "confinement bound" [6] which we state as follows:

Theorem: Let C be a closed, rectangular path on the lattice and $\hat{\chi}$ a character of SU(3), belonging to an irreducible representation of triality different from 0. Then in an SU(3) Yang-Mills theory with sufficiently strong coupling

$$|\langle\hat{\chi}(C)\rangle| \leq \text{const.}\, e^{-m'\cdot \text{area}(C)} . \tag{4.10}$$

Here $m' = \ln g_0^2 - \ln \beta$, see Lemma 1, and (4.7). Area (C) is the area of the rectangle spanned by C.

For a proof one simply has to show that in the cluster expansion (4.4), Q' has to contain a set of plaquettes which span a hypersurface whose boundary is exactly C; terms with Q' not satisfying this condition vanish. A detailed demonstration is contained in [7]; for the cases $\mathcal{G} = U(1)$ and $\mathcal{G} = \mathbb{Z}_2$, see [6] and [10] resp., for weak coupling and $\mathcal{G} = U(1)$, see [21].

Remarks: 1) The condition on the triality of $\hat{\chi}$ appears to be crucial. (Triality can be defined as $\frac{1}{3}\{\#$ of squares in the Young diagram, characterizing the representation, mod $3\}$.) 2) It is not claimed here that, as a consequence of this theorem, quarks would be confined. In [6] Wilson indicates a possible linkage between bounds of the type (4.10) and confinement, but a rigorous treatment of this question cannot exclude the fermions from the discussion and is probably more complicated than the proof of this theorem.

V. THE WEAK COUPLING REGION; THE DUALITY TRANSFORMATION

The duality transformation relates a two dimensional Ising model at high temperature to a dual Ising model at low temperature, see [18], [19] and references given there. There is a generalization of this duality transformation to gauge theories in d dimensions with abelian or nonabelian gauge groups: it turns out to be a Fourier transformation on the group, see e.g. [20]. As in the Ising model, the duality transformation relates weak coupling problems to strong coupling problems. In the case of discrete gauge groups (!), such as $\mathbb{Z}_n$, it appears to be a powerful tool to control the weak coupling region, by applying the methods of chapter IV to the duality transformed model [10]. In more realistic cases (U(1), SU(n)) however, the method hasn't led to rigorous results for the weak coupling region yet, it probably has to be combined with other ideas before it can become effective.

Here we content ourselves with a brief sketch of the method and of the difficulties. The interested reader is referred to [10], [20] and to references given there.

We consider the case of $\mathcal{G} = U(1)$. Let P be a plaquette $x_4 \square x_3$ over $x_1\ x_2$, $g_{x_1x_2} = e^{i\varphi_1}$, $g_{x_2x_3} = e^{i\varphi_2}$, $g_{x_3x_4} = e^{-i\varphi_3}$, $g_{x_4x_1} = e^{-i\varphi_4}$. We choose

$$A_P = \cos\varphi_P, \quad \varphi_P = \varphi_1 + \varphi_2 - \varphi_3 - \varphi_4 .$$

Then for small g_0, $e^{\frac{1}{2g_0}A_P}$ deviates considerably from 1 for some values of φ_P, hence the cluster expansion of Chapter IV will fail. In fact, for g_0 sufficiently small, that factor will be close to zero unless $\varphi_P = 0$, i.e. it will approach a δ function $\delta(\varphi_P)$. This suggests to consider the Fourier transform,

$$e^{\frac{1}{2g_0^2}\cos\varphi} \sim e^{\frac{1}{2g_0^2}(1-\frac{\varphi^2}{2})}$$

$$= \text{Fourier transform of } e^{\frac{1}{2g_0^2}} \cdot e^{-g_0^2x^2} .$$

The Fourier transform thus relates the weak coupling problem to a strong coupling problem, as expected. More precisely

$$e^{\frac{1}{2g_0^2}\cos\varphi} = \sum_{n=-\infty}^{\infty} \lambda_n(g_0)e^{in\varphi} \tag{5.1}$$

where $\lambda_n(g_0) = \sqrt{2\pi}(-i)^n J_n(\frac{i}{4g_0^2})$ as in (3.3). For small g_0 we have $\lambda_n(g_0) \sim e^{-g_0^2n^2}$. For the partition function we find

$$Z = \int e^{\frac{1}{2g_0^2}\sum_P \cos\varphi_P} \prod d\varphi_i$$

$$= \sum_{n_i} \prod_i \lambda_{n_i}(g_0) \int \prod_i e^{in_i\varphi_{P_i}} \prod d\varphi_i . \tag{5.2}$$

The φ_i-integrations can be carried out explicitly and yield δ-functions. Consider a bond b with bond variable φ_b. The bond b occurs in $2k = 2(d-1)$ plaquettes, $P_1, P_2 \cdots P_{2k}$ say. Hence the φ_b integration in (5.2) gives 0 unless

$$n_1 + \cdots + n_k - n_{k+1} - \cdots - n_{2k} = 0 . \tag{5.3}$$

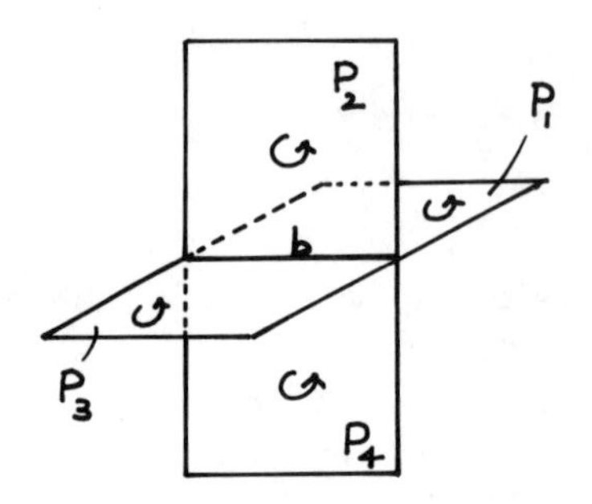

Example: <u>d = 3:</u>

The arrows ↺ indicate that the choice of the sign of φ_P associates a definite orientation with each plaquette.

The constraints (5.3) on the possible values of n_i can be expressed very easily in terms of a dual lattice. E.g., for $d = 3$, we associate with every (oriented) plaquette φ_i a directed bond b_i orthogonal to the plaquette with its middle at the center of the plaquette and directed according to the orientation of the plaquette.

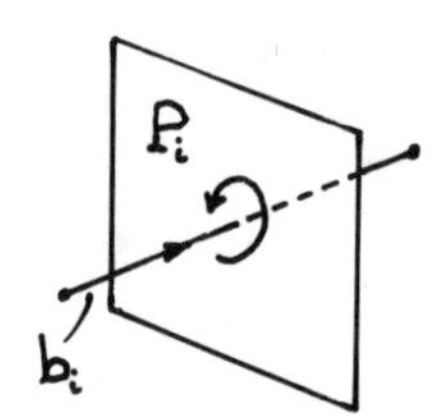

With b_i we associate the "dual lattice bond variable" n_i, the dual lattice being generated by all the b_i. The restriction (5.3) on the variables n_i can now be formulated as follows:

> if we sum up $\pm n_i$ along any closed path on the dual lattice, we obtain 0. The sign in front of n_i has to be chosen depending on whether the path runs parallel or antiparallel with the (directed) bond b_i.

This means that the n_i's are the components of a (discrete) curl free vector field and hence they are the "gradient" of a potential. In other words we can introduce spins x_k at the lattice sites k of the dual lattice and set

$$n_i = x_k - x_{k'}$$

where k and k' are the endpoints of the bond b_i

$$\underset{k}{\bullet}\xrightarrow{\quad b_i \quad}\underset{k'}{\bullet}$$

If we keep one of the spins, x_1 say, at a fixed value and allow all the other x_k's to vary over all integers, then the n_i's will vary exactly over all the values consistent with the constraints (5.3). We define a dual action by

$$\tilde{A} = \sum_{\substack{(k,k') \text{ nearest} \\ \text{neighbors}}} \ln \lambda_{(x_k - x_{k'})}(g_0)$$

$$\sim -g_0^2 \sum_{\substack{(k,k') \text{ nearest} \\ \text{neighbors}}} (x_k - x_{k'})^2, \quad \text{for small } g_0 .$$

The partition function becomes

$$Z = \sum_{x_2, x_3 \cdots \in \mathbb{Z}} e^{\tilde{A}}$$

and correspondingly one can express the expectation of functions of the lattice bond variables as expectations of functions on the dual lattice.

We have thus transformed the (weakly coupled) gauge theory into a lattice spin system with nearest neighbor interaction at high temperature.

In d = 4 dimensions the duality transformation maps the gauge theory into another gauge theory (for d = 4 the dual object of a plaquette is a plaquette), again replacing weak coupling by strong coupling.

In the non-abelian case the Fourier expansion (5.1) becomes

$$e^{\frac{1}{4g_0^2}(\chi(h)+\overline{\chi}(h))} = \sum_\sigma \lambda_\sigma(g_0)\chi_\sigma(h)$$

(compare with (2.6)), and the constraint (5.3) is now that

$$D^{(\sigma_1)} \times \cdots \times D^{(\sigma_k)} \times \overline{D^{(\sigma_{k+1})}} \times \cdots \times \overline{D^{(\sigma_{2k})}}$$

must contain the trivial representation. (σ_i labels the representation that is associated with the plaquette P_i in an individual term in (5.2).) On the dual lattice however we obtain a rather complicated theory, because the dual $\hat{\mathcal{G}}$ of a non-abelian group $\mathcal{G}$ is not a group.

There are troubles in the abelian case already. The character group $\hat{\mathcal{G}}$ of U(1) is discrete but not compact. Compactness, on the other hand, was important in the construction of Section IV.

In order to conclude this lecture on an optimistic note, let me look at an example where the duality transformation works. We choose $d = 4$, $\mathcal{G} = \mathbb{Z}_2$. ($\mathcal{G} = \mathbb{Z}_n$ would work equally well, see [10].) Here the dual group is $\hat{\mathcal{G}} = \mathbb{Z}_2$ and the model is self-dual, as we will see. The group variable h takes just two values $h = \pm 1$, and for $P = n_4 \underset{n_1}{\boxed{n_3}} n_2$

$$A_P = h_1 h_2 h_3 h_4 = h_P \,. \tag{5.4}$$

The two characters on $\mathcal{G}$ are defined by

$$\chi_1(h) \equiv 1, \quad \chi_{-1}(h) = h \,.$$

The Fourier expansion (5.1) yields (with $\beta = 1/2g_0^2$)

$$e^{\beta h_P} = \text{Cosh}\,\beta \cdot \chi_1(h_P) + \text{Sinh}\,\beta \cdot \chi_{-1}(h_P) \,.$$

With a plaquette on the dual lattice we have to associate a variable $\hat{h} = \pm 1$ with a weight $\text{Cosh}\,\beta$ for $\hat{h} = +1$ and $\text{Sinh}\,\beta$ for $\hat{h} = -1$. This weight factor can be rewritten as

$$\frac{1+\hat{h}}{2}\,\text{Cosh}\,\beta + \frac{1-\hat{h}}{2}\,\text{Sinh}\,\beta = c(\beta)e^{\beta^* \hat{h}} \tag{5.5}$$

where β^* is given by

$$\mathrm{Tgh}\,\beta^* = e^{-2\beta} .$$

The constraint for the allowed values of $\hat{h}$ is that

$$\prod_{\substack{\text{surface}\\ \text{of cube}}} \hat{h} = 1$$

(replacing (5.3)). This is guaranteed if we introduce bond variables on the dual lattice and define $\hat{h}$ of a plaquette as the product of the corresponding bond variables. Hence defining $\tilde{A}_P$ as we had defined A_P, (5.4), but replacing all the h_i by dual lattice bond variables $\hat{h}_i$ we find the "action" of the duality transformed model. The Gibbs factor is now $e^{\beta^*\tilde{A}}$, by (5.5). Notice that the constant $c(\beta)$ can be neglected because it drops out if we normalize the Gibbs measure. We have found that

> the $\mathbb{Z}_2$ gauge theory in $d = 4$ dimensions is self-dual with coupling constants related by
>
> $$\mathrm{Tgh}\frac{1}{2g_0^2} = e^{-(g_0^*)^{-2}} .$$
>
> If there is a phase transition in this model, then it must be for the coupling g_c satisfying
>
> $$\mathrm{Tgh}\frac{1}{2g_c^2} = e^{-g_c^{-2}}$$
>
> (see [10]).

This is analogous to the Kramers-Wannier duality for the Ising model in two dimensions [18].

REFERENCES

[1] Yang, C. N., Mills, R. L.; Phys. Rev. 96, 191 (1954).
[2] Higgs, P. W.; Phys. Rev. Letters 12, 132 (1964), and 13, 508 (1964); Phys. Rev. 145, 1156 (1966).
Englert, F., Brout, R.; Phys. Rev. Letters 13, 321 (1964).
Guralnik, G. S., Hagen, C. R., Kibble, T. W. B.;

Phys. Rev. Letters 13, 585 (1964).
Kibble, T. W. B.; Phys. Rev. Letters 13, 585 (1964), and Phys. Rev. 155, 1554 (1967).

[3] Weinberg, S.; Phys. Rev. Letters 19, 1264 (1967).
Salam, A., in Elementary Particle Physics, N. Svartholm (ed.), (Almquist and Wiksell, Stockholm, 1968).

[4] 't Hooft, G.; Nucl. Phys. B33, 173 (1971), B35, 167 (1971).
Lee, B., Zinn-Justin, J.; Phys. Rev. D5, 3121, 3137, 3155 (1972).
Becchi, C., Rouet, A., Stora, R.; Annals of Phys. 98, 287 (1976).
Lowenstein, J., Zimmermann, W.; Nucl. Phys. B86, 77 (1975) and references quoted therein.

[5] Politzer, H. D.; Phys. Rev. Letters 26, 1346 (1973).
Gross, D., Wilczek, F.; Phys. Rev. Letters 26, 1343 (1973).

[6] Wilson, K. G.; Phys. Rev. D10, 2445 (1974).
Susskind, L., Kogut, J.; Phys. Rev. D10, 3468 (1974).
Casher, A., Kogut, J., Susskind, L.; Phys. Rev. D10, 732 (1974), and Phys. Rev. Letters 31, 792 (1973).

[7] Osterwalder, K., Seiler, E.; preprint in preparation.

[8] Osterwalder, K., Schrader, R.; Helv. Phys. Acta 46, 277 (1973).

[9] Osterwalder, K., Schrader, R.; Commun. Math. Phys. 31, 83 (1973) and 42, 281 (1975).

[10] Balian, R., Drouffe, J. M., Itzykson, C.; Phys. Rev. D10, 3376 (1974), 11, 2098 (1975), 11, 2104 (1975).

[11] Berezin, F. A.; The Method of Second Quantization, Academic Press, New York, 1966.

[12] Peter, F., Weyl, H.; Math. Ann. 97, 737 (1927).

[13] Hewitt, E., Ross, K. A.; Abstract Harmonic Analysis II, Springer-Verlag, New York, 1970.

[14] Watson, G. N; Theory of Bessel Functions, Cambridge University Press, 1966.

[15] Ruelle, D.; Statistical Mechanics, Benjamin, New York, 1969.

[16] Glimm, J., Jaffe, A., Spencer, T.; in: Constructive Quantum Field Theory, G. Velo and A. S. Wightman (eds.), Springer Lecture Notes in Physics, Vol. 25, Berlin, Heidelberg, New York, 1973.

[17] Ginibre, J; Commun. Math. Phys. 16, 310 (1970).

[18] Kramers, H. A., Wannier, G. H.; Phys. Rev. 60, 252 (1941).

[19] Wegner, G. J.; J. Math. Phys. 12, 2259 (1971).
Benettin, G., Jona-Lasinio, G., Stella, A.; Lettere al Nuovo Cimento 4, 443 (1972).
Greenberg, Gruber, Merlini, Physica 65, 28 (1973).

[20] Kadanoff, L. P.; The Application of Renormalization Group Techniques to Quarks and Strings, Review of Modern Physics, to appear.

[21] Glimm, J., Jaffe, A.; preprint, to appear in Physics Letters.

CONTINUUM GAUGE THEORIES

R. Stora

Centre de Physique Theorique
CNRS
Marseille, France

I. INTRODUCTION

Maxwell's theory of electromagnetism as well as Einstein's theory of gravitation are without any doubt among the most beautiful classical theories and also the most fundamental since they have to do with the description of elementary interactions. It is remarkable that Maxwell's theory has its quantum counterpart, thanks to the success of renormalization theory, at least in the restricted formal power series sense. Whereas the quantum counterpart of Einstein's theory is yet to be found, quantum electrodynamics has repeatedly served as a model for interactions at the elementary level, in the realm of weak interactions as well as in the realm of strong interactions. Whereas the idea has been put forward that gauge theories are most aesthetical, it has taken a number of theoretical steps before those could be put to work in a way formally at least as satisfactory as quantum electrodynamics.

The proposal by A. Salam and S. Weinberg of a model unifying electromagnetic and weak interactions has taken its full importance through the proof of renormalizability of a whole class of models, on the basis of the Feynman de Witt Faddeev Popov trick.

The subsequent discovery of the "asymptotic freedom" exhibited by some models of this class suggested an elegant explanation of some of the scaling properties of strong interactions.

Whereas these last two pieces of theoretical understanding can be reasonably believed to be correct on the basis of perturbation theory, strong coupling phenomena associated with e.g. quark confinement have been so far mostly attacked through the lattice approach which is infinitely better founded than the continuum perturbative theory in all respects, including geometrical aspects. Actually the mathematical structure of what is known of the continuum theory is at present very loosely grounded, as we shall see, and it is quite likely that even including the smooth classical configurations of the gauge fields which have been so far discovered will only yield part of the correct theory, whether or not it can be reached from the lattice theory.

Since even in the construction of the renormalized perturbation theory most mathematical problems are of a classical nature much of these notes will be devoted to somehow disconnected remarks on the mathematics of gauge fields, which will hopefully belong to a more definite overall picture.

Section II is devoted to a brief discussion of connections on principal fiber bundles pointing to the impression that our present formulation of continuum theories is incomplete.

Section III is devoted to a description of the main ingredients used through the construction of the renormalized perturbation series :

- The Faddeev Popov argument, and the Faddeev Popov Lagrangian
- The Slavnov symmetry and the nature of the Faddeev Popov ghost fields
- The Slavnov identity. An obstruction : the Adler Bardeen anomaly, and its generalizations : the local cohomology of the gauge Lie algebra.

It is concluded by a listing of a number of results and open problems in the framework of renormalized perturbation theory.

Section IV, closer in spirit to Section II, reviews some smooth classical configurations of gauge fields which ought to play a prominent role in the evaluation of the functional integral describing the theory, when it is constructed.

II. GAUGE FIELDS AND RELATED CONCEPTS

a) Yang Mills Fields [1]

Yang Mills potentials $a_\mu^\alpha(x)$ carry a four vector index related to Minkowski (or Euclidean) space R^4 ; $x \in R^4$, and an index α labelling a basis $\{e_\alpha\}$ of the Lie algebra $\mathcal{G}$ of an internal compact symmetry Lie group G . One may as well consider the form

$$a = \sum_{\alpha,\mu} a_\mu^\alpha(x)\, e_\alpha\, dx^\mu \tag{1}$$

with value in $\mathcal{G}$.

Under a gauge transformation $\{ g(x) \in G \}$, a transforms according to

$$^{g}a = g^{-1} a\, g + g^{-1} dg \tag{2}$$
$$\underset{\text{def.}}{=} ad.g\, a + g^{*}\omega$$

In this formula $\{ g(x) \in G \}$ is denoted g, a map from R^4 to G ; $g^*\omega$ is the reciprocal image under this map of the Maurer Cartan form [2] ω on G, with value in $\mathcal{G}$, which fulfills the structure equation

$$d\omega + \frac{1}{2} [\omega, \omega] = 0 \tag{3}$$

In components :

$$\omega = \sum_{\alpha} \omega^{\alpha} e_{\alpha} \tag{4}$$

$$d\omega^{\alpha} + \frac{1}{2} f_{\beta}{}^{\alpha}{}_{\gamma}\, \omega^{\beta} \wedge \omega^{\gamma} = 0 \tag{5}$$

where $f_{\beta}{}^{\alpha}{}_{\gamma}$ are the structure constants of $\mathcal{G}$:

$$[e_{\beta}, e_{\gamma}] = f_{\beta}{}^{\alpha}{}_{\gamma}\, e_{\alpha} \tag{6}$$

and $\wedge$ denotes the exterior product of differential forms. The corresponding Yang Mills field

$$F^{\alpha}_{\mu\nu} = \partial_{\mu} a^{\alpha}_{\nu} - \partial_{\nu} a^{\alpha}_{\mu} + f_{\beta}{}^{\alpha}{}_{\gamma}\, a^{\beta}_{\mu}\, a^{\gamma}_{\nu} \tag{7}$$

defines the curvature (2-) form [2] :

$$F = \frac{1}{2} F^{\alpha}_{\mu\nu}\, e_{\alpha}\, dx^{\mu} \wedge dx^{\nu} \tag{8}$$
$$= da + \frac{1}{2} [a, a]$$

which transforms under a gauge transformation according to

$$^{g}F = ad.g\; F \;. \tag{9}$$

F fulfills the Bianchi identity [2]

$$dF + [a, F] = \nabla F = 0 \tag{10}$$

which summarizes the second set of Maxwell's equations

$$\nabla_{\{\mu} F_{\nu\lambda\}} \tag{11}$$

where ∇_{μ} denotes the covariant derivative associated with the adjoint representation of $\mathcal{G}$.

In mathematical terms [2], a is a connection on the trivial principal bundle $P = R^4 \times G$, and F the corresponding curvature.

b) Connections on Principal. G-Bundles [2]

A more general situation described in the mathematical literature is the following : R^4 is replaced by a connected manifold B, called the base, which admits a locally finite open covering $\{U_i\}$; P is replaced by a manifold $P(B,G)$, such that there is a continuous projection π :

$$\pi : \quad P(B,G) \longrightarrow B \tag{12}$$

$P(B,G)$ is only locally trivial, i.e. $\pi^{-1}(U_i)$ is homeomorphic to $U_i \times G$. G acts on $P(B,G)$ from the right : leaving the base pointwise invariant (in local charts,

$$(x, g)\,\gamma = (x, g\gamma) \quad x \in B \,; g, \gamma \in G)$$

The globalization is given by a set of transition functions

$$g_{ij} : \quad U_i \cap U_j \equiv U_{ij} \longrightarrow G \tag{13}$$

which allow to compare the coordinates of a point of $P(B,G)$ above U_{ij} in the local charts $U_i \times G$ and $U_j \times G$: if (x, γ_i), (x, γ_j) are the two sets of coordinates of such a point,

$$\gamma_i = g_{ij}\,\gamma_j \tag{14}$$

The transition functions fulfill the cocycle condition

$$g_{ij}\, g_{jk} = g_{ik} \qquad \text{in} \quad U_{ijk} = U_i \cap U_j \cap U_k \tag{15}$$

Classes of transition functions are defined by combining refinements of covering and gauge transformations :

$$g_{ij} \Rightarrow h_i\, g_{ij}\, h_j^{-1} \tag{16}$$

Differenciating Eq.(15) [3] we get

$$\omega_{ij} = -\,\mathrm{ad}\, g_{ij}\, \omega_{ki} + \omega_{kj} \qquad \text{in} \quad U_{ijk} \tag{17}$$

where

$$\omega_{ij} = g_{ij}^{*}\,\omega \tag{18}$$

(the reciprocal image of the Maurer-Cartan form ω of G under $U_{ij} \xrightarrow{g_{ij}} G$).
Introducing a partition of unity α_i related to the covering U_i, and defining

$$a_j = \sum_{i \in I} \alpha_i\, \omega_{ij} \qquad \text{in} \quad U_i \tag{19}$$

we get

$$a_j = ad.g_{ij}\, a_i + \omega_{ij} \quad \text{in} \quad U_{ij} \tag{20}$$

The set of 1-forms a_i with values in $\mathcal{G}$, with the transition laws Eq.(20) define a connection on $P(B,G)$. There are many connections on $P(B,G)$, and if a_i is a connection deduced from g_{ij},

$$^h a_i = ad.h_i\, a_i + h_i^* \omega \tag{21}$$

is a connection deduced from $(h_i\, g_{ij}\, h_j^{-1})$.
From the transition laws of the a_i's one can show the existence of a global form a on $P(B,G)$ whose expression in local coordinates above U_i is

$$a\big|_{\pi^{-1}(U_i)} = ad.g\, a_i + \omega \tag{22}$$

The curvature forms locally defined by

$$F_i = da_i + \frac{1}{2}[a_i, a_i] \tag{23}$$

are of course such that

$$F_i = ad.g_{ij}\, F_j \quad \text{on} \quad U_i \cap U_j \tag{24}$$

One can then define on $P(B,G)$ the curvature 2-form

$$F = \delta a + \frac{1}{2}[a, a] \tag{25}$$

where δ is the exterior derivative on $P(B,G)$. Above U_i,

$$F = ad.g\, F_i \tag{26}$$

c) Associated Vector Bundles

Associated vector bundles are similarly defined by replacing each fiber $\pi^{-1}(x) \sim G$, above each point x of B by a vector space V on which acts a representation D of G. If $(x, \varphi_i), (x, \varphi_j)$ are the coordinates of a point above U_i, U_j respectively $(x \in U_{ij})$, then

$$\varphi_i = D(g_{ij})\, \varphi_j \quad \text{in} \quad U_{ij} \tag{27}$$

If $\{x \to \varphi_i(x),\ x \in U_i\}$ is a set of coherent local sections (i.e. fulfilling Eq.(27) for each x) the covariant differential of ϕ_i

$$\mathcal{D}\phi_i = d\phi_i - t(a_i)\,\phi_i \tag{28}$$

fulfills the transition law

$$\mathcal{D}\phi_i = D(g_{ij})\, \mathcal{D}\phi_j \quad \text{in} \quad U_{ij} \tag{29}$$

Here t is the differential of D :

$$t(X) = D(g^{-1})\frac{d}{d\tau}D(g\,e^{\tau X})\Big|_{\tau=0}, \quad X\in\mathcal{G} \tag{30}$$

d) Gauge Invariant Forms

Consider a local form $\mathcal{J}(F_i, \Phi_i, \mathcal{D}\Phi_i \ldots)$ which is invariant under gauge transformation

$$\begin{aligned} F_i &\to \mathrm{ad}\, h_i\, F_i \\ \Phi_i &\to D(h_i)\Phi_i \\ \mathcal{D}\Phi_i &\to D(h_i)\mathcal{D}\Phi_i \end{aligned} \tag{31}$$

Then, clearly

$$\begin{aligned} \mathcal{J}(F_i, \Phi_i, \mathcal{D}\Phi_i \ldots) &= \mathcal{J}(F_j, \Phi_j, \mathcal{D}\Phi_j \ldots) \quad \text{in } U_{ij} \\ &\overset{\text{def.}}{=} \mathcal{J}(F, \Phi, \mathcal{D}\Phi \ldots) \end{aligned} \tag{32}$$

is globally defined on B.

Example.

Let B be endowed with a Riemannian structure such that dual forms may be defined : e.g.

$$a^*_{i_1\ldots i_p}\, dx^{i_1}\wedge\ldots dx^{i_p} = \frac{1}{p!}\varepsilon_{i_1\ldots i_p\, i_{p+1}\ldots i_n}\, a_{i_1\ldots i_p}\, dx^{i_{p+1}}\wedge\ldots d$$

If $\{\ ,\ \}$ is an invariant symmetric quadratic form on $\mathcal{G}$

$\{F, F^*\}$ is globally defined.

Similarly if $(\ ,\ )$ is a bilinear invariant form on V ,

(Φ, Φ^*), $(\Phi, \Phi)(\Phi, \Phi)^*$, $(\mathcal{D}\Phi, \mathcal{D}\Phi^*)$

are globally defined.

These are the main ingredients for coupled gauge fields and matter fields local actions :

$$\int_B -\{F^*, F\} + (\mathcal{D}\Phi, \mathcal{D}\Phi^*) - [\Phi^*, \Phi] - \{[\Phi,\Phi]^*[\Phi,\Phi]\} \tag{33}$$

where the various quadratic forms symbolized by various types of brackets are parametrized in terms of coupling constants masses etc.

Of course the solutions of the corresponding Euler Lagrange equations on a given $P(B,G)$ must be defined locally in each U_i and fit with the transition laws Eqs.(31). This point of view has only been recently advocated by C.N. Yang and T.T. Wu [4] in the formulation of the quantum mechanics of a non relativistic electron in a Dirac monopole field - without strings, and of the classical mechanics of a relativistic electron interacting with a moving relativistic monopole.

In order to investigate such generalized solutions, it is necessary to classify all possible $P(B,G)$'s. This is partially answered by the following :

e) Characteristic Classes. [5]

There is an enormous mathematical literature on the classification of vector bundles which is however not very accessible with elementary knowledge. Somehow one of the first question which comes up is how large the U_i 's can be taken and how they communicate : if one could choose $U_i = B$ one would deal with the trivial principal bundle $P(B,G) = B \times G$ on which there exist global sections

$$B \ni x \to g(x) \in G$$

For vector bundles there always exist global continuous sections but the question is whether one can find k continuously varying vectors in V which stay linearly independent.

Some answers to such problems are provided by the study of globally defined differential forms which are expressible in terms of connections on $P(B,G)$:

Let $P_n(X_1, \dots X_n)$ be a symmetric multilinear form on $\mathfrak{g}$ which is invariant under the adjoint representation :

$$\sum_i P_n(X_1, \dots [Z, X_i], \dots X_n) = 0 \tag{34}$$

Then substitute in $X_1 = \dots = X_n = F_i$ and replace the ordinary product by exterior product (which is meaningful since the F_i 's being 2-forms, commute). Consider the 2n-form :

$$\Omega_{2n,i} = P_n(F_i, \dots F_i) \tag{35}$$

Because of invariance under the adjoint representation and the transition laws for the F_i 's

$$\Omega_{2n,i} = \Omega_{2n,j} \quad \text{in } U_{ij} \tag{36}$$

$$\stackrel{\text{def}}{=} \Omega_{2n} \quad \text{globally defined on } B .$$

Furthermore, from the Bianchi identity and the invariance of P_n, one finds

$$d\Omega_{2n} = 0 \tag{37}$$

i.e. Ω_{2n} is closed.

Now, remember there are lots of connections on $P(B,G)$. In fact, if $\{a_i^{(1)}\}$, $\{a_i^{(2)}\}$ are connections, so are

$$a_i^t = t a_i^{(1)} + (1-t) a_i^{(2)} \qquad 0 \leq t \leq 1 \tag{38}$$

which continuously interpolate $\{a_i^{(1)}\}$, $\{a_i^{(2)}\}$.
The associated curvature is denoted

$$F_i^t = d a_i^t + \frac{1}{2}[a_i^t, a_i^t] \tag{39}$$

One can easily prove that

$$\Omega_{2n}^{(1)} - \Omega_{2n}^{(2)} = d\, Q_{2n-1}^{1,2} \tag{40}$$

where

$$Q_{2n}^{1,2} = n \int_0^1 dt\, P_n(a^{(1)} - a^{(2)}, F^t, \ldots F^t) \tag{41}$$

is also globally defined since $a^{(1)} - a^{(2)}$ transforms under the adjoint representation.

Thus, the cohomology class of Ω_{2n} does not depend on the connection and consequently is attached to the class of $P(B,G)$ (recall that if a_i is a connection attached to g_{ij}, $h_i^{-1} a_i h_i + h_i^* \omega$ is attached to ${}^h g_{ij} = h_i\, g_{ij}\, h_j^{-1}$).

This cohomology class is called a Chern Weil characteristic class for $P(B,G)$.

Formulae (40), (41) can be lifted to $P(B,G)$, namely, in particular, one may consider the form

$$\tilde{\Omega}_{2n} = P_n(F, \ldots F) \tag{42}$$

Then

$$\tilde{\Omega}_{2n} = \delta\, \tilde{Q}_{2n-1} \tag{43}$$

$$\tilde{Q}_{2n-1} = n \int_0^1 P_n(a, F^t, \ldots F^t)\, dt$$

Restricting $\tilde{Q}_{2n-1}$ to any $\pi^{-1}x \simeq G$ yields a form on G defined by a totally antisymmetric invariant P_{2n-1} form of degree $2n-1$ on $\mathcal{G}^{\wedge 2n-1}$, which is associated with P_n. The correspondence

$$P_n \rightarrow P_{2n-1}$$

is called the Cartan map. One can construct inverses to this map. Such inverses are called transgressions. One can prove that both $\wedge \mathcal{G}^*$ and $\vee \mathcal{G}^*$ (resp. the exterior algebra on $\mathcal{G}^*$ and the symmetric tensor algebra of $\mathcal{G}^*$) are generated by "primitive" (i.e. non factorizable) elements whose number is given by the rank of G [6].

Example.

$G = SU3$

$\vee \mathcal{G}^*$ is generated by the Killing form $g^{\alpha\beta}$ and the $d^{\alpha\beta\gamma}$

Gell Mann tensor.
Correspondingly $\Lambda \mathcal{G}^*$ is generated by F, D :

$$F(X,Y,Z) = g^{\alpha\beta} X_\alpha [Y,Z]_\beta$$

$$D(X,Y,Z,T,U) = \underset{XYZTU}{\mathcal{A}} d^{\alpha\beta\gamma} X_\alpha [Y,Z]_\beta [T,U]_\gamma$$

These basic technical facts will be used in section III and section IV in support of the thesis that Yang Mills fields are to be considered as connections on some principal fiber bundle. If so, as supported by the recent ideas of Polyakov and followers, the conventional Yang Mills action, to be hopefully incorporated into the definition of a functional integral defining a vacuum state, is incomplete and misses somehow a measure on classes of fiber bundles, provided such a measure can be defined at all, as it has been possible on some smooth configurations. Since the known measures of euclidean field theory are concentrated on non smooth configurations, the present point of view meets difficulties out of reach which the lattice approach might be able to bypass if the continuum limit comes under control.

III. THE RENORMALIZED PERTURBATION SERIES [7]

We now come back to conventional gauge theories : $P(B,G) = P(M_4,G) = M_4 \times G$, with a gauge invariant action as given in Eq.(33). Because of the degeneracy of the quadratic part due to gauge invariance, this action does not lead directly to the construction of a renormalized perturbation series, as is familiar in quantum electrodynamics. The proper action including a gauge term has been constructed by Faddeev and Popov [7] on the basis of a formal manipulation of the supposedly existing Feynman integral. Further manipulations allowed Slavnov [7] to write some identities formally fulfilled by the Green's functional, whose generalization allowed Becchi and Rouet [7] to discover that at the classical level, the Faddeev Popov action possesses an intriguing symmetry : the Slavnov symmetry. The fulfillment of that symmetry at the quantum level then allows for a reasonable physical interpretation of the theory.

a) The Faddeev Popov Construction

The Faddeev Popov argument consists in replacing the functional integral based on a gauge invariant action of the type given in Eq.(33) by another one such that formally the expectation values of gauge invariant operators be unchanged [8] . Besides, the new action is suited to a perturbative treatment, namely, it is local, it breaks the degeneracy of the quadratic part of the initial action, it is of the renormalizable type.

Such a construction is based on the formal identity [9]

$$\int \mathcal{D}\Omega \, \delta({}^{\Omega}g - C) \det {}^{\Omega}\mathcal{M} = 1 \tag{44}$$

where $\mathcal{D}\Omega$ is the "volume element" of the "gauge group", $g(a,\phi)$ is a gauge function (typically, $g = \partial a + \cdots$), ${}^{\Omega}\mathcal{M} = \frac{\delta\,{}^{\Omega}g}{\delta\Omega}$.
Besides its formal character, this identity assumes that there is one and only one Ω such that ${}^{\Omega}g - C = 0$. In general the integrand must be divided by $\vartheta(g)$, the volume of the stability group of g , which is of course gauge invariant but usually omitted. Eq.(44) is rewritten

$$\int \mathcal{D}\Omega \, \mathcal{D}\bar{\omega} \, \mathcal{D}\omega \, \mathcal{D}\gamma \, e^{i(\bar{\omega}\,{}^{\Omega}\mathcal{M}\,\omega + \gamma\{{}^{\Omega}g - C\})} = 1 \tag{45}$$

where ω belongs to the exterior algebra of the gauge Lie algebra, $\bar{\omega}$ its dual. After simplifying the functional so obtained by the "volume" of the "gauge group" which provides an overall normalization factor cancelling in the expectation values, yields for $C = 0$ the modified action

$$\Gamma_{Landau}(a,\phi,\omega,\bar{\omega},\gamma) = \Gamma_{inv}(a,\phi) + \gamma g + \bar{\omega}\mathcal{M}\omega \tag{46}$$

pertaining to a Landau type gauge.
Integrating Eq.(44) through $\int e^{F(C)} \mathcal{D}C$ yields

$$\Gamma_{Feynman}(a,\phi,\omega,\bar{\omega}) = \Gamma_{inv}(a,\phi) + F(g) + \bar{\omega}\mathcal{M}\omega \tag{47}$$

pertaining to a Feynman type gauge (usually one chooses $F(g) = (g,g)$ for reasons which have to do with power counting).

Note the formal definition of ${}^{\Omega}\mathcal{M}$:

$$\delta\,{}^{\Omega}g = {}^{\Omega}\mathcal{M}\,\tilde{\omega} \tag{48}$$

where δ is the exterior differenciation on the gauge group, $\tilde{\omega}$ the corresponding Maurer Cartan form which fulfills the structure equation

$$\delta\tilde{\omega} + \frac{1}{2}[\tilde{\omega},\tilde{\omega}] = 0 \tag{49}$$

so that

$$\delta^2\,{}^{\Omega}g = 0 \tag{50}$$

This partly explains the Slavnov symmetry [7] . The difference in nature between $\bar{\omega}$ and ω has to be stressed : ω is of a geometric nature comparable to that of the Maurer Cartan form of the gauge group whereas $\bar{\omega}$ - just as γ - serves as a Lagrange multiplier.

b) The Slavnov Symmetry [7]

Define on the Grassmann algebra generated by ω the anti-derivation

$$s\omega = -\frac{1}{2}[\omega,\omega] \qquad (51)$$
$$(s^2 = 0)$$

This antiderivation extends to the algebra generated by a, Φ, according to

$$s a = \frac{\delta^{\Omega} a}{\delta\Omega}\Big|_{\Omega=\text{identity}}^{\omega} \stackrel{\text{def}}{=} \frac{\delta a}{\delta\Omega}\omega \qquad (52)$$
$$s\Phi = \frac{\delta^{\Omega}\Phi}{\delta\Omega}\Big|_{\Omega=\text{identity}}^{\omega} \stackrel{\text{def}}{=} \frac{\delta\Phi}{\delta\Omega}\omega$$

Then Becchi and Rouet [7] made the observation that Γ_{Landau} is invariant under the extension of s, including $\bar{\omega}$ and γ:

$$s\bar{\omega} = -\gamma \qquad (53)$$
$$s\gamma = 0$$

So that

$$s^2 = 0 \qquad (54)$$

Similarly, in the Feynman gauge Eq.(53) is replaced by

$$s\bar{\omega} = -F'(g) \qquad (55)$$

as can be obtained from Eq.(53) by expressing F in terms of its Legendre transform with respect to γ.

This Slavnov symmetry can be extended if one includes in the action Γ terms of the form

$$J.\mathcal{O} + \eta\, s\mathcal{O} \qquad (56)$$

where $\mathcal{O}$ is a local operator, with

$$s\eta = -J \qquad (57)$$
$$sJ = 0$$

For instance, denoting now collectively by Φ both the matter field and gauge field, let

$$\Gamma(\Phi,\omega,\bar{\omega},\eta,K,R) = \Gamma_{inv}(\Phi) - \left(\frac{(g,g)}{2} + \bar{\omega}\mathcal{M}\omega\right) + \eta s\Phi + Rg + Ks\omega \tag{58}$$

The Slavnov symmetry reads

$$\int \frac{\delta\Gamma}{\delta\Phi}\frac{\delta\Gamma}{\delta\eta} + \frac{\delta\Gamma}{\delta\omega}\frac{\delta\Gamma}{\delta K} + \frac{\delta\Gamma}{\delta\bar{\omega}}\left(\frac{\delta\Gamma}{\delta R} + R\right) = 0 \tag{59}$$

Putting

$$\hat{\Gamma} = \Gamma - \frac{1}{2}(R,R)$$

Eq.(58) becomes

$$\int \frac{\delta\hat{\Gamma}}{\delta\Phi}\frac{\delta\hat{\Gamma}}{\delta\eta} + \frac{\delta\hat{\Gamma}}{\delta\omega}\frac{\delta\hat{\Gamma}}{\delta K} + \frac{\delta\hat{\Gamma}}{\delta\bar{\omega}}\frac{\delta\hat{\Gamma}}{\delta R} = 0 \tag{60}$$

with the Slavnov invariance defined by

$$\begin{aligned} s\Phi &= \frac{\delta\Phi}{\delta\Omega}\omega \\ s\omega &= -\frac{1}{2}[\omega,\omega] \\ s\bar{\omega} &= -(g-R) \end{aligned} \tag{61}$$

This is the form of the Slavnov symmetry proposed by J. Zinn Justin [7] .

c) Construction of the Renormalized Perturbation Series

Let us consider a vertex - one particle irreducible - functional $\Gamma(\Phi,\omega,\bar{\omega},\eta,K,R)$ a formal series in $\hbar$:

$$\Gamma(\Phi,\omega,\bar{\omega},\eta,K,R) = \sum_{0}^{\infty} \hbar^{n}\,\Gamma^{(n)}(\Phi,\omega,\bar{\omega},\eta,K,R) \tag{62}$$

where $\Gamma^{(0)}$ coincides with the previously defined classical action, higher terms being given as usual in terms of renormalized Feynman graphs, based on a Lagrangian with counterterms formal series in $\hbar$. The question is whether this Lagrangian can be so defined that the Slavnov identity Eq.(60) holds in the formal power series sense. This turns out to be a purely algebraic problem thanks to

- The renormalized quantum action principle of Y.M.P. Lam [7] and J.H. Lowenstein [7] .
- The implicit function theorem for formal power series.

The action principle is mostly needed in a weak form which ascertains that quantum corrections to its naïve form are limited by power counting [7]. For instance, consider the connected Green functional $Z^c(\mathcal{J}\ldots)$ the Legendre transform of $\Gamma(\phi,\ldots)$, $\mathcal{L}$ the Lagrangian used to construct them, λ a parameter of the theory, T_R a renormalized chronological product.

Then,

$$\frac{\partial Z^c}{\partial\lambda}(\mathcal{J},\ldots) = \left\langle T_R \text{ "}\int \frac{\partial\mathcal{L}}{\partial\lambda}dx\text{"}, \exp\frac{i}{\hbar}\int(\mathcal{L}^{int}+\mathcal{J}\phi)(x)dx\right\rangle \tag{63}$$

with

$$\text{"}\int\frac{\partial\mathcal{L}}{\partial\lambda}dx\text{"} = \left(\int\frac{\partial\mathcal{L}}{\partial\lambda}dx\right)^{naïve} + \hbar\int Q\,dx \tag{64}$$

In Eq.(64) the naïve part involves an assignment of power counting for $\partial\mathcal{L}/\partial\lambda$ under the T_R symbol ; Q is not specified but its power counting index does not exceed that of the naïve part.

Similar results hold upon performing field variations. Consequently, up to $O(\hbar)$ the algebraic problems at hand are of a purely classical nature - in the present case, geometrical, and quantum corrections are controlled by the implicit function theorem [7] :

TH. Let $F_i(x_1,\ldots x_n;y_1,\ldots y_m)$, $i=1\ldots n$ be formal power series in $x_1,\ldots x_n;y_1,\ldots y_m$ without constant coefficients ; $L_x F_i(x_1\ldots x_n;y_1\ldots y_m)$ the parts of the F_i's linear in $x_1\ldots x_n$. Then, if the matrix

$$\frac{\partial L_x F_i}{\partial x_j}(y_1,\ldots y_m)$$

is invertible as a formal power series in $y_1\ldots y_m$, the system of equations

$$F_i(x_1\ldots x_n;y_1\ldots y_m)=0 \tag{65}$$

has a unique solution

$$x_i = \phi_i(y_1\ldots y_m) \tag{66}$$

where the ϕ_i's are formal power series in $y_1\ldots y_m$.

Thus, determining $\mathcal{L}$ so that the Slavnov identity will hold amounts to investigating the stability of some classical algebraic problems, under quantum perturbations.

Applying the action principle to a Slavnov variation of the

fields at the quantum level, given a Lagrangian

$$\mathcal{L} = \mathcal{L}_{classical} + \hbar \Delta \mathcal{L} \tag{67}$$

yields

$$\mathcal{S} Z^c(\mathcal{J}, \dots) = (\Delta_\bullet Z^c)(\mathcal{J}, \dots) \tag{68}$$

$\mathcal{S}$ is now a functional linear differential operator obtained by Legendre transforming Eq. (60) ; $\Delta_\bullet$ is an insertion operator of dimension 5 , ω number + 1 of the form

$$\Delta_\bullet = s \int \mathcal{L} dx + \hbar \int Q dx \tag{69}$$

which obeys a generalized Wess Zumino consistency condition [7] obtained from the algebraic relation

$$\mathcal{S}^2 = 0 \tag{70}$$

Roughly speaking, if

$$s \int Q dx = 0 \tag{71}$$

implies

$$\int Q dx = s \int \Delta' \mathcal{L} dx \tag{72}$$

altering $\Delta\mathcal{L}$ by $-\Delta'\mathcal{L}$ will produce the desired result. So, the problem to be solved is of a cohomological nature. A complete solution has not yet been given. It is however worthwhile mentioning one completely solved subproblem, that concerning the part of the classical Slavnov operator which is the coboundary operator of the gauge Lie algebra :

d) An Obstruction : the Adler Bardeen Anomaly [10]

The question is to solve the equation

$$\delta \Delta(a, \phi, \omega) \equiv \int \left[sa \frac{\delta}{\delta a} + s\phi \frac{\delta}{\delta \phi} - \frac{[\omega, \omega]}{2} \frac{\delta}{\delta \omega} \right] \Delta(a, \phi, \omega) = 0 \tag{73}$$

where

$$\Delta(a, \phi, \omega) = \int \Delta_4(a, \phi, \omega) \tag{74}$$

with $\Delta_4(a, \phi, \omega)$ a 4 form (proportional to d^4x), with coefficient a local polynomial in a, ϕ, ω and their derivatives. Clearly Δ_4 is defined mod $d\Delta_3$. Forms $\Delta(a, \phi, \omega)$ are graded both by their order p and their ω-number g . Their total gradation is $n = p + g$. A component of Δ with such degrees will be denoted

$$\Delta_p^g$$

i.e.

$$^{(n)}\Delta = \sum_{p+g=n} \Delta_p^g \tag{75}$$

and we consider the algebra of the Δ's to be the antisymmetric tensor product of the Δ_0^g and Δ_p^0 algebras, so that it is totally anticommutative. We furthermore define $\hat{d}$ by

$$\begin{aligned} &\hat{d}(^{(m)}\Delta\, ^{(n)}\Delta) = \hat{d}\, ^{(m)}\Delta \cdot {}^{(n)}\Delta + (-)^m\, ^{(m)}\Delta\, d\, ^{(n)}\Delta \\ &\hat{d}\, \Delta_p^0 = d\, \Delta_p^0 \\ &\hat{d}\, \omega = d\omega \end{aligned} \tag{76}$$

The main lemmas and theorems proved by J. Dixon [10] are the following, where all forms are local.

Lemma 1.

If $d\Delta_p = 0$ then $\Delta_p = d\Delta_{p-1} + \chi_p$ (77)

$p \neq 4$ χ_p = constant form.

Δ_p, Δ_{p-1} local forms , $\Delta_{-1} \equiv 0$

In all that follows ϕ is assumed to transform under a fully reducible representation of the compact group G .

Lemma 2.

If $(\hat{d} + \delta)\, ^{4+g}\Delta(a, \phi, \omega) = 0$ (78)

then

$$^{4+g}\Delta(a, \phi, \omega) = (\hat{d} + \delta)\, ^{3+g}F(a, \phi, \omega) + \sum_T T^g(\omega)\, I_4(a, \phi) + \sum_B {}^{4+g}B(a, \omega) \tag{79}$$

$$\begin{aligned} &\delta T^g(\omega) = 0 \quad , \quad T^g(\omega) \text{ involves no derivative.} \\ &\delta\, I_4(a, \omega) = 0 \\ &(\hat{d} + \delta)\, ^{4+g}B(a, \omega) = 0 \end{aligned} \tag{80}$$

Lemma 3.

If $(\hat{d} + \delta)\, ^{4+g}B(a, \omega) = 0$ (81)

then

$$ {}^{4+g}B(a,\omega) = (\hat{d}+\delta)\,{}^{4+g}F(a,\omega) + \sum_{B_{inv}} {}^{4+g}B_{inv}(a,\omega) \qquad (82) $$

where B_{inv} is defined in Lemma 4.

Lemma 4.

Given a ${}^{\tau}T(\omega)$, $\delta_{\tau}\,{}^{\tau}T(\omega) = 0$, which is primitive i.e. cannot be factorized there is a ${}^{\tau}B_{prim}(\omega)$ such that

$$ {}^{\tau}B_{o\,prim}(a,\omega) = {}^{\tau}T(\omega) $$

If ${}^{4+g}T(\omega)$ is a product of primitive factors, the corresponding ${}^{4+g}B_{inv}$ factors out into the corresponding B_{prim}. These results are conspicuously similar to, but different from those pertaining to the cohomology of principal fiber bundles [6]. For $g=1$, they yield the conventional Adler Bardeen anomaly : going back to Eq.(73), (74) yields, through Lemma 1, the existence of Δ_4^1, $\Delta_3^2, \Delta_2^3, \Delta_1^4, \Delta_0^5$ such that

$$ \begin{aligned} \Delta(a,\phi,\omega) &= \int \Delta_4^1(a,\phi,\omega) \\ \delta\Delta_4^1 + \hat{d}\,\Delta_3^2 &= 0 \\ \delta\Delta_3^2 + \hat{d}\,\Delta_2^3 &= 0 \\ \delta\Delta_2^3 + \hat{d}\,\Delta_1^4 &= 0 \\ \delta\Delta_1^4 + \hat{d}\,\Delta_0^5 &= 0 \\ \delta\Delta_0^5 &= 0 \end{aligned} \qquad (83) $$

i.e.

$$ (\delta+\hat{d})\,{}^{(5)}\Delta = 0 \qquad (84) $$

with

$$ {}^{(5)}\Delta = \Delta_4^1 + \Delta_3^2 + \Delta_2^3 + \Delta_1^4 + \Delta_0^5 \qquad (85) $$

Since for all non abelian groups $T^1(\omega)$, Lemma 2 indicates that the only further obstructions to

$$ \Delta = \delta \int \mathcal{L}\,dx \qquad (86) $$

are provided by

$$ \Delta_{Adler\text{-}Bardeen} = \int {}^{5}B^{1}_{prim,4}(a,\omega) \qquad (87) $$

based on

$$ T_0^5 = D(\omega, [\omega,\omega], [\omega,\omega]) \qquad (88) $$

where D is an invariant third rank tensor in $V\mathcal{G}^*$, usually non vanishing for many groups of rank larger than 1 [6].

The most expedient way to carry out in this case the construction indicated in Lemma 4 is to start in six dimensional space, and later on restrict oneself to four dimensions.

Consider the characteristic class [5]

$$T_6^0 = D(F, F, F) = d\, T_5^0 \tag{89}$$

$$T_5^0 = 3 \int_0^1 dt\; D(a, F^t, F^t) \tag{90}$$

$$\begin{aligned} F^t &= t\, da + \frac{t^2}{2}[a,a] \\ &= tF + \frac{t^2 - t}{2}[a,a] \end{aligned} \tag{91}$$

Because of the invariance of D, δT_5^0 has only terms proportional to $\hat{d}\omega$, and since

$$\delta T_6^0 = 0 \qquad \hat{d}\, \delta T_5^0 = 0 \tag{92}$$

There is a T_4^1 such that

$$\delta T_5^0 = \hat{d}\, T_4^1\ . \tag{93}$$

Up to $d(\cdot)$, T_4^1 can be written

$$T_4^1 = \omega\, T_4^0 \tag{94}$$

Going back to Eq.(93)

$$\delta T_5^0 = d\omega\, T_4^0 - \omega\, \hat{d} T_4^0 \tag{95}$$

But we know that δT_5^0 is proportional to $d\omega$. Thus

$$d T_4^0 = 0 \tag{96}$$

hence

$$T_4^0 = d\, T_3^0 \tag{97}$$

The Adler Bardeen anomaly is

$$T_4^1 = \omega T_4^0 = \omega\, d T_3^0 \tag{98}$$

Continuing this chain down would lead to T_0^5 in the last step. The calculation is now straightforward and considerably simpler than previous ones [7] :

$$\delta T_5^0 = 3\int_0^1 dt \left\{ D(d\omega, F^t, F^t) - 2(t^2 - t) D(a, [d\omega, a], F^t) \right\} \quad (99)$$

Selecting the coefficient of $d\omega$ yields

$$T_4^1 = \omega T_4^0 = 3\int_0^1 dt \left\{ D(\omega, F^t, F^t) - 2(t^2 - t) D(a, [\omega, a], F^t) \right\} \quad (100)$$

The coefficient of the $D(\omega, [a,a], [a,a])$ term vanishes as it should and one finds

$$T_4^0 = d\, T_3^0$$

with

$$\omega T_3^0 = D(\omega, a, da) + \tfrac{1}{4} D(\omega, a, [a,a]) \quad (101)$$

In the case where abelian components occur this construction collapses and many more anomalies of the form $\omega^{abel.} I_4(a, \Phi)$ are expected, most of which are hoped to be harmless for renormalization. Formula (101) survives in the form

$$\begin{aligned} \omega^{abel.} d T_3^0 &= \omega^{abel.} T_4^0 = \omega^{abel}(F, F) = \\ &= \omega^{abel} d \left\{ (a, da) + \tfrac{1}{3}(a, [a,a]) \right\} \end{aligned} \quad (102)$$

where (,) is made up with Killing forms of the simple parts and is arbitrary for abelian parts. The fully abelian case was the first one discovered by S. Adler J. Bell R. Jackiw. In that case F is by itself a characteristic class and (F, F) is the relevant object.

e) Discussion and Results

When linear gauges are used, the Slavnov identity simplifies and the Adler Bardeen anomaly is the only obstruction that creates difficulties. When it is absent e.g. in the abelian or SU2 Higgs Kibble model or in the case of quarks coupled to Yang Mills fields, in the absence of pseudo scalar couplings, the Slavnov identity can be proved. In the Higgs Kibble case where a complete particle analysis can be made, and physical degrees of freedom separated out, physical scattering processes can be shown to be gauge independent and obey perturbative unitarity [7].

When massless Yang Mills fields are involved physics has to be defined through gauge invariant local operators. H. Stern and B. Zuber have obtained partial answers, using the background gauge [12] , and S. D.Joglekar and B.W. Lee [11] have proposed a complete solution for pure non abelian Yang Mills fields in the Feynman gauge. In both cases, the dimensional scheme of G. 't Hooft, M. Veltman has been used. The work of J. Dixon [10] , part of which has been exhibited in the previous section considerably streamlines the very involved treatment of S.D. Joglekar and B.W. Lee.

The "academic" problem of non linear renormalizable gauges has so far only found partial answers [12] . The fact which in my opinion has received least attention from a rigorous point of view is the nonrenormalizability theorem for the Adler Bardeen anomaly which states that if the classical Lagrangian is so arranged that this anomaly vanishes at the one loop level, then it vanishes to all orders of perturbation theory. Whereas in electrodynamics compact proofs have been given by A. Zee, J. H. Lowenstein, B. Schroer [7] combining gauge invariance with the known lack of dilatation invariance described by the Callan Symanzik equation, the present version of this theorem, due to W. Bardeen, relies on the mixed use of dimensional and Pauli Villars regularizations at the level of Feynman graphs. Similar proofs also ought to be constructed to cope with possible anomalies associated with abelian parts of the gauge group, in the case of general gauges [7] [12] .

Asymptotic freedom which is one of the most attractive features of sufficiently pure Yang Mills theories will be discussed in the lectures by P.K. Mitter.

The necessary elimination of the Adler Bardeen anomaly is conceptually fundamental in the construction of models and has already led to a number of interesting theoretical ideas : unified models of electromagnetic and weak interactions such as the Salam Weinberg model cannot be made consistent unless either heavy leptons are introduced together with Higgs fields or if they are embedded in a single scheme together with strong interactions, quarks providing the necessary cancellation.

IV. SOME CLASSICAL CONFIGURATIONS OF GAUGE FIELDS

Common experience being based on the repeatedly checked idea that quantum effects act as perturbations on stable classical situations, some efforts have been spent in a search for stable configurations involving gauge fields. The oldest of all is:

a) The Dirac Monopole [4]

It is a connection over a U(1) principal bundle over $R^3\setminus\{0\}$, classified by its first Chern class [2] $F/2\pi = \frac{1}{4\pi} F_{\mu\nu} dx^\mu \wedge dx^\nu$ whose integral on a sphere around 0 is an integer n, which, in terms of the monopole charge at the origin provides the Dirac quantization rule :

$$\frac{2ge}{\hbar c} = n \tag{103}$$

This has been recently put on a geometrical basis by T.T. Wu and C.N. Yang [4] who have also investigated the motion of a quantized electron in a Dirac monopole field, and the classical theory of interacting relativistic charge and monopole, and by J. Madore [13].

b) The 't Hooft Polyakov Monopoles [13] [14]

One looks for finite energy static solutions of the Higgs Kibble model, with $a_0=0$ (Generalizations to $a_0\neq 0$ have also been found). The field equations, in the radiation gauge are elliptic non linear. It is not known whether the Dirichlet problem is (uniquely) soluble for arbitrary data of the Higgs field

$$\phi(\vec{x}) \underset{|\vec{x}|\to\infty}{\longrightarrow} \phi_\infty(\hat{x}) \tag{104}$$

For some boundary values at infinity, solutions have been constructed, by G. 't Hooft, A.M. Polyakov, and followers. Finiteness of the energy implies that

$$\frac{\partial V}{\partial \phi}(\phi_\infty) = 0 \tag{105}$$

where V is the potential of the Higgs field, and

$$\nabla \phi \underset{|\vec{x}|\to\infty}{\longrightarrow} 0 \qquad F \underset{|\vec{x}|\to\infty}{\longrightarrow} 0 \tag{106}$$

where ∇ is the covariant gradient.

Thus

$$a \underset{|x|\to\infty}{\longrightarrow} g^{-1}dg \tag{107}$$

and if

$$\phi_\infty = D(\Omega)\,\phi^0_\infty \tag{108}$$

where ϕ^0_∞ is a space independent minimum of V, Eq.(104) sums up to

$$t\left(\Omega^{-1} g^{-1} \partial_\mu (g\Omega)\right) \phi^0_\infty = 0 \tag{109}$$

Interesting solutions occur when Φ_∞ is not homotopic to the identity on its image. Then, one cannot write Eq.(108) for a continuous Ω. Also the stability group L_Φ of Φ cannot stay isomorphic to itself all over R^3. Wherever this is so (for known solutions, except at the origin), L_Φ defines a sub-bundle of the trivial bundle over R^3 on which a is a connection. This subbundle is in general not trivial. Its first Chern class, in particular, can be expressed in terms of the projected connection deduced from a, and defines a $U(1)$ magnetic monopole provided L_Φ contains a $U(1)$ subgroup. This geometrical interpretation has been given by J. Madore [13] who has made a detailed geometrical study of both the SU2 and SU3 cases.

c) Euclidean Configurations and Classical Vacua (A.A. Belavin, A.M. Polyakov, A.S. Schwartz, Y.S. Tyupkin ; G. 't Hooft ; C.G. Callan, R. Dashen, D. Gross ; R. Jackiw, C. Rebbi.) [15] .

Classically, a vacuum configuration corresponds to a vanishing field, i.e. a potential which is a gauge

$$F = 0 \qquad\qquad a = g^{-1} dg \tag{110}$$

We shall look at fixed time configurations and assume that the time evolution is smooth. For those configurations which are infinite volume limits with prescribed boundary conditions on some large sphere $S^2_R\ (R \to \infty)$, i.e. such that $g(\vec{x},t) \xrightarrow[|\vec{x}| \to \infty]{} g_\infty(\hat{x}\, t)$ we may as well assume that $g_\infty = e$ since g_∞ is homotopic to the identity map, which can be reached by a gauge transformation belonging to the connected component of the identity.

$g(\vec{x},t)$ thus belongs for all t to some homotopy class of mappings from $S^3 \sim R^3 \cup \{\infty\}$ to G. These classes are indexed by integers n for simple groups. Two g's belonging to the same homotopy class can be transformed into each other by a gauge transformation connected to the identity and, so, describe equivalent vacua. This is not so any more for two g's belonging to different homotopy classes. Since the convex combination of two connections is a connection, two inequivalent vacua will be separated by a potential barrier $E(F^t)$ in field configuration space : if

$$a^t = t a_1 + (1-t) a_2 \qquad\qquad F_1 = F_2 = 0$$

$$F^t = \frac{t^2 - t}{2} \left[a_1 - a_2 , a_1 - a_2 \right]$$

$$E(F^t) \propto (t^2 - t)^2 \qquad\qquad E(F^t) > 0$$

It is then argued that at the quantum level, tunnelling will occur between inequivalent vacua, the tunelling amplitude over infinite time being estimated via the Euclidean action of a Euclidean configuration which interpolates between the two vacua at $\pm\infty$ Euclidean time. Such a configuration, which minimizes the Euclidean action and interpolates between two homotopy classes differing by one unit has been found by A.A. Belavin, A.M. Polyakov, A.S. Schwartz, Y.S. Tyupkin. The corresponding characteristic class $\frac{1}{4\pi^2}\int(F,F)$ is unity. It is then conjectured on the basis of these semiclassical arguments that there is a family of gauge invariant vacua parametrized by an angle $0 \leq \theta < 2\pi$ (the dual of $\mathbb{Z} = \pi_3(G)$ which labels the classical vacua), and that these vacua are inequivalent if massive fermions are introduced so that chiral symmetry is broken, a situation analogous to that found in the analysis of the massive Schwinger Thirring model in two dimensions mathematically analyzed by J. Fröhlich.

CONCLUSION

The reader may complain that little physics can be extracted from these notes. Since very little is firmly established within a rapidly changing overall picture which definitely lies outside the framework of formal perturbation theories, the author has found it safer, were it be dull, to compile some type of an algebraic appendix on gauge theories. The conventional perturbative approach is yet crowded with happy coïncidences which, in spite of some efforts, have not been given good explanations, and recent ideas show that a definition of the gauge group, (and eventually of its Lie algebra), which exists in the framework of lattice theories, is just missing in the continuum case. Besides, most studies on the renormalized perturbation theory rely on the dimensional scheme of G. 't Hooft and M. Veltman, whereas mathematicians will only feel like relying on the fairly recent analysis of E.R. Speer [17] which does not yet fill in all details. It was thus essentially impossible to make an exhaustive list of everything that is well known or/and proved by all specialists in this field, to whom I apologize.

ACKNOWLEDGMENT

I wish to thank B. Julia and B. Schroer for clearing up some points brushed over in section IV.

- FOOTNOTES AND REFERENCES -

[1] C.N. YANG, R.L. MILLS,
Phys. Rev. 96 , 191 (1954)

[2] Some notions of differential geometry found here are extracted from

a) S. KOBAYASHI, K. NOMIZU,
"Foundations of Differential Geometry", Vol. I, II,
Interscience, New York 1969

b) W. GREUB, S. HALPERIN, R. VANSTONE,
"Connections, Curvature and Cohomology", Vol. I, II, III,
Academic Press, New York 1976

[3] S.S. CHERN,
"Topics in Differential Geometry", I.A.S. Princeton, 1951

[4] C.N. YANG, T.T. WU,
Phys. Rev. D , 12, 3845 (1975), and Stony Brook
Preprints ITP SB 76-5, 76-11

[5] See, e.g. [2a] Ch. XII
and [2b] , loc.cit. under the title "Weil Homomorphism"

[6] See [2b] , especially Vol. III

[7] Standard references can be found in the lectures by C. BECCHI, A. ROUET, R. STORA in "Renormalization Theory", G. VELO, A.S. WIGHTMAN Ed., NATO Advanced Study Institutes Series C23, D. Reidel Pub. Co. 1976. (Lectures given at the International School of Mathematical Physics, Erice, Italy, 17-31 August 1975).

[8] See, e.g. K. WILSON's Lectures, these proceedings

[9] Spelled out in this form by G. 't HOOFT and M. VELTMAN

[10] J. DIXON,
to be published, private communication

[11] S.D. JOGLEKAR, B.W. LEE,
Ann. Phys.,

[12] J. ZINN JUSTIN,
in Acta Universitatis Wratislaviensis n° 368,
XIIth Winter School of Theoretical Physics in Karpacz

[13] J. MADORE,
I.H.P. Preprint, Paris, June 1976 ; private communication

[14] G. 't HOOFT,
Nucl. Phys. B79, 276 (1974)

[14] A.M. POLYAKOV,
J.E.T.P. 20 , 430 (1974)

J. ARAFUNE, P.G.O. FREUND, G.J. GOEBEL,
J.M.P. 16, 433 (1975)

S. COLEMAN,
"Classical Lumps and their Quantum Descendents",
Ettore Majorana International School, Erice (1975)

[15] a) A.A. BELAVIN, A.M. POLYAKOV, A.S. SCHWARTZ, Yu.S. TYUPKIN,
Phys. Lett. 59B, 85 (1975)

b) G. 't HOOFT,
Harvard Preprints, 1976

c) C.G. CALLAN, R. DASHEN, D. GROSS,
Princeton Preprint COO 2220 75

d) R. JACKIW, C. REBBI,
M.I.T. Preprint, C.T.P. 548

[16] J. FRÖHLICH,
Same volume as [7] , and these proceedings

[17] E.R. SPEER,
Same volume as [7].

MASSLESS RENORMALIZABLE FIELD THEORIES AND THE YANG-MILLS FIELD

P.K. MITTER

Laboratoire de Physique Théorique et Hautes Energies

Université Pierre et Marie Curie, Paris

0. There has been considerable interest in this school in the structure of the massless Yang-Mills field $(Y\text{-}M)_4$ especially in the interplay of ultraviolet (U-V) asymptotic freedom, infrared (I-R) instability and quark confinement. In these notes (which are in the form of remarks) I shall summarise certain results in the renormalised perturbation theory (R.P.T.) of massless $(Y\text{-}M)_4$ and more generally any renormalisable purely massless four-dimensional theory with only dimension 4 terms in the Lagrangian. It is evident that the unsolved problem of central interest is true I.R. behaviour in $(Y\text{-}M)_4$. This is amenable only to non-perturbative analysis, of which lattice methods (Wilson) appear to be of considerable interest. However it is a reasonable hypothesis that the knowledge of short distance behaviour (asymptotic freedom) extracted from R.P.T. via formally consistent renormalisation groups sums will be stable under more constructive analysis. Non-perturbative methods designed for I.R. behaviour, must then be consistent with such knowledge. This justifies our continued interest in results of R.P.T., and related developments in renormalisation group theory which constitute the material of these notes.

These notes are divided into the following sections :

1- I review, informally, rigorous results in I.R. power counting which leads to the proof that the renormalised Feynman integrals (corresponding to each graphical contribution to Green's functions) are tempered distributions. These results, (Lowenstein, Speer) apply to massless $(Y\text{-}M)_4$ and more generally (see above).

2- We review the known structure of massless $(Y\text{-}M)_4$ and the main results of R.P.T. To tie up with section 1 and also for simplicity we focus on dimensional renormalisation, (Speer, Breitenlohner-Maison). For the BPHZ recursive construction of counter-terms to employment Slavnov invariance (Lowenstein), see Stora (these proceedings).

3- The main problem of massless $(Y\text{-}M)_4$ is non-perturbative I.R. behaviour ; it would be useful to disentangle U.V. and I.R. contributions consistent with asymptotic freedom.
(i) We summarise a recent development (Lowenstein-Mitter) in the renormalisation theory of massless fields designed to facilitate the separation of U.V. and I.R. field fluctuations at the renormalised level by separating renormalised Feynman integrals into "hard" and "soft" components. Such decompositions, and the 4-dimensional renormalisation schemes tied to them, lead to renormalisation groups which probe directly regions of integration.
(ii) We give the structure of a hard-soft split of massless $(Y\text{-}M)_4$ /Mitter,Valent/. Hard-soft symmetry is a super-symmetry, intertwined with Slavnov transformations which remain a symmetry of the Lagrangian. Its possible relevance to the problem of reconciling U.V. asymptotic freedom with lattice methods (the idea is due to K.G. Wilson) is pointed out. The material of Section 3 was developed immediately after the Cargèse School. I have included it because of its relevance to the subject under discussion.

1. INFRA-RED POWERCOUNTING AND THE EXISTENCE OF GREEN FUNCTIONS AS DISTRIBUTIONS

We consider, to begin with, an un-renormalised Feynman integral $\mathcal{I}_{\Gamma,\epsilon}$; Γ = 1-vertex irreducible (1 VI) graph without l.o.g. Γ is assumed to have either at least one massive line or not less than 2 external vertices. The propagators are of the form

$$Z_\ell(p_\ell)\left(p_\ell^2 - m_\ell^2 + i\epsilon\left(\vec{p}_\ell^{\,2} + m_\ell^2\right)\right)^{-1}$$

(see / 1,2/), with Z_ℓ a spin polynomial, and m_ℓ the mass of the line 'ℓ'. We allow both massive and massless lines, but later we shall consider only the purely massless case.

- A - All External Momenta are Euclidean and Generic

We explain informally the I.R. power counting (Speer / 3 /), assuming that $\mathcal{I}_{\Gamma,\epsilon}$ is written in parametric (α) space. For the momentum space analysis both at the unrenormalised and renormalised level, see Lowenstein / 1 /. The analysis of U.V. divergences is wellknown.

Let us remark, that the $i\epsilon$ prescription is such that if $m_\ell = 0$ $p_\ell^2 + i\epsilon\vec{p}_\ell^{\,2}$ vanishes when $p_\ell \to 0$. If we can prove that $\mathcal{I}_{\Gamma,\epsilon}$ is free of I.R. singularities, then $\epsilon \to 0^+$ pointwise limit can be taken for generic Euclidean external momenta.

Consider any subgraph $S \subsetneq \Gamma$ such that

i) in S there exists a path wich connects all external vertices of Γ

ii) S contains all the massive lines of Γ

iii) S is minimal in the sense that if S is not 1.V.I, then removing any component of S destroys (i) or (ii).

Such subgraphs in S are called links. We now form the st of all quotients :

$$\mathcal{Q} = \{ Q = \Gamma / S \mid S : \text{link and } Q : \text{IVI} \}$$

Theorem

The Feynman integral $g_{\Gamma,\epsilon}$ is free of I.R. singularities (for generic Euclidean external momenta).

Iff :

$$d(Q) > 0, \forall Q \in \mathcal{Q}$$

Here $d(Q)$ is the naive powercounting dimension of Q considered as a graph. As an immediate application, we consider any purely massless renormalisable theory e.g. to be concrete, massless $(Y\text{-}M)_4$, and consider its Green's functions.

Then

$$d(Q) = d(\Gamma) - d(S) = (4 - B(\Gamma)) - (4 - B(S))$$
$$= B(S) - B(\Gamma)$$

where $B(\gamma)$ is the number of external lines (bosons) of Γ. Since $S \subsetneq \Gamma$, and moreover S contains all the external vertices, it follows that $B(S) > B(\Gamma)$. Thus $d(Q) > 0$ and Green functions of massless $(Y\text{-}M)_4$ has no I.R. divergences. Moreover, this result is stable under suitable renormalisation e.g. for the schemes of Section 2,3. Henceforth we assume such a renormalisation has been carried out e.g. dimensionally /section 2/. The renormalised Feynman integral will be devoted $R_{\Gamma,\epsilon}(P)$. We turn to the problem of taking $\epsilon \to 0^+$ in the sense of distributions. The $\epsilon \to 0^+$ limit in the Euclidean region with generic momenta exists trivially.

- B - Renormalised Feynman Integrals as Tempered Distributions

We restrict ourselves to the purely massless case with $(Y\text{-}M)_4$ as a concrete example. We have to prove that

$$\lim_{\epsilon \to 0^+} \int \prod_i d^4p_i \; R_{\Gamma,\epsilon}(p_1 \cdots p_n) \tilde{\varphi}(p_1 \cdots p_n) \qquad \text{(B.1)}$$

exists, where $\tilde{\varphi} \in \mathcal{S}(R^{4n})$ (space of tempered test functions). The discussion is carried out in two steps, of which the first is to prove (Lowenstein / 1 /)

$$\int \prod_i d^4p_i \; \tilde{\varphi}(\underline{p}) \, R_{\Gamma,\epsilon}(\underline{p}) \quad \text{exists} \qquad \text{(B.2)}$$

with $\epsilon > 0$. This convergence theorem is crucial, and non-trivial since the inverse propagators $(p^2 + i\epsilon \vec{p}^2)$ vanish when the $p_i = 0$. The limit $\epsilon \to 0^+$ follows (second step), although the discussion is delicate (Lowenstein and Speer / 2 /).

We shall sketch the essential idea of the proof of (B_2) by reducing it to the basic theorem of Case A.

Corresponding to Γ, we consider an "augmented" vacuum graph Γ_0 obtained by adding to Γ a new vertex V_0 and joining all external lines at this vertex. No external momentum flows into V_0, hence Γ_0 is a vacuum graph. The old external line ℓ, which is now an internal line ℓ flowing into V_0, is assigned a propagator

$$(p_\ell^2 - M^2 + i\varepsilon(\vec{p}^2 + M^2))^{-\nu}$$

where ν is a positive integer as large as desired and $M \neq 0$, real, and otherwise arbitrary. Lowenstein / 1 / observed that the absolute integrability of $R_{\Gamma_0,\varepsilon}(p)$ i.e. :

$$\int \prod_i d^4p_i \; R_{\Gamma_0,\varepsilon}(p_1 \cdots p_n) \text{ exists} \qquad \text{(B.3)}$$

implies the existence of (B.2). Here $R_{\Gamma_0,\epsilon}$ is the renormalised Feynman integral corresponding to Γ_0 (no new U.V. divergences arises in extending $\Gamma \to \Gamma_0$ since $\nu > 0$ may be chosen as large as necessary). That $(B.3) \Rightarrow B(2)$ follows from the topology of the space $\mathcal{S}$, where $(p_\ell^2 - M^2 + i\varepsilon(\vec{p}^2 + M^2))^{\nu}$ act legitimately as multipliers.

The proof of (B.3) is immediate. Γ_0 is a vacuum graph but it has at least 2 massive lines (since $M \neq 0$), and it remains Iν1. Hence it falls under case (A), and the I.R. powercounting can be used. Now any link S must contain all massive lines. Hence no new quotients Q are generated in $\mathcal{Q}$, besides those already considered for Γ. But for those we already have shown $d(Q) > 0$ for massless $(Y-M)_4$, and indeed for any purely massless theory with anly dimension 4 terms in the Lagrangian. This shows that the integral (B.3), and hence (B.2), with the renormalised parametric representation inserted for $R_{\Gamma,\epsilon}$ are absolutely convergent. The existence of the $\epsilon \to 0^+$ limit then follows from theorem 2 of / 2 /. The discussion at this stage is purely technical (no new graph theoretical powercounting is necessary) and the reader is refered to / 2 / for details.

<u>REMARK</u>

We have seen that renormalised Feynman integrals corresponding to each graphical contributions Γ to Greens functions are tempered distributions. This implies that the <u>absorptive parts</u> exist (graph-wise), and in particular are free of I.R. singularities. In particular the validity of the cutting rules / 4 / in the renormalised theory (axiomatic structure / 5 /) implies that the absorptive part may be expressed as a generalised unitarity sum, which is seen to be free of I.R. singularities.

2. MASSLESS $(Y\text{-}M)_4$

– A –

For the structure of gauge theories the reader is referred to the lectures of Stora / 6 /. Here I shall concentrate on purely massless $(Y\text{-}M)_4$. As is well known the formal Lagrangian is :

$$\mathcal{L} = -\frac{1}{4}\vec{G}_{\mu\nu}\cdot\vec{G}_{\mu\nu} - \frac{1}{\alpha}\left(\frac{1}{2}(\partial_\mu A^\mu)^2 + D_\mu c^+ \cdot \partial^\mu c\right) \qquad (2.1)$$

where we have included the gauge fixing term c^+, c are spinless anticommuting Faddeev-Popov ghosts and, as usual :

$$\vec{G}_{\mu\nu} = \partial_\mu \vec{A}_\nu - \partial_\nu \vec{A}_\mu + g\,\vec{A}_\mu \wedge \vec{A}_\nu \qquad (2.2)$$

$$D_\mu = \partial_\mu + g\,A_\mu \wedge$$

We restrict ourselves to SU(2) for simplicity. The Feynman rules are read off from (2.1). We proceed at first formally. A very important property is the invariance of the corresponding action under global Slavnov formations /7/ discovered by Becchi-Rouet and Stora:

$$\delta A_\mu = D_\mu c^+ \delta\lambda$$

$$\delta c = (\partial A)\,\delta\lambda$$

$$\delta c^+ = g/2\; c^+ \wedge c^+\,\delta\lambda \qquad (2.3)$$

where $\delta\lambda$ is a space-time independent anticommuting c-number. The corresponding Noether current is the conserved Slavnov current :

$$J_\mu = \left(G_{\mu\nu} + \frac{1}{\alpha} g_\mu\,\partial A\right) D^\nu c^+ + g/2\,(c^+ \wedge c^+)\,\partial_\mu c \qquad (2.4)$$

whose conservation leads to the Ward-Slavnov-Taylor identity. (The Slavnov current has dimension 4 and many more renormalisation parts than usual, which leads to complications in application of renormalisation theory).

If we denote by X the formal product :

$$X \equiv \prod_i A_{\mu_i}(x_i) \prod_j c(y_j) \prod_j c^+(z_j) \qquad (2.5)$$

with formal Green functions $\langle T\,X \rangle$, then integrated Slavnov identity can be written as

$$\mathcal{J}\,\langle T\,X \rangle = 0 \qquad (2.6)$$

Where

$$\mathcal{S} \equiv \int dx \left(-D^{\nu} c^{+}(x) \frac{\delta}{\delta A_{\nu}(x)} + \partial A(x) \frac{\delta}{\delta c(x)} + g/_{2}\, c^{+} \wedge c^{+} \frac{\delta}{\delta c^{+}(x)} \right) \tag{2.7}$$

and formal functional derivatives are meant.

An important property of the Slavnov transformation $\mathcal{S}$ is :

$$\mathcal{S}^{2} = - \int dx\, \partial_{\mu} D^{\mu} c^{+}(x) \frac{\delta}{\delta c(x)} = 0 \tag{2.8}$$

by virtue of the ghost field equation

$$\frac{1}{\alpha} \partial_{\mu} D^{\mu} c^{+}(x) = i \frac{\delta}{\delta c(x)} \tag{2.9}$$

in the sense of insertions in Green functions.

The main problem of renormalisation (in perturbation theory) is to construct Green functions (i) as tempered distributions which satisfy (ii) axiomatic structure / 5 / which implies counterterm structure and Lorentz invariance and (iii) the Slavnov identity (2.6). Related to it is the construction of guage invariant composite fields.

\- B -

These problems (i) - (iii) are solved at one stroke if we adopt dimensional renormalisation in the formulation given by Speer / 8 /. See also / 9 /. From the purely perturbative point of view, we shall therefore first expose the consequences of dimensional regularis tion and renormalisation. Dimensional regularisation is wellknown and precise formulations have been given in /8,9/. A length scale (t'Hooft) is introduced by defining the coupling constant $g = \mu^{4-\nu} g_{0}$, g_{0} is dimensionless.

Speer / 3 / has given the complete singularity structure in ν (space time dimension complexified) of dimensionally continued parametric (α-space) Feynman integrals. For purely massless $(Y\text{-}M)_{4}$, for any I V 1 graph Γ the U.V. and I.R. singularities are poles on hyperplanes :

$$\frac{d(H)}{2} = k_{1}, \quad \frac{d(Q)}{2} = - k_{2} \tag{2.10}$$

where k_{1}, K_{2} are non negative integers. The subgraphs $H \subset \Gamma$ are IPI generalised vertices and moreover "saturated" i.e. Γ/H is I V 1. The quotients Q are precisely those encountered in section 1. Finally $d(\gamma)$ is the minimal U.V. powercounting dimension of γ in ν -dimensions, and is obviously linear in ν .

In section 1 we saw that for $\nu = 4$, $d(Q) > 0$. It follows that for any given order n of perturbation theory we can find a neighborhood N_n of $\nu = 4$ in the ν plane, such that $d(Q) > 0$ However this neighborhood shrinks to zero as $n \to \infty$, so that dimensional regularisation is not a non-perturbative regularisation for massless renormalisable theories.

It remains to subtract the U.V. singularities. This is done /8,9/ in the spirit of BPH recursive subtractions /10/ with the following differences, of which (i) is well known and (ii) crucial.

(i) The basic subtraction operation (acting resursively is that of removing the total singular part of the Laurent expansion around $\nu = 4$ which is represented by a Cauchy integral replacing the truncated Taylor operator in external momenta of BPH.

(ii) The recursion on divergent generalised vertices is replaced by recursion on divergent I P 1 subgraphs, as in BPH Z /11/.

Actually this is equivalent to subtractions with recursion on generalised vertices upto fixed (automatically choosen by virture of (i) and (ii)) finite renormalisation /8,9/. Thus dimension renormalisation, so formulated, has counter-term structure. The above formulation of dimensional renormalisation has the following remarkable consequence : Theorem (5.3) of /8/, /9/.

Let P, Q_i represent monomials in elementary fields (and space-time derivatives) with ϕ_j any elementary field (c, c^+, A_μ) then

$$\left\langle T_R \left(\frac{\delta \mathcal{L}}{\delta \phi_j} - \partial^\mu \frac{\delta \mathcal{L}}{\delta \partial^\mu \phi_j} \right)(x) \; P\{\underline{\phi}\}(x) \prod_i Q_i\{\underline{\phi}\}(y_i) \right\rangle$$

$$= \sum_i \delta(y_i - x) \left\langle T_R \; P(x) \frac{\delta Q_i}{\delta \phi_j}(y_i) \prod_{i \neq j} Q_i\{\underline{\phi}\}(y_i) \right\rangle \quad (2.11)$$

Here T_R is the renormalised time ordering i.e. graphwise implementation of the above subtraction procedure. This means that renormalised equations of motion retain canonical structure even when P is a composite field.

We apply this result to the divergence of the Slavnov current j_μ (2.4) inserted into a Green function with T_R ordering. ∂_μ and T_R commute /8/. Applying (2.11) we immediately obtain in the renormalised theory the local Slavnov identity whose integrated form is (2.6, 2.7). Another way of stating this is that (2.11) implies that the Schwinger action principle in its classical form remains valid in the renormalised theory for variations (2.3) engendering (2.6) /9/.

Thus Dimensional renormalisation, as formulated above, is characterised by

(i) Existence of Green functions in R.P.T. as tempered distributions

(ii) Slavnov invariance

(iii) counter-term structure.

(iv) no specific normalisation conditions e.g. at Euclidean symmetry points ad-hoc (the vanishing of 2-point proper functions at zero momentum is an automatic consequence). Of course, the renormalised Green functions can be computed as functions in the Euclidean region.

(i), (ii), (iii) justify the usual procedure renormalising multiplatively using dimensional regularisation and preserving Slavnov invariance. Several well known consequences follow. Thus one can derive renormalisation group equation with $\beta(g)$ independent of the gauge parameter α. (See e.g. /12 /) Local asymptotic freedom is well known / 13 /. It has been shown / 14 /, as a consequence of the definitions of parametric functions and asymptotic freedom that remormalisation constants (as functions of (4 - ε)) are computable i.e. they can be expressed as formal asymptotic series. This fact should play some role in future constructive approaches to $(Y\text{-}M)_4$.

- C -

We have so far discussed dimensional renormalisations. Purely 4-dimensional graphwise renormalisation e.g. Lowenstein / 15 / requires, in general, recursive construction of finite counter-terms (Stora - these proceedings) to implement Slavnov invariance. The main reason is that (2.11) gets modified (when P is a composite field) if one subtracts in external momenta. The modification is simply that individual terms in (2.11) have coefficients which differ from tree graph values. This also implies that the Schwinger action principle acquires non-classical correction terms. This is compensated via finite counter terms using (2.8). Moreover the renormalisation group function is now $\beta(g,\alpha)$ depending on α beyond the lowest order. Actually a reparametrisation is necessary, exchanging g for ρ, the invariant of the guage parameter differential operator $\alpha\,\partial/\partial\alpha + \sigma(g)\,\partial/\partial g$, in order to achieve a gauge invariant coupling strength.

See also Zinn Justin / 16 / and remarks of Symanzik / 17 / for other approaches. It is clear that the main outstanding problem of massless $(Y\text{-}M)_4$ is not renromalisability and U.V. behaviour but non-perturbative I.R. behaviour. Here an intelligent 4-dimensional renormalisation might have a role to play as we discuss in the next section.

3. HARD SOFT RENORMALISATION

-A. We now turn to a simple 4-dimensional renormalisation scheme for (renormalisable) massless theories : hard-soft renormalisation, /18 /. We shall sketch the basic idea on the basis of the simplest example : massless $(\phi^4)_4$. The real reason for developing this scheme is for massless $(Y\text{-}M)_4$ and the motivation will become clearer later. We begin with the formal Lagrangian

$$\mathcal{L} = -\frac{1}{2}\varphi \Box \varphi - \frac{g}{4!}\phi^4 \tag{3.1}$$

and the generating functional of Greens functions is given formally by the path integral

$$Z\{J\} = \int \mathcal{D}\phi \exp i\left[\int d^4x \left(\mathcal{L}(\varphi)(x) + J(x)\varphi(x)\right)\right] \tag{3.2}$$

We add to this an independent Gaussian, without changing anything :

$$Z\{J\} = \text{cnst} \int \mathcal{D}\phi \mathcal{D}\psi \exp i\left[\int d^4x \left(\mathcal{L} - \frac{\Lambda^2}{2}\psi(x)^2 + J\varphi\right)\right] \tag{3.3}$$

with ψ a scalar field. We now make a change of variables :

$$\varphi = \varphi_h + \varphi_s\,,\quad \psi = \varphi_h - \frac{\Box}{\Lambda^2}\varphi_s \tag{3.4}$$

obtaining :

$$Z\{J\} = (\text{cnst})' \int \mathcal{D}\varphi_h \mathcal{D}\varphi_s \exp i\left[\int d^4x \left(\mathcal{L}_{h-s} + J(x)\varphi(x)\right)\right] \tag{3.5}$$

with

$$\mathcal{L}_{h-s} = -\frac{1}{2}\varphi_h(\Box + \Lambda^2)\varphi_h - \frac{1}{2}\varphi_s \Box\left(1 + \frac{\Box}{\Lambda^2}\right)\varphi_s - \frac{g}{4!}\varphi^4 \tag{3.6}$$

with $\varphi = \varphi_h + \varphi_s$. Here h, s refer to hard and soft fields, for obvious reasons. The hard propagator is massive but behaves as $(p^2)^{-1}$ when $p^2 \to \infty$. The soft propagator has the zero-masspole but behaves as $(p^2)^{-2}$ as $p^2 \to \infty$.

The construction of renormalised Greens functions is simple. We consider the Feynman rules from (3.6) in terms of hard and soft fields and use the propagators provided by the bilinear part of (3.6). Then all Feynman integrals can be renormalised by performing minimal. U.V. at zero-momenta, without encountering I.R. divergences The reason is that any (sub)graph which is superficially divergent has at most one soft line, in which case subtraction at zero-momenta is harmless. As in all graphwise renormalisation schemes the contributions to proper functions, superficially U.V. convergent or not, require additional subtractions at zero-momenta /18/ to preserve the zero-mass condition and I.R. convergence. Λ^2 sets the renormalisation mass scale ; it separates infra-red and ultra-violet fluctuations. A convenient way to keep track of the power-counting is to use propagators $i(p^2-\Lambda^2)^{-1}$, and $is^2\Lambda^2(p^2)^{-1}(\Lambda^2-p^2)^{-1}$ where we have introduced a real parameter s . Then all U.V. subtractions are performed at $p = s = 0$, using the U.V. degree of divergence formula $\delta(\gamma) = 4 - n_{ext}$ (independent of field species), and setting $s = 1$ in the end. This automatically enforces the above minimal U.V. renormalisation / 18 /. In contrast the I.R. subtractions for contributions to the two-point proper function are at $p = 0$, $s = 1$. The renormalised theory is free of I.R. and U.V. divergences, the Green's functions existing as tempered distributions. A normal product algorithm is given in / 18

An important question is the consistency of the renormalised theory : one has to show that no indefinite metric ghost has crept in. This follows on observing that the insertion of the field $\psi(x) = \varphi_h - \Box/\Lambda^2 \cdot \varphi_s$, in the Greens functions of (3.5) gives zero in the renormalised theory / 18 /. This provides us with the necessary "gauge invariance" (hard-soft invariance) criterion for the physical theory. Finally we consider the variation $\Lambda^2\,\partial/\partial\Lambda^2$ of $Z\{J\}$. In the unrenormalised theory it is formally zero. However in the renormalised theory it can be proved / 18 / that one obtains the anticipated renormalisation group equation :

$$\left(\Lambda^2\frac{\partial}{\partial\Lambda^2} + \beta(g)\frac{\partial}{\partial g} + \gamma(g)\int d^4x\, J(x)\frac{\delta}{\delta J(x)}\right)Z\{J\} = 0 \tag{3.7}$$

(3.4) is the simplest of a family of transformations, which after renormalisation gives rise to a family of renormalisation groups.

Instead of doing all the integrations in (3.5) in R.P.T., as above, we may choose to do at first only the φ_h integration in R.P.T. We immediately derive the effective action ;

$$\begin{aligned} S_{eff} = \int d^4x \left[-\tfrac{1}{2}\varphi_s\Box\left(1+\frac{\Box}{\Lambda^2}\right)\varphi_s - \frac{g}{4!}\varphi_s^{\,4} + J(x)\varphi_s(x)\right] \\ + \sum_n \int dx_1\cdots dx_n\, J(x_1)\cdots J(x_n)\, G_h^{(n)}(x_1\cdots x_n;\Lambda^2,\varphi_s) \end{aligned} \tag{3.8}$$

$$Z\{J\} = \text{const}\int d\varphi_s \exp S_{eff}\{J\}$$

Here $G_h^{(n)}(x_1 \cdots x_n; \Lambda^2, \varphi_s)$ are renormalised, connected Green functions (according to previous rules) $\langle T\varphi_h(x_1)\cdots\varphi_h(x_n)\rangle_{conn}$ obtained from the Labrangian

$$\mathcal{L}_h = -\tfrac{1}{2}\varphi_h(\Box+\Lambda^2)\varphi_h - g/4!\,(\varphi^4 - \varphi_s^4) \tag{3.9}$$

with φ_s treated as an external field. The $G_h^{(n)}$ can be proved to obey Callan-Symanzik equations for Λ^2 variation. The propagators in (3.8) are highly damped U.V. leading, on the whole, to convergent (U.V.) graphs. Only a (U.V.) logarithmic divengence remaining (by a sharper hard-soft split this too could have been eliminated) in S.E. graphs. The price is all the extra-terms in (3.8). Of course the previous I.R. subtractions at $p = 0$, $s = 1$ for 2-point functions corresponding to "zero-mass" condition are still necessary.

What was the purpose of this exercise? It was to show that the renormalisation scale could be built into the Lagrangian of an U.V. softer but completely equivalent theory, an idea due to K.G. Wilson / 19 /. This would permit a smoother introduction of a lattice than in the bare theory, at least for weak coupling. For massless $(\phi^4)_4$, which is I.R. asymptotic free, Symanzik/ 20 /, one can obtain the necessary I.R. behaviour directly from (3.7) without going to (3.8). But this is precisely not the case for massless $(Y\text{-}M)_4$, to which we turn.

- B - <u>Structure of hard-soft split of (Y-M)4</u>

I shall now point out a possible approach to transforming the orginal massless $(Y\text{-}M)_4$ to an equivalent softer theory where the hard fluctuations have been eliminated. It is based on work with G. Valent / 21 /.

We start with the $(Y\text{-}M)_4$ Slavnov invariant Lagrangian (2.1) of section II to which we add an independent Gaussian, in the sense of path integrals as in part (A). Thus we have :

$$\mathcal{L} = \mathcal{L}_{S\ell}(A_\mu, c, \bar{c}) + \frac{\Lambda^2}{2}\psi_\mu^2 \tag{3.10}$$

with

$$\mathcal{L}_{S\ell} = -\frac{1}{4}G_{\mu\nu}\cdot G_{\mu\nu} + \frac{1}{\alpha}\left(\frac{1}{2}(\partial_\alpha A^\alpha)^2 + \bar{\mathcal{D}}_\mu c^+ \partial_\mu c\right) \tag{3.11}$$

ψ_μ is an independent field, and, as in part (A), this does not change the Y-M theory.

We now make the field transformation :

$$A_\mu = A_\mu^h + A_\mu^s$$
$$\psi_\mu = A_\mu^h - \frac{1}{\Lambda^2}\left(D_\lambda^s G_{\lambda\mu}^s + \frac{1}{\alpha} D_\mu^s (\partial A^s)\right) \tag{3.12}$$

superscript s, means only soft fields involved. In contrast to part (A) we have chosen a non-linear field transformation. Ofcourse we could have used a linear transformation with all covariant derivations D^s in (3.12) replaced by ∂. But then the guage structure of the equivalent soft theory, which we are really after, would be spoil Instead (3.12) has remarkable consequences. Being a non-linear field transformation, the Jacobian determinant in the path integral measure is non-trivial. However, it can be put back into the Lagrangian via new ghosts. We thus obtain :

$$\mathcal{L}_{h-s} = \mathcal{L}_{sl}(A, c, \bar{c}) + \frac{\Lambda^2}{2}\psi_\mu^2 + \Lambda^2 (K_{\mu\nu} B_\nu^+) \cdot B_\mu \tag{3.13}$$

where B_μ, B_μ^+ are anticommuting vector ghosts, and

$$K_{\mu\nu} = \left(\frac{\delta}{\delta A_\mu^h(y)} - \frac{\delta}{\delta A_\nu^s(y)}\right)\psi_\mu(x) \tag{3.14}$$

The fields A_μ, ψ_μ are now <u>defined</u> by (3.12). We could have split the F.P ghosts c, c^+ also, but for the moment we do not do that. Formally, the Lagrangian (3.13) engenders the same theory as (3.10), so long as external sources are coupled to A , c, $\bar{c}$ (the physical sector). From the bilinear part of (3.13), the propagators are obtained : the A_μ, B propagators are massive (U.V. hard) the A^s propagator is U.V. soft, but has zero-mass pole. (3.13) has the following interesting feature : the purely soft gauge field part (obtained by neglecting ghosts and A^h couplings) is

$$-\frac{1}{4} G_{\mu\nu}^s \cdot G_{\mu\nu}^s + \frac{1}{2\Lambda^2}\left(D_\lambda^s G_{\lambda\mu}^s\right)^2 \tag{3.15}$$

which is the gauge invariant Slavnov-regularisation / 23 / of Y-M theory. However in our approach Λ is not a regulator mass but a renormalisation scale !

As in part A we need a symmetry to identify the physical i.e. hard-soft independent sector. Not surprisingly the following nonlinear infinitesimal transformation leaves the Lagrangian invariant:

$$\delta_T A_\mu^h = B_\mu^+ \delta\omega , \quad \delta_T A_\mu^s = - B_\mu^+ \delta\omega$$
$$\delta_T B_\mu = - \psi_\mu \delta\omega \qquad (3.16)$$
$$\delta_T B_\mu^+ = \delta_T c = \delta_T c^+ = 0$$

$\delta\omega$ is an anti-commuting c-number.

We have :

$$\delta_T \mathcal{L} = 0 \qquad (3.17)$$

$\delta_T^2 = 0$ on all fields except B_μ , and there by field equation. The resulting integrated Ward identity enables us to select the physical sector.

The Λ^2 variation gives formally zero in the physical sector, as is seen using the δ_T-Ward identity. This will no longer be true in the renormalised theory, where, as in part (A) we will obtain the (hard-soft) renormalisation group.

We now turn to the Slavnov transformations. These may be extended to the hard-soft level as follows :

$$\delta_\sigma A_\mu^s = \mathcal{D}_\mu^s c^+ \delta\lambda$$
$$\delta_\sigma A_\mu^h = -g\, c^+ \delta\lambda \wedge A_\mu^h$$
$$\delta c = \partial A\, \delta\lambda$$
$$\delta c^+ = g/2\; c^+ \wedge c^+ \delta\lambda \qquad (3.18)$$
$$\delta B_\mu = -g\, c^+ \delta\lambda \wedge B_\mu$$
$$\delta B_\mu^+ = -g c^+ \delta\lambda \wedge B_\mu^+$$

Here $\delta\lambda$ is an infinitesimal constant, anticommuting with all ghosts.

$$\delta_\sigma^2 = 0$$

on all fields exept C. There it vanishes, as usual, only by virtue of equations of motion. Note that the essential gauge content is in A^s_μ . $\mathcal{L}_{s\ell}$ is obviously invariant under (3.18). Moreover ψ_μ is Slávnov covariant except for gauge dependent term. A simple calculation shows :

$$\delta_\sigma \mathcal{L}_{h-s} = \delta_T U \qquad (3.19)$$

Where

$$U = \frac{1}{\alpha} \partial_\nu c^+ \delta\lambda \, D_\nu^s D_\mu^s B_\mu \tag{3.20}$$

Remarkably enough,

$$\delta_\sigma U = 0 \tag{3.21}$$

(3.19 immediately shows that Slavnov invariance is recovered in the h-s independent sector, on using the hard-soft Ward identity.

One can pose the question is there a splitting of the gauge fields such that Slavnov invariance (in a generelised form) holds in general (and not only in the h-s **independent sector)** ?

By a modified Gaussian transformation (with <u>all</u> derivatives in (3.12) replaces by D_μ) and adding F.P. ghosts to the soft gauge fixing one can show (P.K. Mitter and G. Valent (unpublished)) that both hard-soft and generalised Slavnov invariance can be <u>simultaneously</u> maintained.

However, in the B_μ transformation law (3.16) there arrises an additive contribution corresponding to these F.P. ghosts and the $\delta_\sigma c$ transformation (3.18) also changes. Moreover the δ_σ, δ_T symmetries will be preserved on renormalisation The simplest way to see this is via dimensional renormalisation /22/ with Λ as coupling constant scale. The two symmetries δ_T, δ_σ forbid mass-counter terms. As a consequence of /22/, the renormalised action principle has the connonical form i.e. non-symmetric radiative corrections add up to zero in Green functions.

As in part A, $\Lambda^2 \partial/\partial\Lambda^2$ variation leads to the R.G. Let us note that the renormalised hard integration (i.e. of A_μ^h) is gauge invariant because of (3.18). The analogous steps as in part (A) will lead to the Slavnov invariant softened theory which has the form of (3.15) plus correction terms.

In connection with the lattice gauge theory we note the following. The term $-\frac{1}{4} G_{\mu\nu}^s G_{\mu\nu}^s$ has its string variable counterpart in Wilson's basis square, of which it is the leading contribution in the classical continuum limit. Our first correction term is $-\frac{1}{2\Lambda^2} (D_\lambda^s G_{\lambda\mu}^s)^2$ which too has its string variable representation. It is an important task for the future to construct the full lattice gauge representation. Such a lattice gauge theory will have much better U.V. properties consistent with asymptotic freedom.

REFERENCES

1. J.H. LOWENSTEIN : Comm. Math. Phys. 47, 53 (1976).

2. J.H. LOWENSTEIN and E.R. SPEER : Comm. Math. Phys. 47, 43 (1976).

3. E.R. SPEER : Ann.Inst. H. Poincaré 23, 1 (1975).

4. M. VELTMAN : Physica (1963).

5. K. HEPP : Les Houches Lectures (1970).

6. R. STORA : These proceedings.

7. C. BECCHI, A. ROUET, R. STORA : in "Renormalisation theory" ed. A. Wightman and G. Velo, D. Reidel (1976).

8. E.R. SPEER : in / 7 /.

9. P. BREITENLOHNER and D. MAISON : Comm. Math. Phys. 52, 11 (1977) 52, 39 (1977).

10. K. HEPP : Théorie de la Renormalisation, Springer Verlag, Lectures notes in Physics vol.2, (1969).

11. J.H. LOWENSTEIN : in / 7 /.

12. L. BANYAI et al. : Nuovo Cimento Lett. 11 (1974) 151.
W.E. CASWELL and F. WILCZEK : Phys. Lett. 49B (1974) 291.

13. H. POLITZER : Phys. Rep. 14C, 4 (1974).

14. L. BANYAI et al : Nuc. Phys. B93, (1975) 355.

15. J.H. LOWENSTEIN : Nuc. Phys. B96, (1975) 189.

16. J. ZINN-JUSTIN : Bonn Lectures (1974).

17. K. SYMANZIK : these proceedings.

18. J.H. LOWENSTEIN and P.K. MITTER : NYU/TR6/76, Annals of Phys.N.Y. in press.

19. K.G. WILSON : in "New Pathways in High Energy Physics, Vol.II, A. Perlmutter, ed. (Plenum, N.Y., 1976).

20. K. SYMANZIK : Comm. Math. Phys. 34, 7 (1973).

21. P.K. MITTER and G. VALENT : forthcoming pre-prints.

22. P. BREITENLOHNER and D. MAISON : Comm. Math. Phys. 52, 55 (1977).

23. A.A. SLAVNOV : Theor. Math. Phys. 13, 1054, (1972) (trans.).

CONFORMAL INVARIANT QUANTUM FIELD THEORY

G. MACK

Institut für Theoretische Physik

der Universität Hamburg

We shall present here some attempts to analyze quantum field theory (QFT) nonperturbatively. The main tool in this is a partially summed up and supposedly convergent version of Wilson operator product expansions on the vacuum. For realistic quantum field theory with mass and without conformal symmetry we conjecture that they are true. For conformal invariant theories we are able to prove that they hold, given only the existence of Wilson expansions as asymptotic expansions at short distances. Given these vacuum expansions (as we shall call them), arbitrary n-point Wightman functions can also be expanded; the expansion is completely fixed if all the two and three point functions of all the fields (including composite ones) in the theory are known. Given that these two and three point functions satisfy the axioms themselves, the expansion formula provides an Ansatz which satisfies all the Wightman axioms automatically if it converges, except locality. The locality constraint amounts to a crossing relation for the four point functions.

All this discussion is in Minkowski space and makes no reference to a specific Lagrangean. Lagrangean integral equations for conformal invariant Euclidean Green functions can also be analyzed nonperturbatively and a dynamical derivation of Wilson expansions at short distances can be given. This was explained elsewhere[1-3]. In the last lecture of this series we will discuss the connection of some of these results and the Minkowski space approach described before. It turns out that the analytic continuation to Euclidean space requires some care.

1. CONVERGENT OPERATOR PRODUCT EXPANSIONS ON THE VACUUM.

According to Wilson [4], the product of two local fields $\phi^i(x)$ and $\phi^j(y)$ should admit an asymptotic expansion at short distances of the form

$$\phi^i(\tfrac{x}{2})\phi^j(-\tfrac{x}{2})\Omega = \sum_k C^{kij}(x)\phi^k(0)\Omega \tag{1.1}$$

Herein ϕ^k are local fields, and C^{kij} are singular c-number functions. In a scale invariant theory, they are homogeneous functions of x. The expansion is presumably valid on all states Ω in the field theoretic domain $\mathcal{D}$ which is created out of the vacuum by polynomials in smeared field operators. We shall however only consider the special case

$$\Omega = \text{vacuum}$$

Studies in perturbation theory [5] indicate that expansion (1.1) is then valid as an asymptotic expansion to arbitrary accuracy for matrix elements $(\Psi, \phi^i(x)\phi^j(y)\Omega)$, Ψ in $\mathcal{D}$.

Among the fields ϕ^k in expansion (1.1) there are derivatives in other local fields. In general there appears $\partial^\mu\phi$ etc. together with any non-derivative field ϕ.

<u>Example</u> [6]: massless scalar free field $\phi^i = \phi$, $\phi^j = \phi^*$. Then the fields ϕ^k appearing in expansion (1.1) are the unit operator, $\phi^0 = \mathbb{1}$, and normal products $\phi^k = :\phi^*\overleftrightarrow{\partial}_{\mu_1}\cdots\overleftrightarrow{\partial}_{\mu_n}\phi:$, and derivatives thereof, $\partial_\nu\phi^k$ etc. In perturbation theory the situation is similar due to the normal product formalism [5].

According to Ferrara et <u>al.</u> [7], the terms involving derivatives of one and the same field ϕ^k can be formally summed in conformal invariant theories with the result

$$\phi^i(x)\phi^j(y)\Omega = \sum_k \int dz\, \phi^k(z)\Omega\, \mathcal{B}^{kij}(z;xy) \tag{1.2}$$

Herein the sum is only over nonderivative fields, and integration is over Minkowski space. $\mathcal{B}^{kij}(z;xy)$ are singular c-number functions. Because of the spectrum condition for states $\phi^k(z)\Omega$, the Fourier transformed kernel

$$\tilde{\mathcal{B}}^{kij}(p;xy) = \int dz\, e^{ipz}\mathcal{B}^{kij}(z;xy) \tag{1.3}$$

is only physically relevant for momenta p in the spectrum of the theory, $p \in$ sptr. $\subset \bar{V}_+$. ($\bar{V}_+$ = closed forward cone).

Hypothesis: Expansion (1.1) is valid as a convergent expansion for $(\Psi, \phi^i(x)\phi^j(y)\Omega)$, Ψ in $\mathcal{D}$, after smearing with a test function $f(xy)$.

In the second lecture we shall prove that the hypothesis is true in conformal invariant QFT. Therefore all the consequences derived from it in the following are true there. However, we conjecture that the hypothesis is also true in realistic theories with mass and without conformal symmetry[8]. For massive free field theory this was checked by Schroer, Swieca and Völkel [9] .

Let us discuss the connection with Wilson expansions (1.1). In conformal invariant QFT one proves that kernels $\tilde{B}$ can be analytically continued to entire functions of p,

$$\tilde{B}^{kij}(p; \tfrac{x}{2} - \tfrac{x}{2}) = \Sigma_r \, C^{kij,r}_{\alpha_1 \dots \alpha_r}(x) p^{\alpha_1} \dots p^{\alpha_r} \tag{1.4}$$

Let us insert this into (1.2). Writing $\tilde{\phi}^k(p)$ for the Fourier transform of $\phi^k(z)$ one has

$$\begin{aligned} \phi^i(\tfrac{x}{2})\phi^j(-\tfrac{x}{2})\Omega &= \sum_k \int dp \sum_r C^{kij,r}_{\alpha_1 \dots \alpha_r}(x) p^{\alpha_1} \dots p^{\alpha_r} \tilde{\phi}^k(p)\Omega \\ &= \sum_k \sum_r i^r C^{kij,r}_{\alpha_1 \dots \alpha_r}(x) \nabla^{\alpha_1} \dots \nabla^{\alpha_r} \phi^k(0)\Omega . \end{aligned} \tag{1.5}$$

i.e. an expansion of the form (1.1). We have proceeded formally; the argument can however be made precise, and the result is valid as an asymptotic expansion as $x \to 0$.

Let us next discuss how the kernels $\tilde{B}$ can be determined. We will assume for simplicity that all our fields are hermitean. Consider the Wightman two- and three-point functions

$$(\Omega, \phi^i(x)\phi^j(y)\Omega) = \Delta^{ij}(x-y) = \int_{\text{sptr.}} dp \, e^{-ip(x-y)} \tilde{\Delta}^{ij}(p) , \tag{1.6}$$

and

$$(\Omega, \phi^k(z)\phi^i(x)\phi^j(y)\Omega) = W^{kij}(zxy) = \int_{\text{sptr.}} dp \, e^{-ipz} \tilde{W}^{kij}(p; xy) . \tag{1.7}$$

Integration is over the energy momentum spectrum of the theory, in it $p^0 \geq |\mathbf{p}|$.

Let us take the scalar product of the states in (1.2) with $\tilde{\phi}^\ell(p)\Omega$. In view of definitions (1.6), (1.7) the result reads

$$\tilde{W}^{\ell ij}(p;xy) = \sum_k \tilde{\Delta}^{\ell k}(p)\tilde{B}^{kij}(p;xy) \tag{1.8}$$

Because of the spectrum condition for $\tilde{\Delta}$ and $\tilde{W}$, both sides of this equation vanish if $p \notin$ sptr.; on the other hand the value of $\tilde{B}$ for such p is physically irrelevant. For $p \in$ sptr., the two-point matrix $\tilde{\Delta}(p) \geq 0$ and Eq. (1.8) determines the kernels $\tilde{B}$ to the extent that they are physically relevant. We see that they are a kind of amputated Wightman 3-point functions. In conformal QFT, the two-point function $\tilde{\Delta}^{ij}(p)$ vanishes unless ϕ^i and ϕ^j have the same Lorentz spin ℓ and dimension δ. Taking suitable linear combinations of fields one may assume that $\tilde{\Delta}^{ij}(p) \propto \delta_{ij}$, moreover it is a homogeneous function of p; explicitly [10]

$$\Delta^{ij}(x) = const \cdot \delta_{ij}\, \Delta_+^\chi(x) \quad , \quad \Delta_+^\chi(x) = n(\chi)\, D^\ell(i\tilde{x})(-x^2 + i\varepsilon x^0)^{-\delta-\ell_1-\ell_2} \tag{1.9}$$

$D^\ell(\cdot)$ is the representation matrix of the representation $\ell = (\ell_1, \ell_2)$ of SL(2,$\mathbb{C}$), multispinor indices are suppressed. $D^\ell(i\tilde{x})$ is a homogeneous polynomial of degree $\ell_1 + \ell_2$ in coordinates x^μ, and $\tilde{x} = x^0 \mathbb{1} - \underline{x} \cdot \underline{\sigma}$, $\underline{\sigma} = (\sigma^1, \sigma^2, \sigma^3)$ Pauli matrices. $n(\chi)$ is a normalization factor which is not important at present, $\chi \equiv [\ell, \delta]$.

Next we turn to expansion formulae for Wightman functions. Consider finite sequences of test functions $f = f_0, f_1^{i_1}(x_1) \ldots f_N^{i_N \ldots i_1}(x_N \ldots x_1)$. According to the principles of QFT [11], states

$$\Psi(f) = \sum_n \int dx_n \ldots dx_1\, f_n^{i_n \ldots i_1}(x_n \ldots x_1)\, \Psi^{i_n \ldots i_1}(x_n \ldots x_1) \tag{1.10}$$

with

$$\Psi^{i_n \ldots i_1}(x_n \ldots x_1) = \phi^{i_n}(x_n) \ldots \phi^{i_2}(x_2)\, \phi^{i_1}(x_1)\Omega \tag{1.11}$$

form a dense subset of the Hilbert space $\mathcal{H}$ of physical states. The Wightman functions are

$$W^{i_n \ldots i_1}(x_n \ldots x_1) = (\Omega, \phi^{i_n}(x_n) \ldots \phi^{i_1}(x_1)\Omega)$$
$$= (\Psi^{i_{m+1} \ldots i_n}(x_{m+1} \ldots x_n), \Psi^{i_m \ldots i_1}(x_m \ldots x_1)) . \tag{1.12}$$

independent of m, $0 \leq m \leq n$

We can expand the "state" $\phi^{i_2}(x_2)\phi^{i_1}(x_1)\Omega$ in (1.11) according to our hypothesis to obtain

$$\Psi^{i_n \dots i_1}(x_n \dots x_1) = \sum_j \int dz\; \phi^{i_n}(x_n) \dots \phi^{i_3}(x_3) \phi^j(z) \Omega\, \mathcal{B}^{j i_2 i_1}(z; x_2 x_1).$$

The expansion process may now be repeated. Next one expands $\phi^{i_3}(x_3)\phi^j(z)\Omega$, and so on. As a final result one obtains

$$\Psi^{i_n \dots i_1}(x_n \dots x_1) = \sum_k \int dz\; \phi^k(z)\Omega\, \mathcal{B}^{k i_n \dots i_1}(z; x_n \dots x_1) \tag{1.13}$$

with

$$\mathcal{B}^{k_n i_n \dots i_1}(z_n; x_n \dots x_1) = \sum_{k_{n-1}} \dots \sum_{k_2} \int dz_{n-1} \dots dz_2\, \mathcal{B}^{k_n i_n k_{n-1}}(z_n; x_n z_{n-1}) \cdot$$
$$\dots \mathcal{B}^{k_3 i_3 k_2}(z_3; x_3 z_2) \mathcal{B}^{k_2 i_2 i_1}(z_2; x_2 x_1). \tag{1.14}$$

This is an expansion in terms of states $\phi^k(z)\Omega$ which are obtained by applying only the first power of some field to the vacuum.

Thus our hypothesis amounts to the statement that states $\phi^k(f)\Omega = \int dz\, f(z)\phi^k(z)\Omega$ (f = test functions) span the Hilbert space of physical states.
From (1.14) we obtain an expansion for Wightman functions (1.12), viz.

$$W^{i_n \dots i_1}(x_n \dots x_1) = \sum_{k,l} \iint dz\, dz'\, \overline{\mathcal{B}}^{k i_{m+1} \dots i_n}(z; x_{m+1} \dots x_n) \Delta^{kl}(z-z') \mathcal{B}^{l i_m \dots i_1}(z'; x_m \dots x_1) \tag{1.15}$$

independent of m, $0 \leq m \leq n$; with kernels (1.14).

By Fourier transformation in the integration variables, the r.h.s. of (1.15) can also be expressed in terms of kernels $\tilde{\mathcal{B}}^{\dots}(p; xy)$. Formula (1.15) is valid for all m, $0 \leq m \leq n$, if one interprets

$$\tilde{\mathcal{B}}^{kl}(p; x) = e^{ipx} \delta_{kl}, \quad \tilde{\mathcal{B}}^{i}(p) = \delta_{io}, \quad \phi^o = \mathbb{1} \text{ (unit operator).}$$

In the expansions (1.15), (1.13) there are terms coming from the unit operator. The corresponding 2-and 3-point functions are

$$\tilde{\Delta}^{oj}(p) = \delta(p) \langle \phi^j \rangle \quad ; \quad \tilde{\mathcal{B}}^{oij}(p; xy) = \Delta^{ij}(x-y). \tag{1.16}$$

One may assume that the vacuum expectation value $\langle \phi^j \rangle$ vanishes for all fields ϕ^j except the unit operator $\phi^o = \mathbb{1}$.

Let us introduce a graphical notation

$\mathcal{B}^{k i_2 i_1}(z; x_2 x_1) =$ 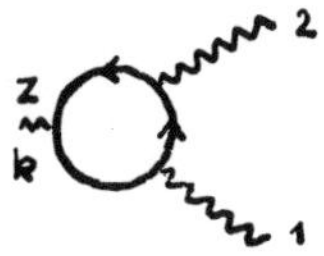

; 

$= \Delta^{ij}(z-z')$ (1.17)

Reversing all arrows will mean complex conjugation.

In this language, expansion (1.15) combined with (1.14) reads

$$(\Omega, \phi^{i_n}(x_n)\dots\phi^{i_1}(x_1)\Omega) = \tag{1.18}$$

$$= \sum_{k_{n-1}}\cdots\sum_{k_2}$$

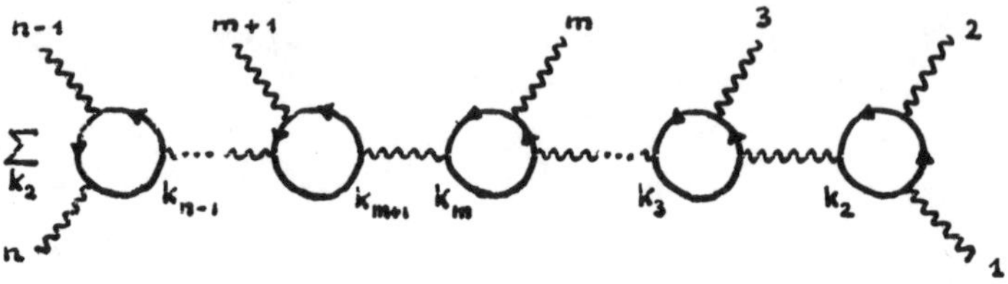

Integration over all internal coordinates is understood; summation is over field labels $k_2 \cdots k_{n-1}$, and m is arbitrary.

Let us now change our point of view and consider expansion (1.18) as an Ansatz for arbitrary n-point functions in a local QFT . Let us ask to what extent it satisfies the Wightman axioms automatically, and what further conditions will have to be imposed.

Let us first remember the axiomatic properties of Wightman two- and three-point functions. (We assume throughout that all our fields are hermitean). They are [11]

i) Lorentz invariance

ii) Positivity and spectrum condition of the two-point matrix $\tilde{\Delta}^{ij}(p) = 0$ for $p \notin \bar{V}_+$ (outside forward light cone)
$\tilde{\Delta}^{ij}(p) = \overline{\tilde{\Delta}^{ji}(p)}$ and $\sum_{i,j} \bar{z}^i \tilde{\Delta}^{ij}(p) z^j \geqslant 0$
for all* sequences of complex numbers $\{z^i\}$ and all p.

iii) The three-point functions W^{kij} resp. Fourier transforms $\tilde{W}^{kij}$ satisfy
hermiticity condition $W^{kij}(z\,x\,y) = \overline{W}^{jik}(y\,x\,z)$
spectrum condition $\tilde{W}^{kij}(p;\,x\,y) = 0$ for $p \notin \bar{V}_+$
locality $W^{i_3 i_2 i_1}(x_3 x_2 x_1) = W^{i_3 i_1 i_2}(x_3 x_1 x_2)$ if $(x_1 - x_2)^2 < 0$.

In addition, $\tilde{W}^{kij}(p;xy)$ must be a measure in p.

---

* $\sum_{i,j \leqslant N} \bar{z}^i \tilde{\Delta}^{ij}(p) z^j$ is allowed to diverge to $+\infty$ as $N \to \infty$ for infinite sequences.

It turns out that it is also necessary to impose a crossing condition on the 4-point function in order to have locality. Because of locality

$$(\Omega, \phi^{i_4}(x_4)\phi^{i_3}(x_3)\phi^{i_2}(x_2)\phi^{i_1}(x_1)\Omega)$$
$$= \pm (\Omega, \phi^{i_4}(x_4)\phi^{i_2}(x_2)\phi^{i_3}(x_3)\phi^{i_1}(x_1)\Omega)$$
$$\text{if } (x_1 - x_2)^2 < 0 ,$$

with $\pm$ depending on the statistics of the fields ϕ^{i_2}, ϕ^{i_3}. Inserting expansion (1.18) this becomes a crossing relation

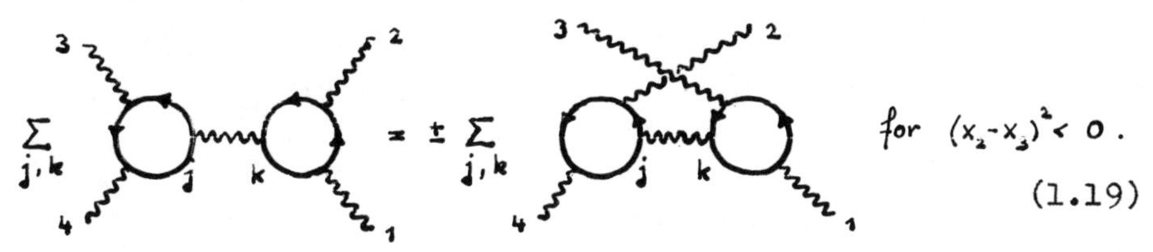

Some attempts at solving such crossing equations in conformal QFT were made by Polyakov [20] . The possible relevance of expansions (1.18) to constructive QFT arises from the following

Assertion: Suppose that one can find a set of two-point functions $\tilde{\Delta}^{ij}(p)$ and amputated three point functions $\tilde{B}^{kij}(p; xy)$ such that axiomatic properties i), ii), iii) of two- and three-point functions listed above are satisfied when W^{kij} are defined by Eqs. (1.8), and such that expansions (1.18) converge and crossing relations (1.19) are satisfied. Then Eqs. (1.18) define a set of Wightman functions with all the axiomatic properties: Lorentz invariance, spectrum condition, positivity and locality.

Let us verify the assertion. First we must be sure that the Ansatz (1.18) is meaningful, i.e. independent of m . This requires merely reshuffling some amputations, and can be verified by straightforward computation, using hermiticity conditions in ii) and iii).

Lorentz invariance is obvious, and so is the spectrum condition, since the propagator between the bubbles with legs m and $m + 1$ can only transmit positive energy, and m is arbitrary.

Positivity is also readily verified. Let $f_0, f_1^{i_1}(x_1) \ldots f_N^{i_N \ldots i_1}(x_N \ldots x_1)$ an arbitrary finite sequence of test functions. Define

$$z^j(p) = \sum_n \sum_{\{i\}} \int \cdots \int dx_n \cdots dx_1 \, \tilde{B}^{j i_n \cdots i_1}(p; x_n \cdots x_1) f_n^{i_n \cdots i_1}(x_n \cdots x_1) \tag{1.20}$$

Then, using (1.15)

$$\sum_{r,s} \sum_{\{i\}} \sum_{\{j\}} \int dx_r \cdots dx_1 dx_1' \cdots dx_s' \, \bar{f}_s^{j_1 \cdots j_s}(x_1' \cdots x_s') W^{j_1 \cdots j_s i_1 \cdots i_r}(x_1' \cdots x_s' x_r \cdots x_1) f_r^{i_r \cdots i_1}(x_r \cdots x_1)$$

$$= \sum_{k,\ell} \int dp \, \bar{z}^k(p) \tilde{\Delta}^{k\ell}(p) z^\ell(p) \geqslant 0 \tag{1.21}$$

as required by axiomatic positivity.

Locality requires that

$$W^{i_N \cdots i_{r+1} i_r \cdots i_1}(x_N \cdots x_{r+1} x_r \cdots x_1) = \pm W^{i_N \cdots i_r i_{r+1} \cdots i_1}(x_N \cdots x_r x_{r+1} \cdots x_1)$$

$$\text{if } (x_r - x_{r+1})^2 < 0 \qquad (r = 1 \ldots n-1). \tag{1.22}$$

We distinguish two cases. If $r=1$ or $r=n-1$, validity of (1.22) follows from locality properties of the three-point functions, $\tilde{W}^{kij}(p; xy) = \pm \tilde{W}^{kji}(p; yx)$ for $(x-y)^2 < 0$, which carry over to the kernels $\tilde{B}^{kij}(p; xy)$ by (1.8). If $r = 2 \ldots n-2$, we put $m = r$ in expansion (1.18). Validity of (1.22) follows then by inserting crossing relation (1.19) for the center piece of the diagram (1.18).

The expansions (1.18) are particularly useful in conformal invariant QFT because in this case all two- and three-point functions are determined up to some constants by conformal symmetry.

2. CONFORMAL INVARIANCE IN MINKOWSKI SPACE

We conjectured that the hypothesis of our first lecture is true for any realistic QFT, and so are therefore all its consequences which we have discussed. From now on we will however restrict ourselves to conformal invariant theories.

Let us first explain what is conformal symmetry. The conformal group is a group of transformations of space-time, and is compounded from the following subgroups

$\tilde{N}$:	translations $\tilde{n}_y$:	$x^\mu \to x^\mu + y^\mu$.	(2.1)
M:	Lorentz transformations m :	$x^\mu \to \Lambda(m)^\mu{}_\nu x^\nu$.	
A:	dilatations a :	$x^\mu \to \lvert a \rvert x^\mu$, $\lvert a \rvert > 0$.	
N:	special conf. transform. n_c:	$x^\mu \to \sigma(x)^{-1}(x^\mu - c^\mu x^2)$ with $\sigma(x) = 1 - 2c\cdot x + c^2 x^2$.	

The special conformal transformations form an abelian four parameter group parametrized by $c = \{c^\mu\}$. The transformation law of the line element ds^2 is $ds^2 \to |a|^{-2} ds^2$ resp. $ds^2 \to \sigma(x)^{-2} ds^2$ under dilatations and special conformal transformations.

There is a special transformation which is not in the conformal group itself but is a reflection, viz. the reciprocal radius transformation R

$$R: \quad x^\mu \to \frac{x^\mu}{x^2} \tag{2.2}$$

If R is multiplied with a time reflection, it becomes an element of the identity component of the conformal group.

Formulae (2.1) can be used both in Minkowski space and in Euclidean space. In Minkowski space, infinitesimal transformations (2.1) form a Lie algebra $g^* = so(4,2)$ in 4 dimensions, resp. so(D,2) in D dimensions. In Euclidean space the Lie algebra is $g = so(5,1)$ resp. so(D + 1,1). Of course, c^μ is a Minkowskian resp. Euclidean 4-vector in the two cases, and the scalar products $c \cdot x$ etc. have to be read with the appropriate metric (+−−−) resp. (++++). Also, the Lorentz group in Minkowski space is SO(3,1)(or SL(2,C)) while in Euclidean space it is SO(4) (or Spin(4)). Special conformal transformations can be generated from translations and the R-operation,

$$n_c = R^{-1} \tilde{n}_{-c} R \quad . \tag{2.3}$$

There are some problems associated with the transformation law (2.1) when one considers global transformations. They come firstly from the fact that $\sigma(x)$ may vanish. That means that the conformal group can take some points to infinity. In the Euclidean domain this is harmless, one simply compactifies the space by adding the point at infinity. The Euclidean conformal group G acts as a group of transformations on the resulting space.

In Minkowski space something like this could also be done. However it is not appropriate because there is a second trouble: One sees by inspecting the transformation law (2.1) that relatively spacelike distances can be transformed into relatively timelike distances. So there seems to be trouble with causality. We will come back to this, because it will be essential later on that we have global conformal symmetry.

Because of this problem, it is costumary to formulate the hypothesis of conformal symmetry in terms of the Euclidean theory. So let us consider the Euclidean Green functions. They have been introduced already in Arthur Jaffe's first lecture; let us remember how it goes. One can make an analytic continuation from Minkowski space to Euclidean space because of the spectrum condition. In fact this analytic continuation can even be done on the states [12]

$$\Psi(x_1 \dots x_n) = \phi(x_1)\dots\phi(x_n)\Omega \tag{2.4}$$

We assume here for simplicity of writing that there exists in the theory a scalar field $\phi(x)$ which is irreducible. One knows that states (2.4) can be analytically continued in x to complex arguments $z_j = x_j + iy_j$; such that all $y_j \in V_+$ and $(z_j - z_k)^2$ are not positive real for $j \neq k$. The Euclidean points have pure imaginary time components and real space components of all arguments z_j. The Euclidean Green functions are*

$$G(\vec{x}_1 \dots \vec{x}_n) = (\Omega, \Psi(z_1 \dots z_n)), \; z_j = (ix_j^4, \underline{x}_j), \; \vec{x}_j = (x_j^4, \underline{x}_j). \tag{2.5}$$

Because of translation invariance the Euclidean Green functions depend only on differences $\vec{x}_j - \vec{x}_k$, and so they are defined for all mutually distinct n-tuples of arguments.

We will assume that the Euclidean Green functions are invariant under the Euclidean conformal group. The transformation law of the Euclidean Greens functions is most easily stated in terms of Euclidean fields. Let us therefore assume for a moment that there are Euclidean fields $\varphi(\vec{x})$ with correlation functions $\langle\varphi(\vec{x}_1)\dots\varphi(\vec{x}_n)\rangle = G(\vec{x}_1 \dots \vec{x}_n)$, cp. A. Jaffe's lectures [13]. Their transformation law (for scalar fields) is

$$\varphi(\vec{x}) \to |a|^{-d}\varphi(a^{-1}\vec{x}) \qquad \text{for dilatations } a \tag{2.6}$$

and

$$\varphi(\vec{x}) \to |x^2|^{-d}\varphi(R^{-1}\vec{x}) \qquad \text{under } R\text{- inversion}$$

d is called the dimension of the field; for free scalar fields $d = 1$, but in the interacting case it is dynamically determined and not a priori known [4]. We assume invariance under time reflection; conformal invariance implies then invariance under the R-operation. Since all conformal transformation can be generated from Poincaré transformations, dilatations, and the R-operation it suffices to know the Lorentz transformation law and (2.6). It follows that under a general conformal transformation $g \in G$,

---

*Throughout these notes we write $\vec{x}$ for Euclidean space arguments and x for Minkowski space arguments.

$$\varphi(\vec{x}) \rightarrow S(g,\vec{x})\,\varphi(g^{-1}\vec{x}) \tag{2.7}$$

This is also valid for fields with spin ℓ, in this case the multiplier $S(g,\vec{x})$ is a matrix *; for Lorentz transformations it has the usual form and is independent of $\vec{x}$.

Conformal invariance of Green functions reads now

$$G(\vec{x}_1 \ldots \vec{x}_n) = S(g,\vec{x}_1) \ldots S(g,\vec{x}_n)\, G(g^{-1}\vec{x}_1 \ldots g^{-1}\vec{x}_n) \text{ for } g \in G. \tag{2.8}$$

A conformal invariant quantum field theory is a QFT which satisfies all the Wightman axioms, locality, spectrum condition and Lorentz invariance, and in addition the Euclidean Green functions of the theory have the symmetry property (2.8) under the Euclidean conformal group $G \approx SO(5,1)$.
There are several reasons for studying conformal invariant QFT.

1. The short distance behavior of more realistic QFT's is described by a conformal invariant QFT if certain conditions are fulfilled (existence of an UV-stable renormalization group fix point, cp. [1]).
2. The long distance behavior of correlation functions in statistical mechanical systems with short range interactions at a critical point (where the correlation length is infinite) can be expected to be described by a conformal invariant Euclidean QFT, assuming that the socalled scaling hypothesis holds true.
3. It can serve as a laboratory for nonperturbative quantum field theory.

In these notes I will skip further discussion of the motivation for studying conformal QFT, it can be found elsewhere (e.g. [1]). Let us turn to the proof of convergence of operator product expansions on the vacuum in conformal QFT instead [14]. (Two dimensional models were studied in[15]).

The essential point in this is the global conformal invariance in Minkowski space. We have assumed that the Euclidean Green functions are invariant under the Euclidean conformal group; how does this carry over to Minkowski space?

---

* Explicitly, let MAN the subgroup defined in (2.1) and let φ transform under Lorentz transformations m according to the representation ℓ by matrices $\mathcal{D}^{\ell}$ in the vector space V^{ℓ}. Define a representation of MAN in V^{ℓ} by $\mathcal{D}^{\chi}(man) = |a|^{-d}\mathcal{D}^{\ell}(m)$, $\chi = [\ell, d]$. Then $S(g,\vec{x}) = \mathcal{D}^{\chi}(\tilde{n}_{\vec{x}}\, g\, \tilde{n}_{\vec{x}'}^{-1})$, $\vec{x}' = g^{-1}\vec{x}$
This formula is only valid in Euclidean space.

The answer is given by a theorem due to Lüscher and the author; it asserts the following [16] .

Let G^* the simply connected group with Lie algebra $g^* \approx so(4,2)$, it is an infinite sheeted covering of the group SO(4,2). Suppose that the theory satisfies the Wightman axioms and Euclidean conformal invariance as stated above. Then the Hilbert space $\mathcal{H}$ of physical states carries a unitary representation of the group G^* .

(There is no suggestion implied that one has a ray representation of a factor group of G^* like SO(4,2). On the contrary, this is probably not true in nontrivial theories in more than two space time dimensions [16]).

Because of the theorem we are sure that we have global conformal invariance in Minkowski space in this sense.

Let me also mention briefly how the causality problem is solved by the group G^* , for this is quite amusing.

It turns out that one can analytically continue the Euclidean Green functions back to Minkowski space; in fact they can be analytically continued into a complex domain which has as a real boundary not just Minkowski space M^4 , but a larger space $\tilde{M}$ which is an infinite sheeted covering of Minkowski space. So it looks as pictured in Fig. 1

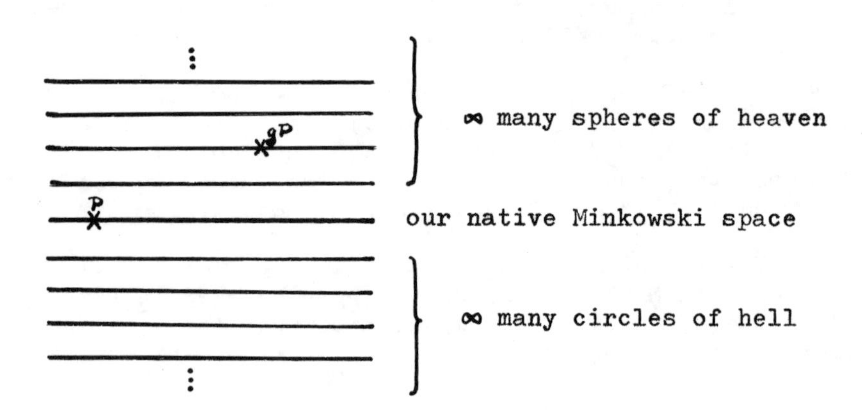

Fig. 1 . Superworld $\tilde{M}$.

Therefore, by uniqueness of analytic continuation, the Wightman functions on Minkowski space determine the Wightman functions on all of superworld $\tilde{M}$. The group G^* can act as a group of transformations on $\tilde{M}$. The causality problem is solved by the fact that $\tilde{M}$ admits a globally G^*-invariant causal ordering which agrees with the usual one on $M^4 \subset \tilde{M}$. Thus there is a conformal invariant notion of relatively spacelike, positive timelike etc. on $\tilde{M}$. Note however, that in general, a transformation g in G^* can take a point P in Minkowski space M^4 into an arbitrary point gP on any sheet of superworld $\tilde{M}$.

Global conformal invariance will enter into our argument in the following way. We shall show that the vacuum expansions (1.2) effect the decomposition of states $\phi^i(x)\phi^j(y)\Omega$ into states which transform irreducibly under the conformal group G^*; so it is just a partial wave decomposition. Now every Hilbert space carrying a unitary representation of a group G^* can of course be decomposed into irreducible subspaces, and if one decomposes any state into the components in these irreducible subspaces a series obtained in this way is of course strongly convergent. This is the basic reason why expansions (1.2) are convergent.

Before we can proceed we need a list of all unitary irreducible representations (UIR's) of G^* with positive energy. This group theoretical problem has been solved completely [17]. The answer is as follows.

The UIR's of G^* with positive energy can be labelled by a pair $\chi = [\ell, \delta]$, where ℓ fixes a finite dimensional irreducible representation of the Lorentz group $M \approx SL(2\mathbb{C})$, acting in a vector space V^ℓ, and δ is a real number, called the dimension, with $\delta \geq \delta_{min}(\ell)$. Finite dimensional irreducible representations ℓ of M can be labelled by a pair of halfintegers $\ell = (j_1, j_2)$ in a standard way. If $j_1 = j_2$ then ℓ is a completely symmetric tensor representation. In this language, $\delta_{min}(\ell) = j_1 + j_2 + 1$ if $j_1 = 0$ or $j_2 = 0$, and $\delta_{min}(\ell) = j_1 + j_2 + 2$ otherwise, except for the trivial 1-dimensional representation which has $\delta = 0, \ell = (0,0)$. All this is for four space time dimensions.

The representation space $\mathcal{E}_\chi$ consists of finite component wave functions $\varphi(x)$ on Minkowski space with values in V^ℓ. Under Lorentz transformations they transform in the standard way $\varphi(x) \to \mathcal{D}^\ell(m)\varphi(m^{-1}x)$, while under general conformal transformations, the transformation law is that of an induced representation like (2.7). Its infinitesimal form was known for a long time [18]. Let us denote the scalar product in V^ℓ by $\langle , \rangle$, we choose it so that $\mathcal{D}^\ell(m^*) = \mathcal{D}^\ell(m)^*$. The scalar product in $\mathcal{E}_\chi$ is then

$$(\varphi_1, \varphi_2) = \int dx_1 dx_2 < \varphi_1(x_1), \Delta_+^\chi(x_1 - x_2) \varphi_2(x_2) > \tag{2.9}$$

where Δ_+^χ is the two point function defined in (1.9).

Let $\mathcal{F}_\chi$ the image of $\mathcal{E}_\chi$ under the intertwining map Δ_+^χ. It consists of vector valued distributions of the form $f(x) = \int dx' \Delta_+^\chi(x-x') \varphi(x')$, φ in $\mathcal{E}_\chi$. The space $\mathcal{F}_\chi$ carries a UIR of G^* which is unitarily equivalent to that on $\mathcal{E}_\chi$. Functions f in $\mathcal{F}_\chi$ satisfy a spectrum condition because the two point function does. Thus their Fourier transforms $\tilde{f}(p) = 0$ for $p \notin \bar{V}_+$. If $\delta = \delta_{min}(\ell)$ there are further constraints. In particular, the representations with $j_1 = 0$ or $j_2 = 0$ and $\delta = j_1 + j_2 + 1$ are zero mass representations. The socalled "canonical dimensions" are precisely $\delta = \delta_{min}$. This completes our discussion of the UIR's.

We will also need information on conformal invariant two- and three-point functions. We shall first discuss the Euclidean case which is simpler. One finds that conformal symmetry alone determines the two- and three-point functions (almost) uniquely. The reason is simple. From Eq. (2.8) we see that the Green function $G(\vec{x}_1 \dots \vec{x}_n)$ (n=2,3) at any one point $(\vec{x}_1 \dots \vec{x}_n)$ $(\vec{x}_i \neq \vec{x}_j$ for $i \neq j)$ determines it for all arguments of the form $(g\vec{x}_1 \dots g\vec{x}_n)$ with g in the Euclidean conformal group G . But it is easy to show that G acts transitively on pairs and triples of noncoinciding arguments $\vec{x}_i$. Thus, the two resp. three point functions are uniquely determined by their value at any one point $(\vec{x}_1 \dots \vec{x}_n)$, n=2,3. In the scalar case this means that they are determined up to normalization. The same consideration applies to the case with spin. In this case, the values of the two resp. three point functions at fixed arguments are matrices. They are however further constrained because they have to be invariant under the subgroup of stability of the pair resp. triple of arguments $(\vec{x}_1, \vec{x}_2)$ resp. $(\vec{x}_1 \vec{x}_2 \vec{x}_3)$ in question. As a result one finds that the two-point function is always unique, while there is in general a finite number of linearly independent three-point functions. When at most one field has spin there is just one at most (none if the spinning field is not a completely symmetric tensor field).

Two- and three-point functions in Minkowski space which satisfy the spectrum condition can be found from the Euclidean result by analytic continuation in time. The result for the two-point function is given in Eq. (1.9). We give an example for the three-point function. (The general result can be found in [14]). Let $\phi_\alpha^a(x)$ a completely symmetric traceless ℓ-th rank tensor field with dimension δ , and $\phi(x)$

a scalar field of dimension d . Then

$$V_\alpha^a(x_3 x_2 x_1) \equiv (\Omega, \phi_\alpha^a(x_3)\phi(x_2)\phi(x_1)\Omega) \tag{2.10}$$
$$= N(\chi)(-x_{12}^2)^{-d+\frac{1}{2}\delta-\frac{1}{2}\ell}[(-x_{13}^2)(-x_{23}^2)]^{-\frac{1}{2}\delta+\frac{1}{2}\ell}(\hat{x}_{\alpha_1}\dots\hat{x}_{\alpha_\ell} - \text{traces})$$

with

$$\hat{x}_\alpha = \frac{(x_{13})_\alpha}{x_{13}^2} - \frac{(x_{23})_\alpha}{x_{23}^2} \quad , (x_{ij})_\alpha = x_{i\alpha} - x_{j\alpha}$$

and

$$(-x_{ij}^2)^\alpha \equiv [-(x_i - x_j)^2 + i\varepsilon(x_i^0 - x_j^0)]^\alpha$$

Let us next consider states $\phi^i(f)\Omega = \int dx\, f_\alpha(x)\phi_\alpha^i(x)\Omega$ which are obtained by applying a smeared field operator (of any spin) to the vacuum. We claim that such states transform irreducibly under the conformal group G^*. Indeed we may consider f as an element of $\mathcal{E}_\chi$ with $\chi = [\ell, \delta]$ given by spin and dimension of the field ϕ^i . The map

$$f \rightarrow \phi^i(f)\Omega$$

is conformal invariant and maps the irreducible representation space $\mathcal{E}_\chi$ into the Hilbert space of physical states. Its image is therefore also an irreducible representation space, and it coincides with the space of states which we consider. In conclusion, the states $\phi^i(f)\Omega$ with f ranging over all test functions f in $\mathcal{E}_\chi$, form a UIR equivalent to χ . $\mathcal{E}_\chi$ contains all Schwartz test functions.

We will now come to the argument why operator product expansions (1.2) on the vacuum are valid and strongly convergent. We recognize the individual terms on the RHS as states which transform according to unitary irreducible representations of G^* . Therefore (1.2) is a partial wave decomposition.

The strategy of its proof is as follows. First one writes down a partial wave decomposition, i.e. one decomposes the states $\phi^i(x)\phi^j(y)\Omega$, smeared with a test function, into a direct sum or integral of states which transform irreducibly.

Then one notes that this expansion is at the same time an asymptotic expansion as $x \rightarrow y$ in homogeneous functions of x - y . This is unusual for harmonic analysis on noncompact groups * ; it comes about because only representations with

---

* Usually one has to decompose into representation functions of the second kind first and shift the path of integration over a representation label into the complex plane [19] .

positive energy are involved here. Their representation functions have good high energy behavior, in contrast with the principal series.

Finally one compares this asymptotic expansion with the Wilson expansion (1.1). Since asymptotic expansions in homogeneous functions are unique, there must be a term by term correspondence. This allows to identify the states in the partial wave decomposition with states that are obtained by applying a smeared local field operator to the vacuum.

Let us go through some of the details of these steps. Let Ψ an arbitrary state in the Hilbert space $\mathcal{H}$ of physical states. We may decompose $\mathcal{H}$ into subspaces which carry copies of one UIR χ of G^*.

$$\mathcal{H} = \int d\mu(\chi)\, \mathcal{H}^\chi \quad \text{whence} \quad \Psi = \int d\mu(\chi)\, \Psi^\chi,\ \Psi^\chi \text{ in } \mathcal{H}^\chi \tag{2.11}$$

Let us simplify the argument by assuming that $\mathcal{H}^\chi$ are actually irreducible. This need not always be true, but we leave it to the reader to work out the necessary generalization for himself. Since $\mathcal{H}^\chi$ carries a UIR which is equivalent to that on $\mathcal{F}_\chi$, there is a (unique) isometric intertwining map

$$\mathcal{B}^\chi : \mathcal{F}_\chi \to \mathcal{H}^\chi \tag{2.12}$$

$\mathcal{B}^\chi$ commutes with the action of the group and preserves the norm. Furthermore, $\mathcal{F}_\chi$ is a function space and $\mathcal{B}^\chi$ is linear. Therefore

$$\mathcal{B}^\chi \varphi = \int dx\, \mathcal{B}^\chi(x)\varphi(x) \quad \text{with} \quad \mathcal{B}^\chi(x) \in \mathcal{H}^\chi \subset \mathcal{H} \quad \text{for } \varphi \text{ in } \mathcal{F}_\chi.$$

Using this we may rewrite the decomposition (3.1) as

$$\Psi = \int d\mu(\chi)\, \mathcal{B}^\chi \varphi^\chi = \int d\mu(\chi) \int dz\, \mathcal{B}^\chi(z)\varphi^\chi(z) \tag{2.13}$$

z-integration is over Minkowski space.

Let ϕ^i and ϕ^j two local fields, and define the kernel

$$\mathcal{B}^\chi(z;xy) = (\mathcal{B}^\chi(z),\, \phi^i(x)\phi^j(y)\Omega) \tag{2.14}$$

Then, from (2.13)

$$(\Psi, \phi^i(x)\phi^j(y)\Omega) = \int d\mu(\chi) \int dz\, \varphi^\chi(z)^* \mathcal{B}^\chi(z;xy) \tag{2.15}$$

with φ^χ in $\mathcal{F}_\chi$

$\varphi^\chi(z)$ and $\mathcal{B}^\chi(z;xy)$ both take values in the vector space V^ℓ.

The scalar product of vectors u, v in V^ℓ is written as $v^* u$.

Eq. (2.15) is valid for arbitrary states Ψ , of course φ^χ depend on Ψ . This completes the first step. Eq. (2.15) is the partial wave decomposition we sought.

Because the map $\mathcal{B}^\chi$ is conformal invariant, i.e. commutes with the action of the group, the kernel $\mathcal{B}^\chi(z;xy)$ is a conformal invariant three-point function in Minkowski space. (It does not satisfy a spectrum condition in z, though). It can be determined by a method similar to that described before for Euclidean three-point functions. It is more complicated though in the following points. For global conformal invariance, arguments x, y must be considered as elements of superworld $\tilde{M}$, of which M^4 is a part, while z is in (compactified) Minkowski space $M^4_c \approx \tilde{M}/\Gamma$, Γ = center of G^*. The group G^* does not act transitively on triples of points (z; xy). It suffices to consider relatively spacelike x , y because of analyticity. Then there are three open orbits under G^*. The values of the kernels $\mathcal{B}^\chi(z;xy)$ on these three orbits are a priori independent.

There is however a further constraint. From the scalar product (2.9) we see that the dual of $\mathcal{F}_\chi$ is the (Hilbert space) $\mathcal{E}_\chi$. Therefore $\int \mathcal{B}^\chi(\cdot\,;xy) f(xy)\,dx\,dy$ must be in $\mathcal{E}_\chi$. It turns out that this fixes the Fourier transform $\tilde{\mathcal{B}}^\chi(p;xy)$ for $p \in \bar{V}_+$. One verifies by inspection that it can be analytically continued in p to an entire function of p . Therefore it admits a power series expansion in p as in Eq. (1.4), viz.

$$\tilde{\mathcal{B}}^\chi(p;\tfrac{x}{2}\,-\tfrac{x}{2}) = \sum_r C_\alpha^{\chi r}(x)\,p^{\alpha_1}\cdots p^{\alpha_r}\ , \quad \alpha = (\alpha_1\cdots\alpha_r) \qquad (2.16)$$

Because of dilatation symmetry

$$C_\alpha^{\chi r}(\lambda x) = \lambda^{-d_i - d_j + \delta + r}\ C_\alpha^{\chi r}(x) \qquad (2.17)$$

$d_{i,j}$ the dimensions of ϕ^i , ϕ^j; $\chi = [\ell,\delta]$. One inserts (2.16) into (2.15) and rewrites the result as after (1.4). Because of (2.17) the result is an asymptotic expansion as $x \to 0$ for $(\Psi, \phi^i(\tfrac{1}{2}x)\phi^j(-\tfrac{1}{2}x)\Omega)$ in homogeneous functions of x, viz.

$$(\Psi, \phi^i(\tfrac{x}{2})\phi^j(-\tfrac{x}{2})\Omega) = \sum_r \int d\mu(\chi)\, i^r C_\alpha^{\chi r}(x)\,\nabla_{\alpha_1}\cdots\nabla_{\alpha_r}\,\varphi^\chi(0)^* \ . \qquad (2.18)$$

Herein $\sum_r \int d\mu(\chi) = \sum_r \sum_\ell \int_{\delta \geq \delta_{min}(\ell)} \mu'(\chi)\,d\delta$

and the summation over r and integration over δ is supposed arranged in order of increasing $\delta+r$. This completes the second step.

Lastly we compare the asymptotic expansion (2.18) with Wilson expansion (1.1). There must be a term by term correspondence, which is only possible if the measure $\mu(\chi)$ is discrete so that

$$(\Psi, \phi^i(x)\phi^j(y)\Omega) = \sum_k \int dz\, \varphi^{\chi_k}(z)^* \mathcal{B}^{\chi_k}(z; xy) \tag{2.19}$$

and

$$\varphi^{\chi_k}(z)^* = (\Psi, \phi^k(z)\Omega)$$

where ϕ^k is a nonderivative field appearing in the Wilson expansion. Since this is valid for all states Ψ , we have completed the derivation of (1.2).

We may look at the argument in another way which makes clear what is involved in the general case.

Let $\mathcal{H}_{\parallel}$ be the subspace of the Hilbert space $\mathcal{H}$ of physical states which is spanned by states of the form $\phi^k(f)\Omega$, when ϕ^k runs over all local fields in the theory, and f runs over all test functions. Write $\mathcal{H}_{\perp}$ for the orthogonal complement of $\mathcal{H}_{\parallel}$. At a heuristic level it is easy to show that the component of the state $\phi^i(x)\phi^j(y)\Omega$ in $\mathcal{H}_{\parallel}$ is given by the right hand side of (1.2). The problem is to show that $\mathcal{H}_{\perp} = (0)$. Let E the projection operator on $\mathcal{H}_{\perp}$ and Ψ an arbitrary state. Consider

$$F(x,y) \equiv (E\Psi, \phi^i(x)\phi^j(y)\Omega)$$

If we insert the Wilson expansion (1.1) on the right hand side, we see that $F(x,y)$ falls off faster than any power of $x - y$ as $x \to y$, because all the terms in the Wilson expansion (1.1) are orthogonal to $\mathcal{H}_{\perp}$. In <u>conformal</u> QFT we were able to show that it follows from this that $F(x,y) \equiv 0$, for arbitrary Ψ . Therefore $\phi^i(x)\phi^j(y)\Omega$ is in $\mathcal{H}_{\parallel}$ (after smearing with a test function)...

3. CONNECTION WITH EUCLIDEAN APPROACH

Our discussion so far has been in Minkowski space. Euclidean Green functions were only considered at some intermediate steps and mainly for pedagogical purposes. There is however also a group theoretical approach to Euclidean conformal invariant QFT. The results of this approach were presented at our earlier lectures in Capri and Bonn which are already published [1,2], they will therefore not be repeated. We will now describe the connection with the Minkowskian vacuum expansions, which are at the same time partial wave expansions on the Minkowskian conformal group G^* as we have seen. A conjecture concerning this connection was stated in our Bonn lectures [2], we will see that it is correct.

Let ϕ_α a symmetric traceless tensor field with Lorentz spin ℓ and dimension δ, and $\phi(x)$ a scalar field with dimension d as in Eq. (2.10). Write $\chi = [\ell, \delta]$ and consider the Euclidean three-point function

$$\Gamma_\alpha^\chi(\vec{x}_1 \vec{x}_2 | \vec{x}_3) = \text{analytic continuation of } (\Omega, \phi_\alpha(x_3)\phi(x_2)\phi(x_1)\Omega). \tag{3.1}$$

Its explicit form is obtained from Eq. (2.10) by analytic continuation to Euclidean space, viz. $(|\vec{x}_i - \vec{x}_j|^2)^\alpha$ should be substituted for $(-x_{ij}^2)^\alpha$.

The socalled elementary representations of the Euclidean conformal group G contain all unitary irreducible ones as subrepresentations and may also be labelled by $\chi = [\ell, \delta]$. The label δ may be complex and ℓ labels a finite dimensional representation of spin (4) or, equivalently (because of Weyls unitary trick) of SL(2$\mathbb{C}$). Γ^χ is at the same time a Clebsch Gordan kernel for such representations, assuming the field is normalized in a special way.

In our Bonn lectures [2] we introduced Clebsch Gordan kernels Q^χ of the second kind as solutions of the integral equations

$$\Gamma_\alpha^\chi(\vec{x}_3 \vec{x}_4 | \vec{x}) = Q_\alpha^\chi(\vec{x}_3 \vec{x}_4 | \vec{x}) + \int d\vec{x}' \Delta_{\alpha\beta}^\chi(\vec{x}, \vec{x}') Q_\beta^{-\chi}(\vec{x}_3 \vec{x}_4 | \vec{x}') \tag{3.2}$$

for $x_3^4, x_4^4 > 0$ with $Q^{\pm\chi}(\cdot\cdot|\vec{x}) = 0$ if $x^4 < 0$.

Here and in the following we write $-\chi = [\ell, 4-\delta]$ when $\chi = [\ell, \delta]$. Δ^χ is the Euclidean two-point function which is obtained from Δ_+^χ defined in (1.9). The normalization factor $n(\chi)$ is assumed to be chosen so that $\Delta^{-\chi}$ is the inverse in the convolution sense of Δ^χ. The $\vec{x}'$- integration is over Euclidean space; because of the support properties of $Q^{-\chi}$ it suffices to extend it over the half space $x^4 \geq 0$.

In [2] it was argued that Eq. (3.2) could be solved by partial wave expansion on the subgroup SO(4,1) of G. Actually there is a simpler way; one can do with the knowledge of the Laplace transform, Eq. (3.7) below, of Q^χ which is related to our old kernel $\tilde{\mathfrak{B}}^\chi$ by analytic continuation as we shall see.

A main result in our Bonn lectures was a convergent expansion, Eq. (5.7) of ref.[2], for the disconnected Euclidean four-point Green function. It reads

$$G(\vec{x}_1 \ldots \vec{x}_4) = \sum_a r_a \int_{x^4>0} d\vec{x}\, \Gamma_\alpha^{\chi_a}(\vec{x}_1\vec{x}_2 | \vec{x})\, Q_\alpha^{-\chi_a}(\vec{x}_3\vec{x}_4 | \vec{x}) \tag{3.3}$$

$$\text{for} \quad x_1^4, x_2^4 < 0 \;, \quad x_3^4, x_4^4 > 0$$

Summation is over all poles of the Euclidean conformal partial wave amplitude $g(\chi)$ at $\chi = \chi_a$. Because of the results of ref.[1] on tensor fields this is the same as a sum over spin and dimension $[\ell_a, \delta_a] = \chi_a$ of all symmetric tensor fields ϕ_α^a in the theory for which the three-point function Γ^χ does not vanish identically. The Clebsch Gordan kernel Γ^χ is identified with the (Euclidean) three-point function of physical fields through these same results. The coefficients

$$r_a = \operatorname*{res}_{\chi = \chi_a} {}_* \, [1 + g(\chi)] \tag{3.4}$$

are, up to a factor, equal to the residue of the Euclidean partial wave amplitude $1 + g(\ell_a, \delta)$ at $\delta = \delta_a$.

Let us temporarily assume that there exist Euclidean fields (see Jaffe's talk [13]) φ, φ_α^a corresponding to the quantum fields ϕ, ϕ_α^a . Then expansion (3.3) reads

$$\langle \varphi(\vec{x}_1) \ldots \varphi(\vec{x}_4) \rangle = \sum_a r_a \int_{x^4>0} d\vec{x} \langle \varphi(\vec{x}_1)\varphi(\vec{x}_2)\varphi_\alpha^a(\vec{x})\rangle\, Q_\alpha^{-\chi_a}(\vec{x}_3\vec{x}_4|\vec{x})$$

$$\text{valid and convergent for } x_3^4, x_4^4 > 0;\; x_1^4, x_2^4 < 0 \;, \tag{3.3'}$$

with some constants r_a . The restriction on the arguments is very important. In view of Euclidean invariance it amounts to requiring that it must be possible to separate the arguments $\vec{x}_1$, $\vec{x}_2$ from $\vec{x}_3$, $\vec{x}_4$ by a hyperplane!

We will now show that expansion (3.3) is equal to the analytic continuation of (and follows from) the expansion of Sect. 1 for the Wightman four-point function in Minkowski space. It can be written as

$$\begin{aligned} W(x_1 \ldots x_4) &= (\Omega, \phi(x_1) \cdots \phi(x_4)\Omega) \\ &= \sum_a \int dp\, \tilde{V}_\alpha^a(x_1 x_2; p)\, \tilde{B}_\alpha^{\chi_a}(p; x_3 x_4) \end{aligned} \tag{3.5}$$

with

$$\tilde{V}_\alpha^a(x_1 x_2; p) = \int dx\, e^{-ipx} (\Omega, \phi(x_1)\phi(x_2)\phi_\alpha^a(x)\Omega) \tag{3.6}$$

Herein $\tilde{B}^\chi$ is the kernel described in Sect. 2. Integration is now over Minkowskian momenta, while in Eq. (3.3) it was over half of Euclidean space.

Consider the Fourier transform $\tilde{V}^a$ of the Wightman three-point function which was defined in Eq. (3.6). It admits analytic continuation to complex arguments $z_1 = x_1 - iy_1$, $z_2 = x_2 - iy_2$, y_1, y_2 in V_+. The Euclidean three-point function is related to it by

$$\Gamma_\alpha^{\chi_a}(\vec{x}_1\vec{x}_2|\vec{x}) = \iint_{E \geq |\underline{p}|} dE\, d^3\underline{p}\; e^{-Ex^4 + i\underline{p}\underline{x}}\, \tilde{V}_\alpha^a(z_1 z_2; p) \tag{3.7a}$$

for $x_1^4, x_2^4 < 0$, $x^4 > 0$, with $p = (E, \underline{p})$, $z_j = (ix_j^4, \underline{x}_j)$ $(j=1,2)$.

For arguments $x^4 < 0$, $x_3^4, x_4^4 > 0$ one has instead

$$\Gamma_\alpha^{\chi_a}(\vec{x}_3\vec{x}_4|\vec{x}) = \iint_{E \geq |\underline{p}|} dE\, d^3\underline{p}\; e^{Ex^4 + i\underline{p}\underline{x}}\, \tilde{V}_\alpha^a(z_1 z_2; p) \tag{3.7b}$$

where $\tilde{V}_\alpha^a(p; x_3 x_4) = \int dx\, e^{ipx}\, (\Omega, \phi_\alpha^a(x)\phi(x_3)\phi(x_4)\Omega)$

and z_j are as before. $\tilde{V}_\alpha^a(p; x_3 x_4)$ admits analytic continuation to $z_3 = x_3 + iy_3$ etc., $y_3 \in \bar{V}_+$.

We insert Eq. (3.7a) into (3.3) and define

$$\int_{x^4>0} d\vec{x}\, e^{-Ex^4 + i\underline{p}\underline{x}}\, Q_\alpha^{-\chi}(\vec{x}_3\vec{x}_4|\vec{x}) = \tilde{B}^\chi(p; z_3 z_4) \tag{3.8}$$

with $z_j = (ix_j^4, \underline{x}_j)$ $(j=3,4)$ and $p = (E, \underline{p})$.

Eq. (3.3) reads then

$$G(\vec{x}_1 \ldots \vec{x}_4) = \sum_a r_a \int d^4p\, \tilde{V}_\alpha^a(z_1 z_2; p)\, \tilde{B}^{\chi_a}(p; z_3 z_4) \tag{3.9}$$

for $x_1^4, x_2^4 < 0 < x_3^4, x_4^4$ with $z_j = (ix_j^4, \underline{x}_j)$ $(j=1\ldots4)$

This looks indeed identical to the conformal partial wave expansion (3.5) of the four-point Wightman function, analytically continued term by term. It only remains to verify that $\tilde{B}^\chi(p; z_3 z_4)$ defined by Eq. (3.8) is the analytic continuation of our old kernel $\tilde{B}^\chi(p; x_3 x_4)$ in x_3, x_4. We show that this follows from the defining equation (3.2) of Q^χ.

Let $\tilde{\Delta}_+^\chi(p)$ the Fourier transform of the (Wightman) two-point function (1.9). The Euclidean two-point function is related to it by

$$\Delta^\chi(\vec{x},\vec{y}) = \iint\limits_{E\geq|\underline{p}|} dE d\underline{p}\ \tilde{\Delta}_+(p) \exp\{-E(y^4-x^4)+i\underline{p}(\underline{y}-\underline{x})\}$$

for $y^4 > 0 > x^4$

It follows from this and definition (3.8) that

$$\int d\vec{y}\ \Delta^\chi_{\alpha\beta}(\vec{x},\vec{y})\, Q^{-\chi}_\beta(\vec{x}_3\vec{x}_4|\vec{y}) = \int dE d\underline{p}\ e^{Ex^4-i\underline{p}\underline{x}}\tilde{\Delta}_+(p)_{\alpha\beta}\tilde{B}^\chi_\beta(p;z_3z_4)$$

for $x^4<0$, z_j as before.

Consider now Eq. (3.2) for $x^4<0$. The first term on its right hand side vanishes then. Because of Eqs. (3.7b) and (3.9) and uniqueness of analytic continuation in x_3, x_4 the equation can be read as

$$\tilde{V}^a(p;x_3x_4) = \int\limits_{\bar{V}_+} dp\ \tilde{\Delta}^{\chi_a}_+(p)\tilde{B}^{\chi_a}(p;x_3x_4)$$

in vector notation.

Thus Eq. (3.8) is equivalent to defining $\tilde{B}^\chi$ as an amputated three-point function as we did earlier, compare Eq. (1.8).

In conclusion, Eq. (3.3), and (3.3') if Euclidean fields exist, are the correct analytic continuation of the vacuum expansion for the conformal invariant Wightman four-point function to Euclidean space.

Remark: We may also conclude from this result that the Euclidean partial wave amplitude $g(\chi)$ is completely determined by the location of its poles and their residues. This is so because they determine the Euclidean four-point function by Eqs. (3.3), (3.4) and uniqueness of analytic continuation (The Euclidean Green functions are real analytic except at coinciding arguments). Conversely, the Euclidean four-point Green function determines $g(\chi)$.

REFERENCES

1. G. Mack, J. de physique (Paris) 34 C1 (1973) 99, and in E.R. Caianello (Ed.), Renormalization and invariance in quantum field theory, Plenum press, New York 1974.

V. Dobrev, G. Mack, V. Petkova, S. Petrova, I. Todorov, Harmonic analysis on the n-dimensional Lorentz group and its application to conformal quantum field theory. Lecture notes in physics (in press), Springer Verlag Heidelberg.

2. G. Mack, Osterwalder-Schrader positivity in conformal invariant quantum field theory, in: Lecture Notes in physics 37 , Springer Verlag Heidelberg 1975.

3. V. Dobrev, V. Petkova, S. Petrova, I. Todorov, Dynamical derivation of vaccum operator product expansions in conformal quantum field theory. Phys. Rev. (in press).

4. K. Wilson, Phys. Rev. 179 (1969) 1499.

5. W. Zimmermann, Commun. Math. Phys. 15 (208) 1969 and in: Lectures on elementary particles and field theory, Brandeis University, vol. I, MIT press 1970.

   J.M. Lowenstein, Phys. Rev. D4 (1971) 2281.

6. B. Schroer, J.A. Swieca, A.H. Völkel, Phys. Rev. D11 (1975) 1509.

7. S. Ferrara, R. Gatto, A.F. Grillo, Springer Tracts in Modern Physics 67, Springer Verlag Heidelberg 1973; Ann. Phys. (N.Y.) 76 (1973) 116.

   S. Ferrara, R. Gatto, A.F. Grillo, G. Parisi, Nucl. Phys. B49 (1972) 77; Lettere Nuovo Cim. 4 (1972) 115.

8. G. Mack, Duality in quantum field theory. DESY 75/44 (1975), and submitted to Nuclear physics.

9. B. Schroer (private communication).

10. W. Rühl, Commun. Math. Phys. 30 (1973) 287, 34 (1973) 149.

11. R.F. Streater, A.S. Wightman, PCT, spin and statistics, and all that. Benjamin, New York 1964.

12. V. Glaser, Commun. Math. Phys.

13. A. Jaffe, lectures presented at this school.

14. G. Mack, Convergence of operator product expansion on the vacuum in conformal invariant quantum field theory. DESY 76/30, to be published in Commun. Math. Phys.

15. M. Lüscher, Commun. Math. Phys. 50 (1976) 23.

W. Rühl, B.C. Yunn, Commun. Math. Phys. (in press).

J. Kupsch, W. Rühl, B.C. Yunn, Ann. Phys. (N.Y.) 89 (1975) 115.

16. M. Lüscher and G. Mack, Commun. Math. Phys. 41 (1975) 203.

17. G. Mack, All unitary representations of the conformal group SU(2,2) with positive energy. DESY 75/50 (submitted to Commun. Math. Phys.).

18. Abdus Salam, G. Mack, Ann. Phys. (N.Y.) 53 (1969) 174.

19. W. Rühl, The Lorentz group and harmonic analysis. Benjamin, New York 1970.

20. A.M. Polyakov, Zh. ETF 66 (1974) 23, engl. transl. JETP 39 (1974) 10.

A.A. Migdal, 4-dimensional soluble models in conformal field theory. Preprint, Landau institute, Chernogolovka 1972.

REGULARIZED QUANTUM FIELD THEORY

K. Symanzik

Deutsches Elektronen-Synchrotron DESY

Hamburg, Germany

0. Purpose of Regularization

Although there are several "axiomatic" formulations of QFT (Wightman, LSZ, Haag-Kastler), most of our ideas about QFT and in particular all successful constructions of examples of such theories derive from working with particular Lagrangian theories. Nontrivial such theories require interaction terms of higher order than quadratic, and even to formulate the Lagrangian is then a nontrivial problem. The possibilities are: A) Choose a finite formulation of the theory (Valatin, Wilson, Brandt) e.g. by writing the field equations in terms of normal products (Zimmermann). The difficulty here is that the theory must essentially already have been "solved" in order to define such normal products adequately by certain (mathematically inconvenient) limiting processes. B) Mutilate the theory (as little as possible) on the Lagrangian level already, and defer the limiting processes that restore the original theory to the level of Green's functions. This method, regularization and its removal, allows a relatively simple analysis of some aspects of also the limiting theory. In fact, the perturbation theoretical proofs of non-renormalization theorems (Adler [1]) and of cancellation of anomalies to all orders (cp. sect. 2) are given simplest in this way.

1. Comparison of Regularizations

Since we have in mind a non-perturbation theoretical analysis of Lagrangian theories, we must use a regularization that is not restricted to perturbation theory or difficult to abstract therefrom.

This rules out dimensional (Bollini-Giambiagi, 't 'Hooft-Veltman) and analytical (Speer) regularization and suggests closer consideration of lattice regularization and of the Pauli-Villars method: See Table 1. Remark thereto: a Hamiltonian lattice regularization using space smearing of the interaction has some features similar to Hamiltonian lattice regularization, but loss of locality; a Hamiltonian formulation of Pauli-Villars regularization (higher space derivatives only) preserves unitarity under loss of covariance. The balance of table 1 lies clearly on the side of lattice regularization, but since this was studied in great depths (see e.g. B. Simon [2]), we will later consider only P.V. regularization in any detail.

Table 1

	lattice	Pauli-Villars
Characteristics	continuous space, or space + imaginary time, replaced by lattice	higher derivatives in kinetic part of Lagrangian (e.g. (2.1)), involving regulator mass Λ
Unitarity (Spectrum condition and pos. metric)	holds, via Hamiltonian acting on Hilbert space, or, via Osterwalder-Schrader positivity condition in the limit theory, respectively	energy spectrum positive, but metric indefinite, positive metric hoped to be recovered if $\Lambda \to \infty$
Poincaré invariance	absent, hoped to be recovered in limit lattice constant $a \to 0$	manifest
locality (commutativity in spacelike distances in real-time theory)	holds in noncovariant sense in Hamiltonian lattice theory, in Euclidean lattice theory hoped to be obtained in $a \to 0$ limit from covariance, spectrum condition, and symmetry of Green's functions via Jost theorem	manifest
tools available	In the Euclidean case: correlation inequalities do allow numerous rigorous proofs, and high-temperature expansions are heuristic non-perturbation-theoretical method	only (formally summed) perturbation theory and abstractions therefrom

2. Large $-\Lambda$ Expansions

To have a concrete regularization in mind, we select the P.V. one and, mainly for notational simplicity, exhibit it for (Euclidean) ϕ_4^4 theory:

$$(2.1)\quad \mathcal{L}_\Lambda = -\frac{1}{2}\sum_{r=0}^{R} c_r^{-1}\left(\partial_\mu\phi_r\partial_\mu\phi_r + \Lambda^{-2}a_r^{-2}\phi_r^2\right) - \\ -\frac{1}{24}g_B\phi_B^4 - \frac{1}{2}\left(m_{B0}^2 + \Delta m_B^2\right)\phi_B^2 + \\ + J(x)\phi_B + \frac{1}{2}K(x)\phi_B^2$$

with c_r such that

$$\sum_{r=0}^{R} c_r\left(-\Delta + \Lambda^2 a_r^2\right)^{-1} = (-\Delta)^{-1}\prod_{r=1}^{R}\left(1-\Lambda^{-2}a_r^{-2}\Delta\right)^{-1}$$

with Δ the 4-dimensional Laplacian, and $\phi_B := \sum_{r=0}^{R}\phi_r$, $a_0 = 0$, $a_1 = 1$. The ϕ_r can also be expressed by ϕ_B acted on by differential operators. $m_{B0}^2 = \Lambda^2 f(g_B)$ is the bare mass squared of the massless theory (we suppress here and later the dependence on the $a_2 \ldots a_R$). We take $g_B > 0$ and, except for a remark in sect. 3, will consider $\Delta m_B^2 \geq 0$ only.

We consider the Green's functions $2^{-\ell}\langle\phi_B(x_1)\ldots\phi_B(x_{2n})\phi_B(y_1)^2\ldots\phi_B(y_\ell)^2\rangle$ and the corresponding vertex functions (connected, and amputated and one-particle irreducible w.r.t. the ϕ-lines) $\Gamma_{\Lambda B}(x_1\ldots x_{2n}, y_1\ldots y_\ell; g_B, \Delta m_B^2)$. Their Fourier transforms (with a factor $(2\pi)^4\delta(\Sigma p + \Sigma q)$ taken out) have for the model (2.1) the formal large-Λ expansions [3]

$$(2.2)\quad \Gamma_{\Lambda B}(p_1\ldots p_{2n}, q_1\ldots q_\ell; g_B, \Delta m_B^2) =$$

$$= \sum_{j=0}^{\infty}\sum_{k=0}^{\infty}\Lambda^{-2j}(\ln\Lambda)^k F_{jk}\left((2n),(\ell); g_B, \Delta m_B^2\right)$$

whereby an L-loop graph contributes only to the F_{jk} with $k \leq L$ ($k \leq L-1$ unless $n + \ell = 2$; we do not consider $n = 0$, $\ell = 0$ or 1). The structure of the r.h.s. of (2.2) can best be elucidated by writing an effective local Lagrangian [3] that generates that r.h.s. upon integration with finiteness-preserving rules; here e.g. Zimmermann's integration rules (if $\Delta m_B^2 > 0$)[4] or rules involving minimal subtraction at Euclidean symmetry points (if $\Delta m_B^2 = 0$)

[5] could be employed.

For other regularizations than P.V., only the j = 0 terms in (2.2) will remain of the same structure that they can be generated by a local effective Lagrangian. $j>0$ terms will be replaced by expansions of rather more involved structure; e.g., nonconvariance of the regularization expresses itself only here. As an example, consider

$$\int dk\, \theta(\Lambda^2 - k^2)\, \theta(\Lambda^2 - (p+k)^2)\, (k^2)^{-1} [(k+p)^2]^{-1} =$$

$$= a \ln(\Lambda^2 |p|^{-2}) + \sum_{j=1}^{\infty} b_j \Lambda^{-j} |p|^j + c$$

involving a sharp momentum cutoff: the presence of Λ-powers not occurring in (2.2) implies that large-Λ expansions in this regularization cannot be generated by a local effective Lagrangian, making the nonlocality of this regularization manifest. For general regularization, then we will write, tentatively,

$$\Gamma_{\Lambda B}((2n),(\ell); g_B, \Delta m_B^2) = \tag{2.3}$$

$$= \sum_{k=0}^{\infty} (\ln\Lambda)^k F_k((2n),(\ell); g_B, \Delta m_B^2) +$$

$$+ O((\ln\Lambda)^{-\infty})$$

meaning the remainder term to vanish for $\Lambda \to \infty$ more strongly than any negative power of $\ln \Lambda$.

The gauge-invariant P.V. regularization of gauge theories has been described by A.A. Slavnov [6] . It involves higher covariant derivatives rather than higher ordinary derivatives. This generates new interaction vertices. Their effect is that one-loop graphs always remain unregularized and require a counter term in the Lagrangian with, however, precisely known coefficient. This coefficient can be defined, e.g., by using temporarily in addition dimensional regularization until the appropriate cancellation has been effected. A consequence hereof is [7] that cancellation of γ_5-anomalies [8], that would inhibit renormalizability of gauge theories, on the one-loop level by choosing appropriate Fermion multiplets suffices to establish renormalizability at also all higher-loop levels. Chiral invariance is hereby effected by parity doublets. Slavnov has also discussed [9] P.V. regularization of the nonlinear sigma-model.

When gauge theories are lattice-regularized in the Hamiltonian (i.e., with still continuous time) formulation [10], the naive "regularized" Hamiltonian appears not to lead to a large-Λ(i.e. small-a, where a is the lattice constant) expansion (2.3) with covariant functions F_k, but that an extra counter term must be supplied [10] that may be interpreted as a renormalization of the speed of light. For a gauge invariant regularization of the Hamiltonian of QED employing space smearing (in terms of a cutoff on three-momentum) this effect was observed by Dirac [11] .

3. Renormalization

We define, for the time being somewhat formally, renormalized vertex functions by omitting the remainder term in (2.3) and setting then $\Lambda = \mu$, kept finite, i.e. by

$$(3.1) \quad \Gamma((2n), (\ell); g, \mu, m^2) = \sum_{k=0}^{\infty} (\ln \mu)^k F_k((2n), (\ell); g, m^2)$$

The "renormalization convention" involved here depends on the manner of regularization employed. The present justification for (3.1) lies in the fact that in all structural relations for vertex functions (like generalized unitary equations) the remainder term in (2.3) cannot interfere with the simply logarithmic terms. (m^2 is not the physical mass, cp. (4.9).)

It can be proven (this is a main result of renormalization theory [12]) that

$$(3.2) \quad \Gamma_{\Lambda B}((2n), (\ell); g_B, \Delta m_B^2) = A(g_B, \Lambda/\mu)^n B(g_B, \Lambda/\mu)^\ell \cdot \Gamma((2n), (\ell); f(g_B, \Lambda/\mu), \Delta m_B^2 B(g_B, \Lambda/\mu), \mu) + \delta_{n0}\delta_{\ell 2} B(g_B, \Lambda/\mu)^2 C(g_B, \Lambda/\mu) + O((\ln \Lambda)^{-\infty})$$

where A,B,f, and C are double power series in g_B and $\ln \Lambda/\mu$. (3.2) is a consequence of the existence of a local Lagrangian, as discussed before, that generates the r.h.s. of (2.3) apart from the remainder term, and will be further commented upon later. Consistency of (3.2) with (3.1) requires

(3.3) $A(g_B,1)=B(g_B,1)=1,\ f(g_B,1)=g_B,\ C(g_B,1)=0.$

The relation expressed by (3.2) can be inverted (at least in the sense of formal power series in g_B and g) to the more usual form

(3.4)
$$\Gamma((2n),(\ell);\ g,\mu,m^2)=$$
$$=Z_3(g,\Lambda/\mu)^n Z_2(g,\Lambda/\mu)^\ell \cdot$$
$$\cdot \Gamma_{\Lambda B}((2n),(\ell);\ g_B(g,\Lambda/\mu), m^2 Z_2(g,\Lambda/\mu))-$$
$$-\delta_{n0}\delta_{\ell 2} Z_2(g,\Lambda/\mu)^2 K(g,\Lambda/\mu)+O((\ln\Lambda)^{-\infty})$$

where Z_2, Z_3, g_B, and K and double power series in g and $\ln\Lambda/\mu$, with, corresponding to (3.3)

(3.5) $Z_2(g,1)=Z_3(g,1)=1,\ g_B(g,1)=g,\ K(g,1)=0.$

From (3.4) with (3.5) and (2.3), (3.1) is reobtained.

That the renormalized functions are the $\Lambda\to\infty$ limit of the regularized ones with appropriate substitutions for the parameters, and factors supplied, means in terms of (3.4): The explicit $\ln\Lambda/\mu$ occurrence cancels all logarithmic terms stemming from the Λ-dependence of the propagators; only terms $O((\ln\Lambda)^{-\infty})$ remain unaccounted for and are discarded in the limit.

The two-masses form (3.2) may be less familiar than the one-mass form obtained by setting in (3.1) $\mu=m$ and in (3.2) μ constrained such that $\mu^2=\Delta m_B^2 B(g_B,\Lambda/\mu)$. The validity of (3.2) for general μ can be proven by starting from this special case, and then changing Δm_B^2 at fixed Λ and g_B, and computing the change using the Schwinger action principle, formulae of sect. 4, and a simple argument resting on regularity in g. - The l.h.s. of (3.4) is not expandible in a powers series in m^2, but the logarithmic terms encountered can all be "computed" (an example being formula (III.9) of ref. [13]) on the basis of "asymptotic freedom in the infrared" in ϕ_4^4 theory.

So far we kept $\Delta m_B^2 \geq 0$. Formula (3.4) applies, in a sense, also for $\Delta m_B^2<0$ such that $m^2<0$. In this case one must, however, use the Lagrangian with fields shifted: One can prove [14] that the shift $\phi=\psi+(-6m^2g^{-1})^{1/2}$ determined by the "classical" (or "tree graph") Lagrangian (and the corresponding shift for $\phi_B=Z_3^{1/2}\phi$) leads to a Lagrangian in ψ of $\psi^4+\psi^3$ type with precisely those counter terms necessary to effect cancellation of all divergences as $\Lambda \to \infty$. (3.3) remains unchanged (apart from replacement of 2n by an arbitrary positive integer) if $\ell = 0$, while for $\ell > 0$ the precise form depends on a convention.

(3.1) leads to a simple integral representation for the renormalized functions if we use P.V. regularization. For simplicity, we set $m^2 = 0$ and $a_2 = \cdots = a_R = 1$. Define $\Gamma_G((2n),(\ell);\mu,g,\lambda_1 \ldots \lambda_L)$ as the contribution to the vertex function from an L-line graph G computed with the Feynman rules: propagator $\Gamma(1+\lambda_j)\mu^{2\lambda_j}(p^2)^{-1-\lambda_j}$ for the jth line, vertex -g. Then the renormalized contribution is

(3.6) $$\Gamma_{G\,ren}((2n),(\ell);\mu,g) =$$

$$= \oint \frac{d\lambda}{2\pi i} \prod_{j=1}^{L-1} \left[\int_{\uparrow} \frac{d\lambda_j}{2\pi i} \frac{\Gamma(R-\lambda_j)}{\lambda_j \Gamma(R)} \right] \cdot \frac{\Gamma(\lambda_L)}{\lambda_L \Gamma(R)} \cdot$$

$$\cdot \Gamma_G((2n),(\ell);\mu,g,\lambda_1 \ldots \lambda_L)$$

with the constraint $\lambda_L = \lambda - \sum_{j=1}^{L-1} \lambda_j$. Hereby the λ-integration path encircles $\lambda = 0$ with $|\lambda| < 1$. The λ-integrand is to be obtained by evaluating the L-1 fold integral for $Re\,\lambda > 0$ and analytically continuing from there; it is a meromorphic function of λ with poles only at the real integers. The L-1 fold integral is hereby to be evaluated with $0 < Re\,\lambda_j < 1$, $j = 1 \ldots L$. Furthermore, whenever the lines $i_1 \ldots i_s$ constitute a self-energy part, $\sum_{\sigma=1}^{s} Re\,\lambda_{i_\sigma} < 1$ is to be observed in the L-1 fold integration. (3.6) follows from (3.1) by use of the integral representation

$$(p^2)^{-1}[\mu^2/(p^2+\mu^2)]^R =$$

$$= \int_{\substack{\uparrow \\ 0 < Re\lambda < R}} \frac{d\lambda}{2\pi i} \frac{\Gamma(R-\lambda)}{\lambda \Gamma(R)} \frac{\Gamma(1+\lambda)\mu^{2\lambda}}{(p^2)^{1+\lambda}}$$

The subtraction of self energy parts at zero momenta, implied by the term prop. m_{B0}^2 in (2.1), is secured by the evaluation prescription given: the self energy parts then vanish at zero momenta for homogeneity reasons. (3.6) displays one particular evaluator in the sense of Speer [15]. - Replacing in the propagators for (3.6)

p^2 by $p^2 + m^2$ yields a massive-theory function with, however, different convention for m^2 than the one implied by (3.1).

4. PDEs and Applications

From (3.4), using the independence of $\Gamma_{\Lambda B}$ on μ for fixed Λ, g_B, and Δm_B^2 one derives (see, e.g., [3])

(4.1a) $$Op_{2n,\ell}\,\Gamma((2n),(\ell);g,\mu,m^2) + \delta_{n0}\delta_{\ell 2}\,\kappa(g) = 0$$

where

(4.1b) $$Op_{2n,\ell} = \mu\frac{\partial}{\partial\mu} + \beta(g)\frac{\partial}{\partial g} - 2n\gamma(g) + \ell\eta(g) + \eta(g)m^2\frac{\partial}{\partial m^2}$$

with

(4.2a) $$\beta(g) = b_0 g^2 + b_1 g^3 + \cdots \qquad b_0 = \frac{3}{16\pi^2}$$

(4.2b) $$\gamma(g) = c_0 g^2 + \cdots \qquad b_1 = -\frac{17}{2^8 3\pi^4}$$

(4.2c) $$\eta(g) = \frac{1}{3}b_0 g + \cdots$$

(4.2d) $$\kappa(g) = \frac{1}{3}b_0 + \cdots \qquad c_0 = \frac{1}{2^{10}3\pi^4}$$

The coefficients written out are independent of the manner of regularization. (4.1) is complemented by

(4.3) $$\frac{\partial}{\partial m^2}\Gamma((2n),(\ell);g,\mu,m^2) = -\Gamma((2n),(\ell)0;g,\mu,m^2).$$

Define

(4.4) $$\rho(g) = \int^g dg'\,\beta(g')^{-1}.$$

Then in (3.4)

(4.5a) $$g_B(g,\Lambda/\mu) = \rho^{-1}(\rho(g) + \ln\Lambda/\mu) = \exp[\ln\Lambda/\mu\cdot\beta(g)\frac{\partial}{\partial g}]g$$

(4.5b) $$Z_3(g,\Lambda/\mu) = \exp[2\int_{g_B}^{g}dg'\,\beta(g')^{-1}\gamma(g')]$$

(4.5c) $$Z_2(g,\Lambda/\mu) = \exp\left[-\int_{g_B}^{g} dg'\beta(g')^{-1}\eta(g')\right]$$

(4.5d) $$K(g,\Lambda/\mu) = \\ = \int_{g_B}^{g} dg'\beta(g')^{-1}\kappa(g')\exp\left[2\int_{g_B}^{g'} dg''\beta(g'')^{-1}\eta(g'')\right]$$

wherefrom expansions in $\ln\Lambda/\mu$ are obtained using for any F(·)

$$F(g_B) = \exp\left[\ln\Lambda/\mu\cdot\beta(g)\frac{\partial}{\partial g}\right]F(g).$$

Differentiating (3.4) w.r.t. Λ keeping g, m^2, μ fixed gives, using (4.5), the Zinn-Justin PDE for regularized unrenormalized vertex functions

(4.6a) $$Op_{2n,\ell B}\Gamma_{\Lambda B}((2n),(\ell);g_B,\Delta m_B^2) + \\ + \delta_{n0}\delta_{\ell 2}\kappa(g_B) = O((\ln\Lambda)^{-\infty})$$

where

(4.6b) $$Op_{2n,\ell B} = \Lambda\frac{\partial}{\partial\Lambda} + \beta(g_B)\frac{\partial}{\partial g_B} - 2n\gamma(g_B) + \\ + \ell\eta(g_B) + \eta(g_B)\Delta m_B^2\frac{\partial}{\partial\Delta m_B^2}$$

The analog to (4.3) is

(4.7) $$\frac{\partial}{\partial\Delta m_B^2}\Gamma_{\Lambda B}((2n),(\ell);g_B,\Delta m_B^2) = \\ = -\Gamma_{\Lambda B}((2n),(\ell)0;g_B,\Delta m_B^2).$$

The functions appearing in (3.2) are trivially related to those in (3.4) given in (4.5), whereby

$$F(g) = F(\rho^{-1}(\rho(g_B) - \ln\Lambda/\mu)) = \\ = \exp\left[-\ln\Lambda/\mu\cdot\beta(g_B)\frac{\partial}{\partial g_B}\right]F(g_B).$$

In equs. (4.1) and (4.6) one can let $m^2 \searrow 0$ resp. $\Delta m_B^2 \downarrow 0$ at non-exceptional momenta [16], since at these the r.h. sides of (4.3) resp. (4.7) have only (due to "asymptotic freedom in the infrared", even computable [13]) logarithmic singularities in m^2 resp. Δm_B^2.

At $\mu = m$, (4.1) with (4.3) becomes the inhomogeneous PDE [17]

(4.8) $$\left[m\frac{\partial}{\partial m} + \beta(g)\frac{\partial}{\partial g} - 2n\gamma(g) + \ell\eta(g)\right]\cdot \\ \cdot\Gamma((2n),(\ell);g,m,m^2) + \delta_{n0}\delta_{\ell 2}\kappa(g) = \\ = -(2-\eta(g))\Gamma((2n),(\ell)0;g,m,m^2).$$

In (3.1) the physical mass is defined by the vanishing of $\Gamma(p(-p),;g,\mu,m^2)$ at $p^2 = -m_{phys}^2$. Inserting

$$\Gamma(p(-p),;g,\mu,m^2) =$$
$$= -Z(g,m/\mu)(p^2 + m^2_{phys}) + O((m^2 + m^2_{phys})^2)$$

into (4.1) gives $\mathcal{O}p_{00}\, m^2_{phys} = 0$ and $\mathcal{O}p_{20} Z = 0$, with the solutions

(4.9a) $m^2_{phys} =$
$$= m^2 \exp\Big[-\int_{\hat g(g,m/\mu)}^{g} dg' \beta(g')^{-1} \eta(g')\Big] \mathcal{R}(\hat g(g,m/\mu))$$

where

(4.9b) $\hat g(g,m/\mu) = \hat\rho^{-1}(\hat\rho(g) + \ln m/\mu)$

with

(4.9c) $\hat\rho(g) = \int^{g} dg' \beta(g')^{-1}[1 - \tfrac{1}{2}\eta(g')]$

and $\mathcal{R}(\ )$ is a (in perturbation theory regular) function definable by $m^2_{phys}(\mu^2 = m^2) = m^2 \mathcal{R}(g) = m^2(1 + r_1 g + \cdots)$, and

(4.10) $Z(g,m/\mu) =$
$$= \exp\Big[2\int_0^{g} dg' \beta(g')^{-1} \gamma(g')\Big] Z(\hat g(g,m/\mu))$$

with $Z(\cdot)$ a regular function in the same sense. For $m^2 \searrow 0$, (4.9) gives m^2_{phys} prop. $m^2 |\ln m|^{-1/3}$ (cp. Sect. 5).

The vertex functions as defined in (3.1) can be related to the functions characterized by some more usual type of renormalization conditions (e.g., by prescriptions at zero momenta or on the mass shell, or some Euclidean symmetry points) via reparametrization formulae that are straightforward to derive; only the $\beta, \gamma, \eta, \kappa$ functions of the two renormalization conventions are needed (apart from an easily determined constant). The relevant formulae are a straightforward extension of those e.g. in [13] sect. II.3. The determination outside of perturbation theory of the parametric functions for any particular manner of regularization will be discussed in sect. 5.

Of the equations (4.5) we only discuss (4.5a) from the speculative point of view. Assuming $\beta(g)$ to be a continuous function connecting smoothly with the behaviour near g = + 0 indicated by (4.2a), consider the section of the g axis adjoining the origin where $\beta(g)$ is positive. There $\rho(g)$ of (4.4) is monotonically increasing, and thus $g_B(g,\Lambda/\mu)$ is monotonically increasing in both g and Λ/μ. Λ is allowed to increase indefinitely if $\rho(g)$ increases indefinitely with g in that section. If $\rho(g)$ does not do so, we have a "Landau catastrophe" where Λ cannot increase indefinitely without forcing $g \searrow 0$, as Landau [18] observed using effectively the approximation $\beta(g) = b_0 g^2$. The behaviour of g_B for $\Lambda \nearrow \infty$, fixed g, depends on assumptions; e.g., if $\beta(g)$ has a first positive zero at g_∞ and $-\beta'(g_\infty) > 0$, then [19]

(4.11)
$$g_B(g, \Lambda/\mu) \approx g_\infty - (\mu/\Lambda)^{-\beta'(g_\infty)} (g_\infty - g) \cdot \exp\left\{-\frac{1}{\beta'(g_\infty)} \int_g^{g_\infty} dg' \left[\frac{1}{\beta(g')} - \frac{1}{(g'-g_\infty)\beta'(g_\infty)}\right]\right\}.$$

Thus, g depends on how fast one lets $g_B \nearrow g_\infty$ when $\Lambda \nearrow \infty$. For further discussions, see [19]. - Note that the behaviour of $g_B(g, \Lambda/\mu)$ as described before applies only to the fictitious g_B of (4.5a) and not to the true bare coupling constant, since the remainder term in (3.4) is neglected and thus (4.5a) approximates the true bare coupling constant only if Λ/μ is large. Also, g_∞ will depend on the manner of regularization!

Formulae (4.1), (4.3) have been derived from (3.4), and (3.4) in turn is an abstraction from perturbation theory. One must not, however, consider these equations as so restricted. E.g., (4.8) with some more "physical" renormalization convention [17] is simply a response equation: it is the infinitesimal form of the statement that change of the bare mass (defined in the regularized theory) in the Lagrangian leaves the Lagrangian in the family of regularized ones renormalizable, as $\Lambda \to \infty$, by mass-, coupling constant-, and amplitude renormalization, and from (4.8), (4.1) is derived by integrating (4.8) along the characteristic and change of notation. That the parametric functions in (4.1) have a certain behaviour, in particular that they connect smoothly with the behaviour at small g as suggested by the formulae (4.2), is where speculation comes in.

5. Large-Λ-Behaviour of Unrenormalized Green's Functions

If in (3.2) in

(5.1)
$$f(g_B, \Lambda/\mu) = \rho^{-1}(\rho(g_B) - \ln \Lambda/\mu)$$

Λ becomes large, f can be computed accurately provided g_B lies in the section of the g-axis from to origin to g_∞ in (4.11). Using (4.2a) we write

$$\bar{\varrho}(g) = \int_0^g dt\left[\frac{1}{\beta(t)} - \frac{1}{b_0 t^2} + \frac{b_1}{b_0^2 t}\right] = O(g)$$

such that

$$\varrho(g) = -b_0^{-1} g^{-1} - b_0^{-2} b_1 \ln g + \bar{\varrho}(g)$$

and for λ large

$$(5.2) \quad \varrho^{-1}(\varrho(t) - \ln\lambda) =$$
$$= \left[b_0 \ln\lambda + b_0^{-1} b_1 \ln(b_0 t \ln\lambda) + t^{-1} - b_0 \bar{\varrho}(t)\right]^{-1} +$$
$$+ O((\ln\lambda)^{-3} \ln\ln\lambda).$$

For brevity, we discuss only the simplest cases of (3.2):

1) $\Delta m_B^2 = 0$, $n = 2$, $\ell = 0$, $p_1 \ldots p_4$ nonexceptional:

$$(5.3) \quad \Gamma_{\Lambda B}(p_1 \ldots p_4; ; g_B, 0) =$$
$$= \exp\left[4\int_f^{g_B} dg' \beta(g')^{-1} \gamma(g')\right] \Gamma(p_1 \ldots p_4; ; f(g_B, \Lambda/\mu), 0, \mu)$$
$$+ O((\ln\Lambda)^{-\infty}) =$$
$$= \exp\left[4\int_0^{g_B} dg' \beta(g')^{-1} \gamma(g')\right] \cdot \Big\{ -f(g_B, \Lambda/\mu) +$$
$$+ f(g_B, \Lambda/\mu)^2 \left[-\frac{1}{6} b_0 \sum_{j=2}^{4} \ln\frac{(p_1 + p_j)^2}{\mu^2} + c\right]\Big\}$$
$$+ O(f(g_B, \Lambda/\mu)^3)$$

where the constant c is regularization-manner-dependent, and the r.h.s. is independent of μ. From (5.1-3) we learn: $\Gamma_{\Lambda B}(p_1 \ldots p_4, ; g_B, 0)$ for large Λ, with g_B fixed in the range described, possesses an "asymptotic expansion", of which the first two terms prop. $(\ln\Lambda)^{-1}$ and prop. $(\ln\Lambda)^{-2} \ln\ln\Lambda$ are known up to the first factor on the r.h.s. of (5.3); the coefficient of $(\ln\Lambda)^{-2}$ involves in addition the function $\bar{\varrho}(g_B)$. Thus, one obtains both functions $\beta(g_B)$ and $\gamma(g_B)$ (which depend on the manner of regularization) if that "asymptotic expansion" could (necessarily nonperturbatively) be constructed.

2) $\Delta m_B^2 > 0$ fixed, $n = 2$, $\ell = 0$, zero momenta:

$$(5.4)\quad \Gamma_{\Lambda B}(0000,;g_B,\Delta m_B^2) =$$

$$= \exp\left[4\int_f^{g_B} dg'\beta(g')^{-1}\gamma(g')\right]\cdot$$

$$\cdot\Gamma(0000,;f(g_B,\Lambda/\mu),\Delta m_B^2 \exp\left[\int_{g_B}^{f} dg'\beta(g')^{-1}\eta(g')\right],\mu) +$$

$$+ O((\ln\Lambda)^{-\infty}) =$$

$$= \exp\left[4\int_0^{g_B} dg'\beta(g')^{-1}\gamma(g')\right]\cdot\Big\{-f(g_B,\Lambda/\mu) +$$

$$+ f(g_B,\Lambda/\mu)^2\left[-b_0 \ln\frac{\Delta m_B}{\mu} + \frac{1}{2}b_0\int_f^{g_B} dg'\beta(g')^{-1}\eta(g') + c'\right]\Big\} +$$

$$+ O(f(g_B,\Lambda/\mu)^3)$$

wherefrom now also $\eta(g_B)$ could be determined as described before. c' is another regularization-manner-dependent constant.

3) $m^2_{phys} > 0$ fixed, n = 2, ℓ = 0, zero momenta:
From (5.4), (4.9), (3.2), and (4.5c) we find

$$(5.5)\quad \Gamma_{\Lambda B}(0000,;g_B,\Delta m_B^2)\Big|_{m^2_{phys}\ \mathrm{fixed}} =$$

$$= \exp\left[4\int_0^{g_B} dg'\beta(g')^{-1}\gamma(g')\right]\cdot\Big\{-f(g_B,\Lambda/\mu) +$$

$$+ f(g_B,\Lambda/\mu)^2\left[-b_0\ln\frac{m_{phys}}{\mu} + c'\right]\Big\} + O(f(g_B,\Lambda/\mu)^3)$$

6. Conclusion

The formulae (5.1), (5.4-5) rest on rather weaker assumptions than e.g. those discussed in connection with (4.11), and their verification would not imply the existence of a renormalized theory. However, the weaker assumptions are formally related to those on which the idea of "asymptotic freedom" for gauge theories [20] rests. The analoga of the formulae in sect. 5 in that case, however, concern the large-momenta behaviour in the deep-Euclidean region of the renormalized theory, or the manner in which the bare coupling constant has to go to zero in order to have a nontrivial "computable" effect when the cutoff becomes large, and these situations might not be easily accessible constructively. Failure of (5.3) or (5.4) or (5.5) even for small g_B, on the other hand, would indicate that the regularization employed is not a suitable one for possibly arriving at a renormalized theory.

Consideration of ϕ_4^4 theory as the limit theory of $\phi_{4-\varepsilon}^4$ as $\varepsilon \downarrow 0$, certain high-temperature-expansion results (discussed in sect. 13 of [21]), and in particular the trend of recent rigorous work by R. Schrader [22] shed doubt on the existence of renormalized ϕ_4^4 theory, and this desease might befall also other non-asymptotically-free formally renormalizable theories, with QED having possibly the escape route of being imbedded in a nonabelian gauge theory. - I am unable to comment on the recurring idea that gravity be a universal regulator (also for itself) [23].

References and Footnotes

1. S.L. Adler, in: Lectures on Elementary Particles and Fields, Eds. S. Deser, M. Grisaru, H. Pendleton; Cambridge: MIT Press 1970
2. B. Simon, "The $P(\ )_2$ Euclidean (Quantum) Field Theory", Princeton University Press 1974
3. K. Symanzik, Comm. math. Phys. 45, 79 (1975), GIFT lecture 1975 (DESY 75/24)
4. W. Zimmermann, Ann. Phys. (N.Y.) 77, 536, 570 (1973)
5. P.K. Mitter, these Proceedings
6. A.A. Slavnov, Theor. Math. Phys. 13, 1054 (1972) (transl.)
7. I thank J. Zinn-Justin for this remark
8. D. Gross, R. Jackiw, Phys. Rev. D6, 477 (1972)
9. A.A. Slavnov, Nucl. Phys. B31, 301 (1971)
10. J. Kogut, L. Susskind, Phys. Rev. D11, 395 (1975); L. Susskind, PTENS 76/1. The author thanks L. Susskind for a discussion.
11. P.A.M. Dirac, in "Fundamental Interactions at High Energy II", Eds. A. Perlmutter, G.J. Iverson, R.M. Williams, New York: Gordon and Breach 1970
12. E.G. "Renormalization Theory", Eds. G. Velo, A.S. Wightman, Dordrecht: Reidel 1976
13. K. Symanzik, Comm. math. Phys. 34, 7 (1973)
14. K. Symanzik, unpublished

15. E.R. Speer, "Generalized Feynman Amplitudes", Princeton University Press 1969

16. K. Symanzik, Comm. math. Phys. 23, 49 (1971)

17. K. Symanzik, Comm. math. Phys. 18, 227 (1970); C.G. Callan Jr., Phys. Rev. D2, 1541 (1970)

18. L.D. Landau, in "Physics in the Twentieth Century", Eds. R. Jost, V. Weisskopf

19. K.G. Wilson, Phys. Rev. D3, 1818 (1971)

20. D. Politzer, Physics Reports 14C, 129 (1974)

21. K.G. Wilson, J. Kogut, Physics Reports 12C, 75 (1974)

22. R. Schrader, Comm. math. Phys. 49, 131, 50, 97 (1976), and FUB preprints

23. C.J. Isham, A. Salam, J. Strathdee, Phys. Letts. 46B, 407 (1973)

ON NON-RENORMALIZABLE INTERACTIONS

G. PARISI

(I.H.E.S. - BURES-sur-YVETTE)

I. Introduction

Nonrenormalizable interactions have always been the black sheep of field theory. Long time ago (1) it was supposed that non-renormalizable interactions are characterized by having Green functions which are not C^{∞} in the coupling constant : if this interpretation is correct, the ultraviolet divergences found in the perturbative expansion arise from the non existence of the quantities which are computed in the standard approach (i.e., the coefficients of the Taylor expansion for zero coupling constant).

The first attempts in this direction were done using or the ξ-limiting procedure (2) or the peratization technique (3). However they were mainly inconclusive ; the full understanding of the problem required a better non perturbative knowledge of quantum field theory which is now given by the modern theory of second order phase transitions (4).

The purpose of these lectures is to study the existence and the properties of non-renormalizable interactions at the light of the knowledge gathered in the study of critical phenomena. We do not prove any rigorous theorem and we are able to control our results only in favorable cases where a divergence-free perturbative expansion can be used. We also do serious attempts to understand the general case which is outside the range of perturbation theory. Our interest is more concentrated on general structural properties than on specific numerical computations.

These lectures are divided in nine sections :

In section II we explain the relation between critical phenomena and non-renormalizable interactions and we present a general method for computing the Green functions of a non-renormalizable theory.

In section III we study two different concrete applications of the methods described in the previous section.

In section IV we discuss an explicit example of a non-renormalizable interaction for which a new divergence-free perturbative expansion has been constructed.

In section V we show that the Green functions constructed in the previous section are not C^∞ in the coupling constant at the origin. The general structure of the irregular terms is found.

In section VI we write the C.S. (Callan Symanzik) equation and we use the properties of its solutions to recover the general structure of zero coupling singularities found in the previous section.

In section VII we see how the C.S. equation can be used as a self consistency requirement, to compute the values of the counter-terms which are arbitrary in the standard perturbative approach.

In section VIII we show how the C.S. equation may be used as a substitute of the equations of motion.

Finally in section IX we present our conclusions.

II. The Relation with Critical Phenomena

It is not well known that non-renormalizable interactions can be regarded as superrenormalizable interactions with infinite bare coupling constant (5). Although this statement may sound very deep it is trivial ; its only utility is due to the fact that changing the language we transform an unsolved problem (the infinite cutoff limit in non-renormalizable interaction) in a solved problem (the infinite bare coupling limit in superrenormalizable interactions). The reader must realise that contrary to the appearence there are many theories in which the infinite coupling limit can be controlled. Indeed the celebrated Kadanoff scaling law for critical phenomena (6) (which it is considered to be well understood) is equivalent to the existence of the renormalized Green functions of a superrenormalizable interaction in the infinite coupling limit (7).

The connection between critical phenomena (massless theory) and infinite coupling theories is quite simple : if the coupling constant has positive mass dimension, the dimensionless coupling constant (on which dimensionless quantities do depend) goes to infinity when the mass goes to zero.

Let us now see some example of NRI (non-renormalized interactions) which are obtained as infinite coupling limit of SRI (superrenormalizable interactions), some of them are well known. Let us consider a vector field of mass m interacting with a conserved current with a coupling constant, λ , the space time dimension D being less than 4 . When we send the mass and the coupling constant to infinity at fixed λ^2/m^2 , we recover the local Fermi current-current interaction. Similarly the $\lambda\Phi^4$ interaction in the infinite coupling limit becomes the non linear σ model (8) which is characterized by the measure $\delta(\Phi^2-1)$. This last result can be formally proved using the identity

$$\lim_{\lambda\to\infty} \exp-\lambda(\Phi^2-1)^2 \alpha \delta(\Phi^2-1) \tag{2.1}$$

A perturbative check of the identity of the two theories can be found in ref.(8) and in Zinn-Justin's lectures at this school.

These results are not surprising ; indeed the coupling constant of the SRI plays the same rôle of the cutoff of the NRI. The familiar divergences in the infinite cutoff limit of NRI are traded with the more familiar divergences of the standard perturbative expansion when $\lambda \to \infty$.(The value at infinity of a non trivial polynomial is always infinity!) However, the modern theory of second order phase transitions teaches us that these divergences are spurious : the renormalized coupling constant (g) acquires a finite value g_c when the bare coupling becomes infinite ; the renormalized Green functions can be computed in this limit : one uses their expansion in powers of the renormalized coupling constant to evaluate them at $g = g_c$ (7), (the so called infrared stable fixed point).

Using these definitions the statement at the beginning of this chapter reads "non-renormalized interactions are superrenormalizable interactions computed at the infrared stable fixed point". It is now clear which is the general procedure which we can follow to construct a NRI : we put a cutoff in the interaction, we interpret the cutoffed theory as a SRI and we find its infrared stable fixed point : this can be done looking for the zeros of a well defined function (9) $(\beta(g_c)=0)$. The renormalized Green functions of the SRI at the infrared unstable fixed point are also the renormalized Green functions of the NRI.

The weak point of this approach is that the infrared stable fixed point may not exist (e.g. the system undergoes always a first order and never a second order transition) or if it exists, it cannot be easyly found . A general classification of NRI can be done using as a criterion the control that we have on the position and on the existence of the zero. There are essentially three different cases:

a) g_c is very small ; it may be of order ϵ in $4-\epsilon$ (4) or $2+\epsilon$ (8) dimensions or it may be of order $1/N$ in the theories invariant under the $O(N)$ group (10). In this case the value of g_c is quite well known : there are asymptotic expansions in $1/N$ or ϵ for small values of $1/N$ of ϵ which allow the evaluation of g_c with very great accuracy.

b) g_c is of order one ; however the function $\beta(g)$ has a zero also if one includes only the contributions coming from one loop diagrams. Taking care of higher order diagrams the zero does not disappear and its position can be extimated with improved accuracy. This situation is realized in the three dimensional $\lambda\Phi^4$ theory (7), (11). Also in this case very precise results may be obtained.

c) The function $\beta(g)$ does not have any non trivial zero in the one loop approximation and no one has computed the contribution from a number of loops to high enough to see if improving the approximation, a stable zero appears.

In this last case no conclusions can be drawn ; unfortunately this is the situation for a local current-current interaction in 4 dimensions ; consequently the problem of constructing a finite non renormalizable realistic model for weak interaction will not be solved in these lectures. (My personal opinion is that such a construction can hardly be obtained without giving up the unification among weak and electromagnetic interactions, which may be a too high price to pay).

At this stage the reader which is not too familiar with the theory of critical phenomena will be amazed by the existence of case a . He would think that the author must have done a terrific trick to transform an infinite coupling theory in a theory with infinitesimal coupling and maybe he will doubt on the ability of the author to cope with such difficult problems. Consequently it may be useful to spend some words to explain the origin of case a ; the existence of a computable expansion for the critical exponents in powers of ε has a similar origin.

Let us consider a $\lambda\Phi^4$ interaction in $4-\varepsilon$ dimension, ε being a small but non zero number. Given an expansion for a quantity in powers of the bare or the renormalized coupling, we will call it "good" or "bad" if the coefficients of the Mc Laurin expansion have respectively a finite or an infinite limit when $\varepsilon\to 0$ It is clear that although the differente between a "good" and a "bad" expansion is defined only in the limit $\varepsilon\to 0$ this distinction will play a crucial rôle also for small ε . The renormalizability of the theory in $4-\varepsilon$ dimensions implies that the expansion of the Green functions in powers of the bare coupling constant is "bad", while the expansion of the renormalized Green functions in powers of the renormalized coupling constant is "good". The function $\lambda(g)$ which gives the bare coupling λ as function of the renormalized one (g) has also a "bad" expansion. It is a crucial observation to note that the function $\lambda(g)$ has the representation (7), (9).

$$\lambda(g) = g \exp \int_0^g [-\varepsilon/\beta(g')-1/g']dg' \tag{2.2}$$

where the $\beta(g)$ has a "good" expansion! Moreover the point $\lambda\to\infty$ corresponds to the point g_c such that $\beta(g_c) = 0$. At the one loop level we find

(2.3) $\beta(g) = -\varepsilon+g^2+O(g^3)$.

For small ε the high order terms are negligible when $g = O(\varepsilon)$ (the expansion for β is "good"). We finally get $g_c=\varepsilon+O(\varepsilon^2)$ which is the first term of a systematical expansion of g_c in powers of ε .

Roughly speaking the physical picture is that for small ε the ultraviolet divergences nearly shield the interaction so that a theory with a large bare coupling constant is always reduced to a theory with a renormalized coupling constant of order ε .

Having explained the general ideas laying behind our approach to the study of NRI , we think that it is better to consider some specific applications of them. This will be the subject of the next section.

III. Two Examples

In the first part of this section we study a current-current interaction among Fermions in dimension $D(2 < D < 4)$; the Lagrangian density is :

(3.1) $\mathcal{L} = \sum_{1}^{N}{}_i\bar{\Psi}_i(\partial\!\!\!/+m)\Psi_i+\frac{1}{2}GJ_\rho J^\rho$; $J_\rho = \sum_{1}^{N}{}_i\bar{\Psi}_i\gamma_\rho\Psi_i$

where Ψ_i is a N component spin 1/2 field.

This interaction is non-renormalizable : the expansion in powers of G can be defined only after the introduction of a cut-off and at any finite order in G , divergent results are obtained in the infinite cutoff limit. Our goal is to show that these divergences disappear using non pertubative techniques (12).

We will use the following cutoffed Lagrangian density :

(3.2) $\mathcal{L}_c = \sum_{1}^{N}{}_i\bar{\Psi}_i(\partial\!\!\!/+m)\Psi_i+uA_\rho J^\rho+\frac{1}{2}A_\rho(\mu^2-Z\Box)A^\rho$.

Where $G = u^2/\mu^2$ and Z plays the rôle of the inverse of the cutoff ; in the limit $Z \to 0$ we recover the local Lagrangian (3.1) : this fact can be easily proved integrating over the A_ρ field. The dimensionless bare parameters which are invariant under a redefinition of the A_ρ field are :

$$(3.3) \qquad g_B = u/[\mu^{(4-D)/2} . Z^{(D-2)/2}] \; ; \; r = u^2 m^{(D-2)}/\mu^2 \quad .$$

When Z goest to zero, the coupling constant g_B goes to infinity if $D > 2$; consequently the limit $Z \to 0$ can be controlled directly only when the interaction is renormalizable.

Renormalized fields, masses and coupling constant can be defined in the usual way (e.g. the propagator of the A field is $1/(p^2+\mu^2)+O(p^4)$. The bare quantities can be computed as functions of the renormalized ones ; using the CS equation and the Ward identities, we find :

$$(3.4) \quad \mu_R = \mu \; ; \; u_R = u \; ; \; Z(g) = \exp \int_0^g [(D-4)/(2\beta(g')) - 1/g'] dg'$$

where $\beta(g) = (D-4).g/2 + Ag^3 + O(g^5)$ $(A>0)$ and g is the dimensionless renormalized coupling constant $(g = u/\mu^{(4-D)/2})$. When D is near to 4 , the function β has a zero at $g_c = [(4-D)/2A]^{1/2} + O(4-D)$. At $g = g_c$; $Z = 0$; the Green functions of the NRI (3.1) concede with those of the SRI (3.2) computed at a particular value of the coupling constant, which is small and where perturbation theory can be successfully applied. The non-renormalizable Fermi interaction turns out to be an intermediate boson theory in which the value of the dimensionless renormalized coupling constant is fixed by the condition $\beta(g_c)=0$; the Fermi coupling is proportional to the inverse of the mass (μ) of the Bose field : the weak coupling limit is obtained when μ goes to infinity.

If $4-D$ is not small (e.g. $D = 3$) , $g_c = O(1)$: it is not clear if we can get sensible results using the perturbative expansion in powers of g . We do not know the answer in the case of the Lagrangian (3.1), however we will show that in a different case using a similar procedure one can estimate the Green functions of the NRI with a few per cent accuracy also when g_c is not small.

As we have discussed in the previous section, the non linear σ-model can be regarded as an infinite coupling $\lambda\Phi^4$ theory ; consequently its Green functions coincide with those of the $\lambda\Phi^4$ theory computed at the infrared stable fixed point ; in particular the off shell scattering amplitude at zero momenta in the single phase region $(\langle\Phi\rangle=0)$ is equal to the renormalized coupling constant (g_c) satisfying the condition $\beta(g_c) = 0$. One can try to estimate the position of the zero of the function $\beta(g)$ using perturbation theory in g (the renormalized coupling constant of the $\lambda\Phi^4$ interaction). An explicit computation shows that (7), (11) :

$$(3.5) \qquad \beta(g) = -g+g^2-.42g^3+.35g^4-.38g^5+.50g^6+0(g^7) \quad .$$

We know from general theorems that the expansion of β in powers of g has zero radius of convergence (13) and it is an asymptotic expansion (14). If the series is summed using the powerful Padé-Borel technique (15), taking as input only the first 2,3,4,5,6 terms we find (11) respectively g_c=1,1.60,1.42,1.43,1.42 . There are few doubts that the true value of g_c is 1.42 with an error of a few per cent (the precise amount of the error may be a matter of debate).

This example shows that, if an enough high number of diagrams is computed, the approach we propose may be able to produce accurate results for a NRI also when the coupling constant is not small. Having succeeded to construst non trivial NRI's, we would like to understand some structural properties (e.g. the nature of the singularities at zero coupling constant, the validity of the standard divergent pertubative expansion). In principle this can be done without changing the technique, however for simplicity we prefer to study these problems using an explicit renormalizable expansion in which general properties can be verified order by orber. The construction of such an expansion will be the subject of the next section.

IV. The Large N Expansion

We are now going to see that in some non-renormalizable interactions the $1/N$ expansion is renormalizable (16). More precisely there are NRI's, symmetric under the action of the $O(N)$ group, in which the Green functions can be exactly computed in the limit $N\to\infty$ and a systematic expansion in powers of $1/N$ can be constructed. At each order in $1/N$ the only divergences present can be absorbed by mass, wave function and coupling constant renormalization. The $1/N$ expansion may not be suited for pratical computations, however each order can be written in a closed form ; this technique is therefore quite useful to derive general properties.

For example let us consider the quadrilinear interaction of a scalar field in a D dimensional space $(4 < D < 6)$; similar considerations can be extended to a quadrilinear interaction of Fermions for $2 < D < 4$. The theory is non-renormalizable ; its Lagrangian density is :

$$\mathcal{L} = 1/2 \sum_{1}^{N} {}_i (\partial_\mu \Phi_i) + 1/2 (m^2 + (g/N)^{1/2} \sigma) \sum_{1}^{N} {}_i \Phi_i^2 + 1/2 \sigma^2 \tag{4.1}$$

where Φ_i is an N component scalar field ; σ is an auxiliary field which can be eliminated reproducing the usual $\lambda\Phi^4$; the Lagrangian is invariant under the group $O(N)$. Notice that our definition of coupling constant has the opposite sign of the conventional one : in our notations, if the coupling constant (g) is positive, the hamiltonian is formally unbounded from below ; the theory is well defined only for negative g , however the $1/N$ expansion can be constructed only for positive g . This contradiction is peculiar of this model and it is absent in the slightly more complicated $1/N$ expansion for a current-current interaction or for the nonlinear σ-model. For simplicity we disregard the problem and we consider here only the Lagrangian (4.1) ; we think that the study of the essentially selfadjointness of the hamiltonian and the construction of a formal perturbative expansion are disconnected at this level of sophistication.

In the limit $N \to \infty$ the Green functions are those of the free field theory, the only exception being the renormalized propagator of the σ field, which is equal to :

$$D(p^2) = 1/(1-g\Pi(p^2))$$

(4.2)

$$\Pi(p^2) = \int [q^2+m^2)^{-1}((q+p)^2+m^2)^{-1}-(q^2+m^2)^{-2}]d^Dq \quad .$$

Standard dimensional arguments imply that $\Pi(p^2)$ is finite for $D < 6$. The lack of convergence of the integral (4.2) forbids the application of the $1/N$ expansion for $D \geq 6$. In the large p region one finds that $\Pi(p^2) \simeq Ap^{(D-4)}$ $(A < 0)$. For negative g the $D(p^2)$ propagator has a pole in the Euclidean region $(p^2 > 0)$; to avoid the presence of this unwanted singularity we are bound to take g positive.

At the first order in $1/N$ the elastic scattering amplitude for the process $i+ \to j+m$ ‖ i , ℓ, j, and m are indexes which refer to the internal degrees of freedom) is

$$(4.3) \quad A_{i,\ell;jm}(s,t,u)=(\delta_{i\ell}\delta_{jm}D(s)+\delta_{i'}\delta_{\ell m}D(t)+\delta_{im}\delta_{j\ell}D(u))/N$$

s,t and u being the Mandestam variables. Elastic unitarity is satisfied at the order 1/N . At high energies s-wave amplitude goes to a constant and the differential cross section $d\sigma/dt$- at wide angles scales like $s^{(D-2)/2}F(\theta)$, as suggested by naive scale invariance (17). There is a striking difference among these results and those obtained from the first order in pertubation theory, where the unitarity bounds and asymptotic scaling invariance are both violated. At the first non trivial order in 1/N all the Green functions are asymptotically scale and conformal invariant. The field Φ has canonical dimension (D-2)/2 while the dimension of the field σ has jumped from its canonical value (D-2) to 2 . The behaviour in the large momentum region is g independent, the only dependence from the coupling constant is in the point where asymptotics set in.

The diagrammatic rules for constructing the 1/N expansion are very simple and they will not described here (10). The main difference with standard perturbation theory consists in the systematic use of the renormalized σ-propagator $D(p^2)$. Using simple power counting arguments one can show that the 1/N expansion is renormalizable in the usual sense (16) : only a finite number of superficial divergent diagrams are present. All the divergences disappear after mass, wave function and coupling constant renormalization. As far as ultraviolet divergences are concerned we find the same situation as in quantum electrodynamics.

This result is not unaspected. Using general arguments it can be shown that in an asymptotically scale invariant theory, if the dimensions of the fields are not too small or too high, the number of superficially divergent diagrams is finite and the theory can be consequently renormalizable (18). Notice that the shift in the dimensions of the field σ makes the effective dimensions of the interaction Lagrangian equal to the space time dimensions

D . The effective coupling constant in the large momentum region is dimensionless as in the standard renormalizable theories, its value is not arbitraty but fixed (it is g independent). Each order of the $1/N$ expansion is no more a polynomial in the coupling constant, indeed one finds a very complicated structure at $g = 0$. The study of this structure will be the main subject of the remaining part of these lectures.

V.- The Structure around $g = 0$

The presence of ultraviolet divergences in the standard perturbative approach and the absence of these divergences in the $1/N$ expansion suggest that the Green functions are not C^{∞} in the coupling constant. We will prove that at each order in $1/N$, terms proportional to non-integer powers of g are present ; their origin is clear : as suggested by unitarity something like an effective cutoff is present at momenta of order $g^{1/(D-4)}$. A term which in the cutoffed perturbation theory is proportional to $g^n \Lambda^{\nu}$, becomes $g^{n+\nu/(D-4)}$; the conjecture of T.D . Lee (2) done in the framework of the ξ-limiting procedure finds here its explicit realization.

A greater insight on the structure of the irregular terms maybe given by studying in detail a simple example ; we consider the contribution to the six points function at zero momentum coming from only one diagram (i.e. the one looking like a hexagon) (16) ; it can be written as :

$$(5.1)\quad \Gamma^6(g)=g^3 I(g)=g^3\int d^D k(k^2+m^2)^{-3}[1-g\Pi(k^2)]^{-3} \quad .$$

This integral exists for any positive value of g , however a divergent integral is obtained if we perform enough derivatives respect to g at the point $g=0$. In order to understand the precise nature of the singularities of the function $I(g)$ at $g=0$, we study the positions of the pôles of its Mellin transform (16, 19-21) :

$$(5.2)\qquad M(s) = \int_0^{\infty} dg\, g^{s-1} I(g)$$

$$I(g) = (2\pi i)^{-1} \int_{-i\infty}^{+i\infty} g^s M(s) ds \quad .$$

One easily finds :

$$(5.3)\qquad M(s)=\Gamma(D/2)\Gamma(s)\Gamma(3-s)/2. \int_0^{\infty} \Pi(X)^{-s} X^{(D-2)/2}/(X^2+m^2)^3 dX \quad .$$

The pôles of $M(s)$ have two different origines : some arise from the pôles of the Γ functions in front of the integral, others are produced by divergences in the integral itself. If the theory is superrenormalizable $(D < 4)$, the integral is a regular function in the negative s plane ; there are pôles on the positive s plane but they are not interesting for small g : the integration path can always be shifted to the left. When the theory becomes non-renormalizable $(D > 4)$, these pôles migrate via the point at infinity from the positive to the negative s plane and they become relevant in determining the expansion of $I(g)$ in broken powers of g. These extra pôles are located at $s = 2i/(4-D)-j \ (i,j \in Z^+)$. For irrational dimensions none of these pôles collides and one obtains the following double expansion :

$$(5.4)\qquad I(g)=\Sigma_i f_i(D) g^i + \Sigma_k \Sigma_i h_{i,k}(D) g^{i+k/(D-4)} \quad .$$

This analysis can be extended to any Feynman diagram of the $1/N$ expansion ; a distinction must be done between the pôles that come form the last integration and pôles of the integrand itself ; the difference between the two cases is similar to that between only superficial divergent diagrams and divergent subdiagrams. In the general case one obtains a double expansion similar to (5.4).

The functions f and h have pôles when two or more of them multiply the same power of g (D must be rational) ; finite results are obtained : the divergences cancel out, and integer powers of $\ln(g)$ appear. In other words when two or more pôles of the Mellin transform collide, a higher order pôle is produced.

If D is irrational, dimensional regularization maybe used to define the divergent integrals of the standard perturbative expansion ; it is not difficult to check (16) that this procedure gives correctly the functions $f(D)$ (the functions $h(D)$ are obviously missing). This fact implies that the functions $f(D)$ have a closed representation for fixed N ; such a representation does not exist for the functions $h(D)$, which must be considered as the substitutes of the divergent counterterms of the old perturbative expansion. Information on their behaviour at finite N can be obtained imposing the cancellation of their singularities with those of the functions $f(D)$.(This situation has many points in common with the one described by Symanzik) (22).

If we want to use the expansion (5.4) for small values of the coupling constant and not small $1/N$ (e.g. $N=1$) , it is imperative to get information on the irregular terms independently from the $1/N$ expansion. This cannot be done unless we have under control the large momenta behaviour of the theory : the functions $h(D)$ comes from the integration region where the momenta are of order $g^{1/(4-D)}$. Although we can cope with this problem using the techniques described in sections II and III, it would be nice to get direct estimates in the infinite cutoff limit. The rest of these lectures will be devoted to the study of this problem.

VI. - The Solution of the Callan Symanzik Equation

The proof of the validity of CS (Callan Symanzik) equation (9), (23) can be extended to NRI, provided that the renormalized Green functions depend only on the renormalized mass and coupling constant. For example in the case of the $\lambda\Phi^4$ interaction for $D < 6$ we find :

$$(6.1) \quad [-\omega\frac{\partial}{\partial\omega} + \beta(g)\frac{\partial}{\partial g} + d_n - n\gamma(g)]\Gamma_n(\omega p,g) = \Delta\Gamma_n(\omega p,g) = m^2\Gamma_{n,\Phi_R^2}(\omega p,g)$$

$$\beta(g)=(D-4)g-(g^2+O(g^3);C=(N+8)/N\Gamma(3-D/2),\gamma(g)=O(g^2) ;$$

$$d_n=2D-n(D-2) \quad ,$$

where g is the renormalized coupling constant (we use the same sign convention as in section IV), Γ_n are the one particle irreducible Green functions and Γ_{n,Φ_R^2} is the zero momentum insertion of the renormalized Φ^2 field. If the higher order terms are neglected $\beta(g)$ has a zero at $g_u = (D-4)/C$; the corrections to the position of the zero are small if $1/N$ or $D-6$ are small.

Let us consider eq.(6.1) as a differential equation in Γ ($\Delta\Gamma$ is supposed to be known) ; a family of solutions can be found using the method of the characteristic curves (24) (the solution of a first order differential equation is not unique if we do not specific the boundary conditions). However, if $\exists g_u$ such that $\beta(g_u)=0, \beta'(g_u) < 0$ (g_u is the ultraviolet stable fixed point), there is only one solution which remains finite at $p = 0$ when $g \to g_u$; it is given by (25) :

$$(6.2) \quad \Gamma_n(p,g) = \int_{g_u}^{g} dg'/\beta(g').F_n(g)/F_n(g')\Delta\Gamma_n(R(g)/R(g')p,g')$$

$$R(g) = g^{1/(D-4)}\exp\int_0^g [1/\beta(g')-(D-4)^{-1}g'^{-1}]dg'$$

$$Z(g) = \exp\int_0^g \gamma(g')/\beta(g')dg' \; ; \; F_n(g) = R(g)^{-d_n}Z(g)^n \quad .$$

If the interaction is superrenormalizable or renormalizable but asymptotically free (26), $g_u=0$; if the interaction is non renormalizable or renormalizable but not asymptotically free, $g_u \neq 0$. There are no doubts that in the first case the Green functions are regular near g_u , however the condition of regularity near g_u seems very plausible also in the second case (24). The integrated form (eq. 6.2) of the CS equation may be used to investigate the large momentum behaviour of the Green functions : if $\lim_{\lambda\to\infty} \Delta\Gamma(\lambda p,g)/\Gamma(\lambda p,g) = 0$ as it happens at any order of perturbation theory, after some manipulations which are described in ref.(25), we find that asymptotic scale invariance is satisfied :

$$(6.3) \quad \Gamma_n(\lambda p,g) \xrightarrow[\lambda\to\infty]{} a^n(g)\lambda^{-[d_n-n\gamma(g_u)]} \int_0^\infty dx\, x^{[d_n-n\gamma(g_u)-1]} \Delta\Gamma_n(xp,g_u).$$

An interesting feature of eq.6.2 in the case of NRI is that Γ_n is not C^∞ at $g=0$ also if the functions $\Delta\Gamma_n$, β and γ are analytic around $g=0$. For sake of simplicity let us condiser the case $p=0$: eq.(6.2) can be written

$$(6.4) \quad \Gamma_n(g) = \int_0^g dg'/\beta(g').F_n(g)/F_n(g')\Delta\Gamma_n(g') +$$

$$+ F(g) \int_{g_u}^0 dg'/[\beta(g')F_n(g')\Delta\Gamma_n(g') \quad .$$

After the split both integrals in eq. 6.4 are divergent, however, if D is irrational, they can be defined using analytic regularization ; the first term is C^∞ in g (always for irrational dimension) while the second term is proportional to $g^{-d_n/(D-4)}$ (16), (25) . Comparing eq. 6.4 with eq. 5.5 we see that we have obtained an explicit representation for the functions $h(D)$ as integrals from 0 to g_u . This fact explains why the computation of $h(D)$ is not simple (it involves the knowledge of the Green functions up to $g = g_u$) , however it may be used to give rough estimates of $h(D)$. If the same argument is used to estimate the singularities at $g=0$ of the function Γ_{n,Φ_R^2}, assuming that $\Delta\Gamma_{n,\Phi_R^2}$ is regular, we find that $\Delta\Gamma_n = g^{(-d_n+2)/D-4)} +$ regular terms, i.e. the power of the irregular term of $\Delta\Gamma$ is greater than the power of the irregular term of Γ.

In conclusions, if the functions β, γ and $\Delta\Gamma$ are reasonable, eq. 6.2 has the virtue to implement automatically asymptotic scale invariance and to generate functions which are not C^∞ starting from a C^∞ input.

VII. - The Counterterms

In the standard perturbative approach to NRI finite results are obtained adding counterterms to the Lagrangian, whose number increase with the order of perturbation theory ; the infinite part of the counterterms is fixed but their finite part is arbitrary : for a given value of the coupling constant we obtain an infinite class of theories. We will argue that the true NRI (which is uniquely defined by the $1/N$ expansion) belongs to this class and it is the only one in which the CS equation is satisfied and the Green functions are finite at the ultraviolet stable fixed point.

As discussed in Symanzik's lectures the introduction of counterterms corresponds to the use of the Lagrangian :

$$\mathcal{L}_E = \mathcal{L}_0 + \Sigma\, F_i(g,m,D) O_i \tag{7.1}$$

where $\mathcal{L}_0$ is the Lagrangian without counterterms for NRI the sum runs over all possible local operators O_i, the functions F have singularities at rational dimensions which cancel out with those coming from the perturbative expansion of $\mathcal{L}_0$.

Improving the analysis of section V on the structure of singular terms in the framework of the $1/N$ expansion, we find that the irregular terms $h(D)$ can be generated by a suitable form of the counterterms. Indeed the singularities of the Mellin transform of the Green functions arise either from the superficial divergence of a diagram, or from the divergence of a subdiagram in the large momenta region ; the divergent term is a polynomial in the external momenta and corresponds to the insertion of a local operator. The final result is :

$$F_i(g,m,D) = g^{-d_i/(D-4)} f_i(gm^{(D-4)}, m^2 g^{2/(D-4)}, D) \quad ,$$

where d_i is the mass dimension of the operator O_i and for irrational dimension the functions f_i are C^∞ in both variables ; they can explicitly be computed in the $1/N$ expansion.

Let us see now how we can fix the functions f_i without using the $1/N$ expansion. If extra counterterms are introduced, we can compute in perturbation theory the Green function of the renormalized Φ and Φ^2 fields. A CS equation can be written for these Green functions : it will be satisfied in the correct theory but it will not be satisfied for an arbitrary choice of the counterterms. At a finite order (k) in g only a finite number (ℓ) of counterterms is needed. The finite part of the ℓ counterterms can be reconstructed from the knowledge of the value of ℓ different Green functions (usually computed at zero external momenta). Let us concentrate our attention on these ℓ Green functions. In the conventional approach their values are arbitrary and can be treated as ℓ independent parameters ; however they can be computed using the integrated form of the CS equation (6.2). As far as we are not using the exact form of the Green functions, but only an approximated one, we cannot pretend that eq.(6.2) is exactly satisfied ; a reasonable requirement is that the discrepance among the l.h.s. and the r.h.s. must be of order g^{k+1}. It easy to check that this will not happen for an arbitrary choice of the counterterms, imposing such a requirement we find a set of

ℓ coupled non linear equation, whose solution fix the value of all the ℓ Green functions and consequently of the counterterms. The simplest case is realized when no counterterms are needed to define $\Delta\Gamma$ and the counterterms appear only in Γ : eq. 6.4 gives the correct values of the counterms and there is no system of non linear equations to be solved. In the general case also the function $\Delta\Gamma$ will depend on the counterterms.

Unfortunately this technique to find the counterterms is not stable respect to the increase in the order in perturbation theory, in the sense that the value we obtain for a fixed counterterm depends on the order k of perturbation and it is not at all clear what happens when k goes to infinity. (We are also unable to prove the existence of a solution for large k). We think however that these results are interesting because the counterterms are computed using a procedure which remain internal to the theory : it does not involve the introduction of any specific cutoff (we can use dimensional regularization) ; moreover asymptotically ASI is satisfied at any stage : we pick that particular solution of the CS equation (eq. 6.2) which remains finite at the ultraviolet stable fixed point.

VIII. - The Callan Symanzik Equation as a Substitute of the Equations of Motion

It is well known that the equations of motion for the field Φ (let us restrict to the case of a scalar interaction) are equivalent to an infinite system of coupled integral non-linear equations for all the connected Green functions. Performing formal manipulations one obtains a system of a finite number of equations for only the low degree Green functions (27) (the so-called Dyson equations). The standard perturbative expansion can be generated by the iterative solution of these equations. Although there are solutions of the Dyson equations which are asymptotically scale invariant, (28) (29), ASI (asymptotic scale invariance) is not automatically implemented : if we try to solve the Dyson equations by iteration and we start from a zero order approximation which violates ASI, the amount of violations of ASI increases after each iteration up to the point where ultraviolet divergences appear. This is the origin of the divergence present in the standard treatment of NRI interactions ; it also explains the lack of convergence of the perturbative expansion in the large momenta region for a renormalizable interaction.

Longtime ago it was noticed that in the case of SRI the CS equation can be used as a substitute of the equations of motion : (24), (30), $\Delta\Gamma_n$ can be written as an integral over the first n+2 Γ_n functions ; if both sides of eq. 6.1 are expanded in powers of g , we obtain the k-order of perturbative theory for Γ_n as function of the first (k-1)-orders of perturbation theory for $\Gamma_i(i=2,N+2)$. If we write the CS equation at the infrared stable fixed point $g_c(\beta(g_c)=0$, i.e. the bare coupling constant is infinite), we obtain a system of integral equations for the Green functions (6) which is the equivalent of Wilson's fixed point condition for the Hamiltonian (4).

In the same spirit we can consider eq. 6.2 as an infinite set of non-linear equations for the Γ_n functions ; the requirement that in the limit $g\to 0$ we recover the first non zero order of perturbative theory must be imposed as boundary condition. This boundary condition is automatically satisfied if we solve eq. 6.2 by iterations using as a zero order approximation the first order of perturbation theory. This approach becomes particulary interesting when the interaction is no more superrenormalizable, because ASI is built in the formalism. If we apply the same procedure to NRI we discover that no ultraviolet divergence is present and that terms no C^{∞} in the coupling constant appear in the Green functions (eq. 6.4).

Although also in the case of SRI it is not known if the solution of this kind of infinite coupled non linear equations can be found by iteration (i.e. if the iterative solution converges)we think that the methods described in this section maybe used to perform approximated computations of the Green functions and, perhaps, to obtain non formal results on non-renormalizable and renormalizable interactions (the case of asymptotically free interactions seems to be the most promising).

In this discussion we have been cavalier with two major points; the definition of the renormalized Φ^2 field (Φ_R^2) and the existence of a zero of the $\beta(g)$ functions.

The first problem is due to the divergence of the Green functions of the renormalized field (Φ_R) at two coinciding points, the remedy is well known and consistes in the introduction of the field $\Phi_R^2 = Z_2(\Phi_R)^2$ (Z_2 is an infinite constant). Divergence free equations can written (23), (24), at the price of introducing the whole machinery of Beth-Salpeter kernels, two particle irreducible Green functions... However this is only a technical complication. The second problem may give us serious troubles : eq. 6.2 makes sense only if g_u exists. If the interaction is asymptotically free, $g_u=0$, however if the interaction is not asymptotically free or it is not renormalizable, g_u is the first zero of the $\beta(g)$ function. If we start from a zero order approximation in which the $\beta(g)$ does not have a zero, eq. 6.2 cannot be written for absence of a candidate for g_u . If the $\beta(g)$ does not have a zero at the one loop level, it is not clear which should be the starting points of our iterative procedure (maybe the methods described in the previous section can be useful in this case). This difficulty is much more serious and goes to the heart of the construction of NRI. In this framework there is practically no difference between renormalizable non asymptotically free and non-renormalizable interactions:

if we would be able to construct in this way a $\lambda\Phi^4$ interaction in 4 dimensions with the good sign of the coupling constant, we should have no serious problems to construct the nearby NRI in $4+\epsilon$ dimensions). We note en passant that the conformal invariant self consistency conditions for the propagator and the vertex (29) do not show any pathology (31) when we go from a renormalizable to a non-renormalizable interaction provided that a non-trivial solution exists in the renormalizable case : the values of the coupling constant and of the anomalous dimensions seem to be regular functions of the dimension of the space (5).

IX.- Conclusions

We hope to have convinced the hypothetical reader of all the previous sections that the concept of a finite non-renormalizable interaction is not contradictory and that using an appropriate perturbative expansion accurate results maybe obtained. We consider quite gratifying the validity of asymptotic scale invariance. We stress that our results are valid only for a limited class of interactions and there are many interesting cases which have not been the object of systematic investigations (e.g. a non abelian current-current interaction and the Einstein theory of gravity), however we think that we have settled a general scheme in which the properties of a particular non-renormalizable interaction can be investigated.

References

1) P.J. Redmond and J.L. Uretski, Phys. Rev. Letters 1, 145 (1958), Ann. of Phys. 9, 106 (1960).

2) T.D. Lee, Phys. Rev. 128, 899 (1962).

3) G. Feinberg and A. Pais, Phys. Rev. 132, 2724 (1963).

4) K. Wilson and J. Kogut, Phys. Reports 12, 75 (1974).

5) G. Parisi, "Some considerations on nonrenormalisable interactions", on the Proceedings of the "Colloquium on Lagrangian Field Theory", Marseille 1974.

6) L.P. Kadanoff, et al. Rev. Mod. Phys. 39, 395 (1967).

7) G. Parisi, Lectures given at the Cargese Summer School, July 1973, Columbia University, Preprint.

8) E. Brezin and J. Zinn-Justin, Phys. Rev. Letters 36, 639 (1976) and Phys. Rev. B (October 1976).

9) K. Symanzik, Comm. Math. Phys. 18, 227 (1970).

10) E. Brezin and D.J. Wallace, Phys. Rev. B7, 1967 (1973).

11) G.A. Baker, B.G. Nickel, M.S. Green and D.I. Meiron, Phys. Rev. Letters 36, 1351 (1976).

12) G. Parisi (in preparation).

13) A. Jaffe, Comm. Math. Phys. 1, 127 (1965).

14) J.T. Eckan, J. Magnen and R. Sénéor, Comm. Math. Phys. 39, 251 (1975).

15) S. Graffi, V. Grecchi and B. Simon, Phys. Letters 32B, 631 (1970).

16) G. Parisi, Nucl. Phys. B100, 368 (1975).

17) K. Wilson, Phys. Rev. D4, 2911 (1973).

18) G. Mack and L. Todorov, Phys. Rev. D8, 1764 (1973).

19) V. de Alfaro and E. Predazzi, Nuovo Cimento 39, 235 (1965).

20) G. Parisi, Nuovo Cimento Lettere 6, 450 (1973).

21) K. Symanzik, Nuovo Cimento Lett. 8, 771 (1973), Cargese Lectures 1973, DESY preprint 73158 .

22) K. Symanzik, Comm. Math. Phys. 45, 79 (1975).

23) J. Callan, Phys. Rev. D1, 1541 (1971).

24) K. Symanzik, Comm. Math. Phys. 23, 61 (1971).

25) G. Parisi, Nuovo Cimento 21A, 179 (1974).

26) K. Symanzik, Nuovo Cimento Lettere 6, 420 (1973).

27) J.D. Bjorken and S.D. Drell, Relativistic Quantum Field, Theory (New York, McGraw-Hill, 1965).

28) A.M. Poliakov, Soviet Physics JEPT 28, 533 (1969).

29) G. Parisi and L. Peliti, Nuovo Cimento Lettere 2, 623 (1971).

30) G. Parisi, Phys. Letters 39B, 643 (1972).

31) K. Symanzik, Nuovo Cimento Lettere 3, 734 (1972).

THE CLASSICAL SINE GORDON THEORY

K. POHLMEYER

INSTITUT FÜR THEORETISCHE PHYSIK

DER UNIVERSITÄT HEIDELBERG

I. INTRODUCTION

The organizing committee of this school had invited Ludwig Fadeev as one of the lecturers for the tutorials. Unfortunately, Fadeev was prevented from coming. Thus I was asked to step into his place. I do not know the programme which Fadeev had planned to follow. Since in the seminars by M. Lüscher, A. Luther, and A. Neveu the quantum sine Gordon theory (or, equivalently, the massive Thirring model) will play a central rôle, I finally decided to give an introduction to the classical sine Gordon theory. The classical theory instructs us about the elementary excitations of which the system is capable. Moreover, it provides a variety of explicit analytic results used in the WKB approximation for the mass spectrum of the corresponding quantum theory.

The sine Gordon theory is a theory of a single scalar field $\phi(t,x)$ in one time and one space dimension. Its dynamics is determined by the local Lagrangian density

$$\mathcal{L}=\mathcal{L}(t,x)=\frac{1}{2}\phi_t(t,x)^2-\frac{1}{2}\phi_x(t,x)^2-\frac{m^4}{\lambda}\left[1-\cos\left(\frac{\sqrt{\lambda}}{m}\phi(t,x)\right)\right] \quad (1.1)$$

Here the subscripts t and x stand for partial derivatives with respect to t and x. The velocity of light as well as an action quantity, say $\hbar$, have been set equal to one. The remaining parameters m and λ play the rôle of a mass and a coupling constant.

We pass to the new dimensionless space, time and

field coordinates

$$x' = mx\ ,\ t' = mt\ ,\ \check{u}(t',x') = \frac{\sqrt{\lambda}}{m}\,\phi(t,x)$$

in terms of which the Lagrangian density reads

$$\mathcal{L} = \mathcal{L}'(t',x') = \frac{1}{2\gamma}\left[\check{u}_{t'}(t',x')^2 - \check{u}_{x'}(t',x')^2 + 2(\cos\check{u}(t',x') - 1)\right]$$

where

$$\gamma = \frac{\lambda}{m^4}\ .$$

Once and forever we treat the theory in these new coordinates and drop primes

$$\mathcal{L} = \frac{1}{2\gamma}\left[\check{u}_t^2 - \check{u}_x^2 + 2(\cos\check{u} - 1)\right]. \tag{1.2}$$

The corresponding equation of motion - the sine Gordon equation - is

$$\Box\check{u} + \sin\check{u} = 0 \tag{1.3}$$

where the symbol $\Box$ denotes the d'Alembertian

$$\Box = \partial_t^2 - \partial_x^2 \tag{1.4}$$

This equation is invariant under Lorentz transformations L

$$L_\zeta : (t,x) \longrightarrow (\alpha_+ t + \alpha_- x, \alpha_- t + \alpha_+ x) \quad \alpha_{\pm} = \frac{\zeta \pm \zeta^{-1}}{2}\ ,\ \zeta \in \mathbb{R}^1 \setminus \{0\}$$

Under L_ζ, a solution $\check{u}(t,x)$ transforms into another solution $\check{u}^{(\zeta)}(t,x) = \check{u}(\alpha_+ t - \alpha_- x, -\alpha_- t + \alpha_+ x)$.
The solutions of the sine Gordon equation with minimal energy, the vacuum solutions, are constant in space and time. The values of these constants coincide with the positions of the minima of $(1 - \cos\check{u})$, i.e.

$$\check{u}(t,x) \equiv \check{u}_n = 2\pi n \qquad n = 0, \pm 1, \pm 2, \cdots \tag{1.5}$$

Whereas the Lagrangian (1.2) is invariant under the discrete symmetry operations

$$\check{u} \to -\check{u}\ ,\ \check{u} \to \check{u} + 2\pi j \qquad j = 0, \pm 1, \pm 2, \cdots \tag{1.6}$$

the vacuum solutions are not. Thus, these discrete symmetries are spontaneously broken. The mass of the "ordinary" vacuum excitations is equal to 1.

Before we discuss the structure of the classical sine Gordon theory, let me briefly mention that there are other branches of physics besides relativistic local field theory and elementary particle physics [1], where this non-linear theory plays a major rôle, e.g. Solid State Physics (Frenkel theory of dislocations in crystals

[2]; propagation of Bloch walls in magnetic crystals [3]), Superconductivity (macroscopic theory of Josephson junctions [4]), Optics (propagation of ultra-short optical pulses)[5].

The sine Gordon theory can be realized by first taking the infinite "volume" limit $L \to \infty$, $N/L = a^{-1} = \text{const}$ and then the continuum limit $a \downarrow 0$ of the following mechanical model:
N identical masses are fixed at the ends of N identical, practically massless rigid sticks. These sticks are attached in equal distances (a) to a straight rod of length L such that they point perpendicular to the axis of the rod and run parallel to each other. The rod is assumed to be inflexible, but to possess ideal torsional properties. The whole arrangement is placed into the gravitational field of the earth in such a way that the rod lies in the horizontal plane.

In mathematics, more precisely in differential geometry, the sine Gordon equation (1.3) was studied in connection with the two-dimensional Riemannian surfaces of constant negative curvature [6]. The fact that the differential geometers contented themselves with particular solutions may be taken as an indication that the, say, Cauchy initial value problem of the sine Gordon equation cannot be solved in a closed form [7].

In 1973 Ablowitz, Kaup, Newell and Segur "solved" the characteristic initial value problem [8]. In 1974 Fadeev and Takhtadzhyan handled the Cauchy initial value problem. They showed that the sine Gordon theory defines a completely integrable Hamiltonian system [9]. The work of either group relies on the inverse scattering method for the "solution" of certain non-linear partial differential equations. This method was developed by Gardner, Green, Kruskal and Miura in 1967 in the context of the Korteweg-de Vries equation, a non-relativistic equation which describes in a certain approximation the propagation of wave fronts in shallow water [10].

The idea of the inverse scattering method will be explained at the end of this section.

Turning back to the sine Gordon equation, we realize that in light-cône variables, i.e. in characteristic coordinates

$$\xi = \frac{t+x}{2} \, , \, \eta = \frac{t-x}{2} \tag{1.7}$$

in which the d'Alembertian factorizes

$$\Box = \frac{\partial^2}{\partial\xi\partial\eta} \tag{1.8}$$

it takes the simple form

$$\check{u}_{\xi\eta} + \sin\check{u} = 0 \, . \tag{1.9}$$

Let us now enumerate some of the remarkable features of the sine Gordon theory.

1) Finite energy solutions must adopt integer multiples of $2\pi : n_+ \cdot 2\pi$ and $n_- \cdot 2\pi$ as their values at $x = +\infty$ and $x = -\infty$ respectively (n_+ not necessarily equal to n_-). An infinitely high energy barrier lies between solutions with different values of n_+ and n_-, in particular with different values of $N = n_+ - n_-$.
There exists a non-trivial topological conserved charge Q which has the meaning of a winding number

$$Q = \int_{-\infty}^{+\infty} dx \, j_0(t,x) \tag{1.10}$$

with

$$j_\mu(t,x) = \frac{1}{2\pi} \epsilon_{\mu\nu} \partial^\nu \check{u}(t,x) \, , \quad \partial^\mu j_\mu(t,x) = 0 \tag{1.11}$$

i.e.

$$Q = \frac{1}{2\pi} \left(\check{u}(t,+\infty) - \check{u}(t,-\infty) \right) = N \tag{1.12}$$

Note that the charge density $j_0(t,x)$ does not contain the momentum canonically conjugate to the field $\check{u}(t,x)$. Hence the Poisson bracket of Q and u vanishes; Q is not the infinitesimal generator of a continuous symmetry group of the theory.

2) The sine Gordon system has non-trivial (indifferent) equilibrium configurations, i.e. the sine Gordon equation possesses stable static solutions of finite energy "solitons"/"antisolitons" with their center of gravity x_0 at rest with respect to the chosen frame of reference

$$\begin{aligned} \check{u}_S(x;0,x_0) &= 4 \operatorname{arctg}(e^{x-x_0}) + 2\pi n_- \\ \check{u}_A(x;0,x_0) &= 4 \operatorname{arctg}(e^{-(x-x_0)}) + 2\pi n_+ \\ n_-, n_+ &= 0, \pm 1, \pm 2, \ldots \end{aligned} \tag{1.13}$$

e.g. for $n_- = 0, x_0 = 0$ $n_+ = 0, x_0 = 0$

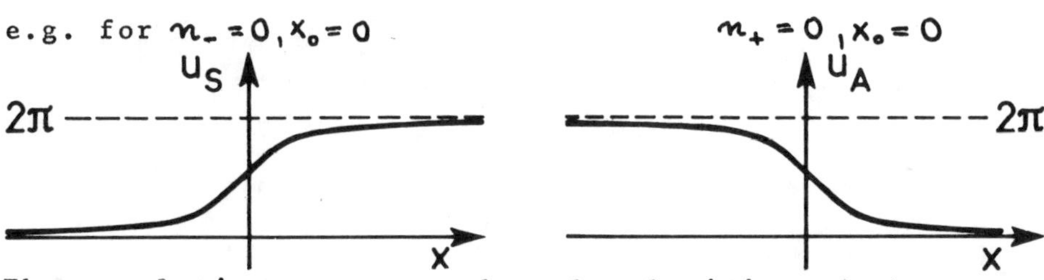

These solutions correspond to local minima of the potential energy in the chosen frame of reference or rather to valleys of the potential energy which stretch in one direction at constant elevation. The energy of the solitons/antisolitons at rest is localized in space and amounts to $M = 8/\gamma$. The charge Q of the soliton is equal +1, that of the antisoliton equal to -1. The soliton (antisoliton) is the lowest energy state in the Q = +1 (Q = -1) sector. If the center of gravity of the soliton (antisoliton) is pushed very far to the left (right), the state of the system locally looks like one of the (physically equivalent) ground states described by $\check{u} = \check{u}_n = 2\pi n$.

From now on we identify all solutions with each other which differ by an additive integer multiple of 2π only.

3) There exist periodic time-dependent solutions (in closed form) of period τ and energy

$$M_\tau = \frac{16}{\gamma} \frac{\sqrt{\tilde{\tau}^2 - 1}}{\tilde{\tau}} \quad , \quad \tilde{\tau} = \frac{\tau}{2\pi} > 1 \tag{1.14}$$

so-called breathers, with their center of gravity x_0 at rest and with "phase" δ at time 0

$$\check{u}_B(t,x;\tau,\delta,0,x_0) = 4 \operatorname{arctg}\left(\sqrt{\tilde{\tau}^2-1} \cdot \frac{\sin(t/\tilde{\tau} + \delta)}{\cosh\left(\frac{\sqrt{\tilde{\tau}^2-1}}{\tilde{\tau}}(x-x_0)\right)}\right). \tag{1.15}$$

These breather solutions can be thought of as bound states of one soliton and one antisoliton with binding energy

$$\Delta E = \frac{16}{\gamma} \frac{1}{\tilde{\tau}(\tilde{\tau} + \sqrt{\tilde{\tau}^2 - 1})} \quad , \quad 0 < \Delta E < \frac{16}{\gamma} . \tag{1.16}$$

Their charge Q is equal to zero.

Lorentz boosts provide solitons, antisolitons and breathers, whose center of gravity - at position x_0 at time 0 - moves with constant velocity v

$$\begin{aligned} \check{u}_S(t,x;v,x_0) &= 4 \operatorname{arctg}\left(e^{\frac{(x-x_0)-vt}{\sqrt{1-v^2}}}\right) \\ \check{u}_A(t,x;v,x_0) &= 4 \operatorname{arctg}\left(e^{-\frac{(x-x_0)-vt}{\sqrt{1-v^2}}}\right) \end{aligned} \tag{1.17}$$

$$u_B(t,x;\tilde{\tau},\delta,v,x_0) = 4\,\mathrm{arctg}\left(\sqrt{\tilde{\tau}^2-1}\,\frac{\sin\left(\frac{t-v(x-x_0)}{\tilde{\tau}\sqrt{1-v^2}}+\delta\right)}{\cosh\left(\frac{\sqrt{\tilde{\tau}^2-1}}{\tilde{\tau}}\frac{(x-x_0)-vt}{\sqrt{1-v^2}}\right)}\right).$$

4) There exist solutions (in closed form) which both in the remote past and distant future can be interpreted in terms of freely moving solitons, antisolitons and breathers. The charge Q being conserved guarantees that the difference of the number of solitons and the number of antisolitons in the remote past and distant future are the same.

Actually, the number of solitons, the number of antisolitons and the number of breathers of a given binding energy are separately conserved; there is no soliton-antisoliton pair creation, there is neither creation nor excitation of breathers.

Moreover, the set of momenta of the incoming solitons (antisolitons, breathers of a given type) is the same as the set of momenta of the outgoing solitons (antisolitons, breathers of a given type): no creation of new momenta! The classical scattering can be described as a sequence of subsequent two-body collisions [11], in which in particular solitons and antisolitons attract and penetrate each other while solitons repel and reflect their likes, which is true for the antisolitons as well. The whole effect of the interaction is to give the solitons, antisolitons and breathers a certain time delay.

As an example consider the time asymptotics of the soliton-antisoliton scattering solution with center of gravity at rest at x = 0, with soliton velocity with respect to the center of gravity -v and with hypothetical positon of the soliton $-v\frac{\Delta}{2}$ at time 0 if the free motion of the soliton at $t=-\infty$ is freely extrapolated to $t=0$:

$$\check{u}_{SA}(t,x;-v,v,-v\tfrac{\Delta}{2},v\tfrac{\Delta}{2}) = 4\,\mathrm{arctg}\left(\frac{\sinh\left(t\frac{v}{\sqrt{1-v^2}}\right)}{v\cosh\left(x\frac{1}{\sqrt{1-v^2}}\right)}\right) \qquad (1.18)$$

$$\Delta = 2\frac{\sqrt{1-v^2}}{v}\ln v < 0 .$$

In the remote past

$$\check{u}_{SA}(t,x;\cdots)\xrightarrow[t\to-\infty]{}\check{u}_A(t,x;v,v\tfrac{\Delta}{2})+\check{u}_S(t,x;-v,-v\tfrac{\Delta}{2}) \qquad (1.19)$$

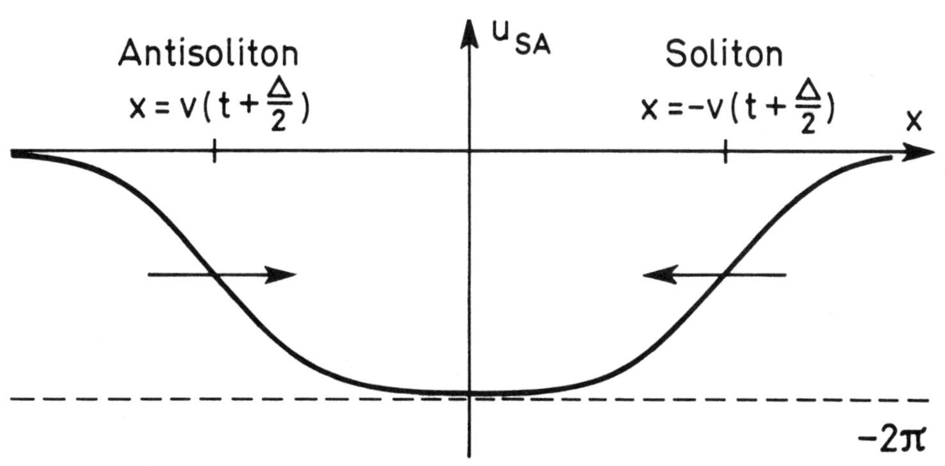

In the distant future

$$\check{u}_{SA}(t,x;\cdots)\xrightarrow[t\to+\infty]{}\check{u}_S(t,x;-v,+v\tfrac{\Delta}{2})+\check{u}_A(t,x;v,-v\tfrac{\Delta}{2})\quad(1.2o)$$

u_{SA}

2π

$x=-v(t-\frac{\Delta}{2})$ $x=v(t-\frac{\Delta}{2})$ x

Soliton Antisoliton

Δ has the meaning of a time "delay".

At the bottom of this extreme stability of the sine Gordon solitons and breathers is a denumerably infinite set of local covariant (explicitly known) conservation laws and their associated conserved charges, e.g.

$$\begin{aligned}&\frac{\partial}{\partial t}\left\{\frac{1}{2}(2\check{u}_{xx}+2\check{u}_{tx}-\sin\check{u})^2-\frac{(\check{u}_t+\check{u}_x)^2}{2}\cos\check{u}-\frac{1}{8}(\check{u}_t+\check{u}_x)^4\right\}\\&-\frac{\partial}{\partial x}\left\{\frac{1}{2}(2\check{u}_{xx}+2\check{u}_{tx}-\sin\check{u})^2+\frac{(\check{u}_t+\check{u}_x)^2}{2}\cos\check{u}-\frac{1}{8}(\check{u}_t+\check{u}_x)^4\right\}\quad(1.21)\\&=0.\end{aligned}$$

4) There exists a non-linear Bäcklund transformation T_1 of the set of finite energy solutions onto itself defined (modulo boundary conditions) by the two ordinary differential equations

$$T_1 \begin{cases} \left(\dfrac{\check{u}(\cdot;1)+\check{u}}{2}\right)_\xi = \sin\left(\dfrac{\check{u}(\cdot;1)-\check{u}}{2}\right) \\ \left(\dfrac{\check{u}(\cdot;1)-\check{u}}{2}\right)_\eta = -\sin\left(\dfrac{\check{u}(\cdot;1)+\check{u}}{2}\right) \end{cases} \tag{1.22}$$

with the consistency requirements

$$\check{u}_{\xi\eta}(\cdot;1) + \sin\check{u}(\cdot;1) = 0 \ , \quad \check{u}_{\xi\eta} + \sin\check{u} = 0 \tag{1.23}$$

The Bäcklund transformation T_1 does not commute with the Lorentz transformations L_ζ. Hence by forming

$$T_\zeta = L_\zeta T_1 L_\zeta^{-1} \begin{cases} \left(\dfrac{\check{u}(\cdot;\zeta)+\check{u}}{2}\right)_\xi = \zeta^{-1}\sin\left(\dfrac{\check{u}(\cdot;\zeta)-\check{u}}{2}\right) \\ \left(\dfrac{\check{u}(\cdot;\zeta)-\check{u}}{2}\right)_\eta = -\zeta\sin\left(\dfrac{\check{u}(\cdot;\zeta)+\check{u}}{2}\right) \end{cases} \tag{1.24}$$

we obtain a whole one-parameter family of Bäcklund transformations.

If we expand the Bäcklund transformed solutions $\check{u}(\cdot;\zeta)$ around $\check{u}$ in an asymptotic series in powers of ζ and insert this expansion into the conservation law

$$\zeta\left\{\cos\left(\frac{\check{u}(\cdot;\zeta)+\check{u}}{2}\right)\right\}_\xi + \zeta^{-1}\left\{\cos\left(\frac{\check{u}(\cdot;\zeta)-\check{u}}{2}\right)\right\}_\eta = 0 \tag{1.25}$$,

collect terms of the same power in ζ and equate their sums separately to zero, we find the expressions for all the conserved currents in terms of the field $\check{u}(t,x)$ and its derivatives. Thus, T_ζ serves as a generating functional for the conserved currents.

Actually, the T_ζs commute and have an inverse

$$T_{\zeta_2} T_{\zeta_1} = T_{\zeta_1} T_{\zeta_2} \tag{1.26}$$

$$T_\zeta^{-1} = T_{-\zeta} \quad \text{for appropriate boundary conditions} \tag{1.27}$$

However, the product $T_{\xi_2} T_{\xi_1}$ is not equal to a Bäcklund transformation T_{ξ_3}. Hence, the family $\{T_\xi / \xi \in \mathbb{R}^1 \setminus \{0\}\}$ does not form a group. Only after extending it to

$$\left\{ \prod_{k=1}^{<\infty} T_{\xi_k} / \xi_k \in \mathbb{R}^1 \setminus \{0\} \quad k = 1, 2, \cdots \right\} \tag{1.28}$$

it does so. It yields an infinite parameter abelian group which can be decomposed into abelian one-parameter subgroups.

In the following lectures extensive use will be made of the inverse scattering method for solving certain non-linear partial differential equations, e.g. the sine Gordon equation. This method reduces the non-linear, say, characteristic initial value problem to a chain of linear problems. It only works if one can find a one-parameter family of linear differential operators, say, $L(\eta)$ into which the characteristic initial data for fixed coordinate η - $\{ \mathring{u}_\xi(\eta, \xi) \ -\infty < \xi < +\infty \}$, $\mathring{u}(\eta, -\infty) = 0$ by virtue of finite energy momentum - enter as potentials such that the spectra of all operators $L(\eta)$ coincide on account of the non-linear differential equation. The "linear" chain consists of the following steps:

1) Determine the potential of the linear differential operator $L(0)$ from the characteristic initial data.

2) Compute the scattering data s(0) of the operator L(0) (direct scattering problem). They consist of the reflection coefficient $r(\varkappa, 0) \ -\infty < \varkappa < +\infty$ and the discrete eigenvalues $\varkappa_j$ together with complex numbers $m_j(0)$ which for the corresponding bound state wave functions determine the ratio of the coefficients in front of the exponential damping factors for $\xi \to +\infty$ and $\xi \to -\infty$

$$s(0) = \{ r(\varkappa, 0) \ -\infty < \varkappa < +\infty \ ; \varkappa_j, m_j(0) \quad j = 1, \cdots, N \} .$$

3) Integrate explicitly the equations for the η-evolution of the scattering data using the finiteness of the energy of the characteristic data and the isospectrality of the operators $L(\eta)$.

4) Construct the potential of the linear differential operator $L(\eta)$ from the scattering data

$$s(\eta) = \{ r(\varkappa, \eta) \ -\infty < \varkappa < +\infty \ ; \varkappa_j, m_j(\eta) \quad j = 1, \cdots, N \}$$

by means of the Gelfand-Levitan-Marchenko equation (inverse scattering problem).

5) Finally, extract the characteristic initial data $u_\xi(\eta,\xi)$ $-\infty < \xi < +\infty$ from the potential of the operator $L(\eta)$

The linear eigenvalue problems

$$L(\eta)\,\psi(\cdot\,;\varkappa,\eta) = \varkappa\,\psi(\cdot\,;\varkappa,\eta) \tag{1.29}$$

need not be self-adjoint. It is, however, essential that the spectrum of the operators $L(\eta)$ is η-independent.

The isospectrality of the operators $L(\eta)$ is equivalent to the existence of an operator $M(\eta)$ - which also involves the characteristic initial data for fixed η - such that

$$\frac{\partial L}{\partial \eta} = [L, M] \tag{1.30}$$

To see this, define the operator M by

$$\frac{d}{d\eta}\psi(\cdot\,;\varkappa,\eta) = -M\psi(\cdot\,;\varkappa,\eta) \tag{1.31}$$

and differentiate the eigenvalue equation once with respect to η .

To derive the isospectral family of linear differential operators for the sine Gordon theory, we start from the first of the two differential equations defining T_ζ :

$$\left(\frac{\check{u}(\cdot\,;\zeta) + \check{u}}{2}\right)_\xi = \zeta^{-1}\sin\left(\frac{\check{u}(\cdot\,;\zeta) - \check{u}}{2}\right)$$

and reduce its transcendental non-linearity to a quadratic one by substituting for $\check{u}(\cdot\,;\zeta)$

$$\begin{gathered} \Gamma = \mathrm{tg}\left(\frac{\check{u}(\cdot\,;\zeta) - \check{u}}{4}\right) \\ \Gamma_\xi + \frac{\check{u}_\xi}{2}(1+\Gamma^2) = \zeta^{-1}\Gamma . \end{gathered} \tag{1.32}$$

These Riccati equations can be linearized by the following ansatz:

$$\Gamma = \frac{\psi_1}{\psi_2} . \tag{1.33}$$

The resulting differential equation is satisfied if ψ_1 and ψ_2 solve the following linear system of first order differential equations

$$i\psi_{1\xi} + i\frac{\check{u}_\xi}{2}\psi_2 = i(2\xi)^{-1}\psi_1$$
$$-i\psi_{2\xi} + i\frac{\check{u}_\xi}{2}\psi_1 = i(2\xi)^{-1}\psi_2 \tag{1.34}$$

This is a linear eigenvalue problem for each value of η

$$L(\eta)\psi = \varkappa\psi \tag{1.35}$$

where

$$\psi = \begin{pmatrix}\psi_1\\ \psi_2\end{pmatrix}$$
$$L(\eta) = i\begin{pmatrix}1 & 0\\ 0 & -1\end{pmatrix}\frac{d}{d\xi} + \frac{i}{2}\check{u}_\xi\begin{pmatrix}0 & 1\\ 1 & 0\end{pmatrix} \tag{1.36}$$

and

$$\varkappa = \frac{i}{2\xi} \ .$$

The η -evolution of ψ can be determined from the second of the two differential equations defining T_ξ . The result is

$$\frac{d\psi(\cdot;\varkappa,\eta)}{d\eta} = -M\psi(\cdot;\varkappa,\eta) \tag{1.31}$$

with

$$M = -\frac{1}{4i\varkappa}\begin{pmatrix}\cos\check{u}, & \sin\check{u}\\ \sin\check{u}, & -\cos\check{u}\end{pmatrix} . \tag{1.37}$$

One can check that the family of linear eigenvalue problems (1.36) with the deformation parameter η is isospectral:

$$\frac{\partial L}{\partial\eta} = [L, M] \ .$$

We want to show that the evolution equation for the scattering data can be integrated explicitly.

The main point is the following: The η-evolution of the scattering data involves only the asymptotic form of M. Since we are interested in solutions with finite energy momentum: $\cos\check{u}(\eta,\xi) \xrightarrow[|\xi|\to\infty]{} 1$, $\sin\check{u}(\eta,\xi) \xrightarrow[|\xi|\to\infty]{} 0$.
Thus, for the solutions of interest the asymptotic form of the operator M is known

$$M \approx -\frac{1}{4i\varkappa}\begin{pmatrix}1 & 0\\ 0 & -1\end{pmatrix} \tag{1.38}$$

although M itself is not known. After all, it involves the unknown function $\tilde{u}(\eta,\xi)$ which we set out to determine. Inserting the asymptotic forms of M and ψ

$$\psi(\xi;\varkappa,\eta) \underset{\xi\to-\infty}{\approx} a(\varkappa,\eta)e^{i\varkappa\xi}\binom{0}{1} + b(\varkappa,\eta)e^{-i\varkappa\xi}\binom{1}{0}$$

$$\underset{\xi\to+\infty}{\approx} e^{i\varkappa\xi}\binom{0}{1} \qquad (1.39)$$

$$\psi(\xi;\varkappa_j,\eta) \underset{\xi\to-\infty}{\approx} b_j(\eta)\, e^{-i\varkappa_j\xi}\binom{1}{0}$$

$$\underset{\xi\to+\infty}{\approx} e^{i\varkappa_j\xi}\binom{0}{1}$$

into the equation (1.31) and taking the normalization for ψ into account, we find the following η-evolution for the scattering data:

$$r(\varkappa,\eta) = \exp\left\{\frac{\eta}{2i\varkappa}\right\} r(\varkappa,0) \qquad -\infty<\varkappa<+\infty$$

$$\varkappa_j(\eta)=\varkappa_j(0)=\varkappa_j\,, \quad m_j(\eta) = \exp\left\{\frac{\eta}{2i\varkappa_j}\right\} m_j(0) \quad j=1,\dots,N \qquad (1.40)$$

Here we have used the expression for r in terms of a and b, and the expression for m_j in terms of b_j and a:

$$r = \frac{b}{a}\,, \quad m_j = \frac{b_j}{i\frac{d}{d\varkappa}a(\varkappa)/_{\varkappa=\varkappa_j}}\,. \qquad (1.41)$$

Multi-soliton, multi-antisoliton scattering solutions correspond to reflectionless potentials: $r(\varkappa,\eta)\equiv 0$

$$a(\varkappa,\eta) = a(\varkappa,0) = \prod_{j=1}^{N}\left(\frac{\varkappa-\varkappa_j}{\varkappa-\varkappa_j^*}\right) \qquad (1.42)$$

$\varkappa_j$ purely imaginary.

The sign of (ib_j) decides whether the eigenvalue $\varkappa_j$ corresponds to a soliton or an antisoliton. The eigenvalue $\varkappa_j$ itself specifies the soliton or antisoliton momentum.

Multi-breather solutions also correspond to reflectionless potentials. Here the zeros of $a(\varkappa,\eta)=a(\varkappa,0)$ lie symmetrically about, but off the imaginary axis.

There is a 1:1 correspondence between multi-soliton, multi-antisoliton, and multi-breather solutions on the one hand and reflectionless potentials on the other hand.

II. THE SINE GORDON THEORY AS A COMPLETELY INTEGRABLE HAMILTONIAN SYSTEM

The remaining lectures shall be used to explain the work by Fadeev and Takhtadzhyan on the integrability of the sine Gordon theory[9]. In the framework of Hamiltonian mechanics the authors have shown that the inverse scattering method provides a canonical transformation to action-angle variables. In these variables the elementary excitations of the system are readily determined and the equations of motion easily integrated.

We take $\check{u}$ and $\frac{1}{\gamma}\check{u}_t$ as the generalized coordinates to start with. The Hamiltonian (energy) of the system in these coordinates is

$$P_0 = \frac{1}{2\gamma}\int_{-\infty}^{+\infty} dx\,\{\check{u}_t^2 + \check{u}_x^2 + 2(1-\cos\check{u})\}, \tag{2.1}$$

the momentum

$$P_1 = -\frac{1}{\gamma}\int_{-\infty}^{+\infty} dx\,\check{u}_x\check{u}_t \tag{2.2}$$

and the symplectic form

$$\Omega = \frac{1}{\gamma}\int_{-\infty}^{+\infty} dx\,\{d\check{u}_t(x)\wedge d\check{u}(x)\}\,. \tag{2.3}$$

Canonical transformations are by definition those transformations which leave the external differential form Ω invariant. In order to establish the integrability of the system we have to find a complete set of constants of motion which are in involution, i.e., whose Poisson brackets vanish. Moreover, we would like to determine the canonical transformation which maps the old variables into those new ones whose generalized momenta are just the above constants of motion or certain functions of them.

Actually, we shall proceed the other way round, namely as follows. In a first section we shall derive an isospectral family of linear differential operators L(t) - with the time t as the deformation parameter - adequate for the solution of the Cauchy initial value problem. In a second section we shall express the symplectic form Ω in terms of the scattering data of L(t). This gives us an idea which combinations of the scattering data could be introduced as new canonical variables Then we deduce so-called trace identities which furnish a complete set of constants of motion in involution, and which allow in particular to express P_0 and P_1 in terms of the new canonical variables. These expressions for P_o and P_1, finally, suggest an interpretation of the field theory in terms of particles.

II.1. The "Solution" of the Cauchy Initial Value Problem

In order to apply the inverse scattering method to the Cauchy initial value problem for the sine Gordon equation we must find an isospectral family of linear differential operators L(t) whose "potentials" are in a one-to-one correspondence to the Cauchy data at time t. In addition we must determine how the Cauchy data enter into the operator M which describes the time evolution of the bound state and continuum wave functions.

We start from equations (1.36) and (1.31) supplemented by equation (1.37). By taking appropriate linear combinations and changing the basis according to

$$\Psi = \frac{1}{\sqrt{2}}\begin{pmatrix} 1, & -i \\ -i, & 1 \end{pmatrix}\psi \tag{2.4}$$

we arrive at the linear eigenvalue problem

$$-i\sigma^2\frac{d}{dx}\Psi + \left\{\frac{i}{4}w\sigma^1 + \frac{B^2}{\lambda}\right\}\Psi = \lambda\Psi \tag{2.5}$$

and the corresponding time evolution equation

$$\frac{\partial}{\partial t}\Psi = \frac{d}{dx}\Psi + 2i\sigma^2\frac{B^2}{\lambda}\Psi\,. \tag{2.6}$$

Here we have introduced the following notation:

$$w = \breve{u}_t + \breve{u}_x$$

$$B = \frac{1}{4}\begin{pmatrix} e^{i\frac{\breve{u}}{2}}, & 0 \\ 0, & e^{-i\frac{\breve{u}}{2}} \end{pmatrix} \tag{2.7}$$

$$\lambda = \frac{\varkappa}{2}$$

σ^i $i = 1,2,3(,4)$ the 3 Pauli (2x2 unit) matrices.

Equation (2.5) is equivalent to the following family of degenerate linear eigenvalue problems

$$L(t)\begin{pmatrix}\Psi \\ \chi\end{pmatrix} = \lambda\begin{pmatrix}\Psi \\ \chi\end{pmatrix}, \tag{2.8}$$

$$L = L(t) = \begin{pmatrix} -i\sigma^2, & 0 \\ 0, & 0 \end{pmatrix}\frac{d}{dx} + \begin{pmatrix} \frac{i}{4}w\sigma^1, & B \\ B, & 0 \end{pmatrix} \tag{2.9}$$

equation (2.6) equivalent to the time evolution

$$\frac{\partial}{\partial t}\begin{pmatrix}\Psi\\ \chi\end{pmatrix} = -M\begin{pmatrix}\Psi\\ \chi\end{pmatrix} \tag{2.10}$$

$$M = M(t) = \begin{pmatrix}-\sigma^4, & 0\\ 0, & \sigma^4\end{pmatrix}\frac{d}{dx} + \begin{pmatrix}0 & ,-2i\sigma^2 B\\ -2iB\sigma^2, & 0\end{pmatrix}. \tag{2.11}$$

It is an easy exercise to verify the relation

$$\frac{\partial}{\partial t}L = [L,M] \tag{2.12}$$

which guarantees the isospectrality of the family of operators L(t).

We consider the eigenvalue problems (2.8) in the Hilbert space of vector functions $\begin{pmatrix}\Psi\\ \chi\end{pmatrix}$ with

$$\int_{-\infty}^{+\infty} dx\,\{\Psi^\dagger(x)\Psi(x) + \chi^\dagger(x)\chi(x)\} < \infty\,. \tag{2.13}$$

Here the dagger denotes the Hermitian adjoint. We impose the following condition on the Cauchy data

$$\int_{-\infty}^{+\infty} dx\,\{|w(x)| + |\check{u}_x(x)| + |\sin(\tfrac{\check{u}(x)}{2})|\} < \infty. \tag{2.14}$$

Still, the operators L(t) are neither self-adjoint nor Hermitian with respect to the scalar product

$$\int_{-\infty}^{+\infty} dx\,\{\Psi^{(2)\dagger}(x)\Psi^{(1)}(x) + \chi^{(2)\dagger}(x)\chi^{(1)}(x)\}\,. \tag{2.15}$$

However, they are Hermitian with respect to the following scalar product

$$\int_{-\infty}^{+\infty} dx\,\{\Psi^{(2)T}(x)\Psi^{(1)}(x) + \chi^{(2)T}(x)\chi^{(1)}(x)\} \tag{2.16}$$

where the letter T denotes transposition. This scalar product leads to an indefinite "norm". For solutions of the eigenvalue equation (2.8) the scalar product (2.16) reads

$$\int_{-\infty}^{+\infty} dx\,\Psi^T(x,\mu)\Big[\sigma^4 + \frac{B^2(x)}{\lambda\mu}\Big]\Psi(x,\lambda) \tag{2.17}$$

Though we shall work in the Hilbert space equipped with the scalar product (2.15), in the orthonormality relations below we shall recognize the scalar product (2.16) (or (2.17)).

The Cauchy initial value problem for the sine Gordon equation consists of determining the Cauchy data at time t: $\{\check{u}(t,x),\, \check{u}_t(t,x) \quad -\infty < x < +\infty\}$ from the Cauchy data at time zero: $\{\breve{u}(0,x), \breve{u}_t(0,x) \; -\infty < x < +\infty\}$. With the help of the inverse scattering method and the isospectral family of associated linear eigenvalue problems (2.8)-(2.12) this non-linear problem can be decomposed into the following chain of linear problems:

1) Determine the "potentials" $\frac{i}{4}w(0,x)\sigma^1$ and $B(0,x)$ of the linear differential operator $L(0)$ from the Cauchy initial values $\{\breve{u}(0,x), \breve{u}_t(0,x) \; -\infty < x < +\infty\}$.

2) Compute the scattering data $s(0)$ of the operator $L(0)$ (direct scattering problem).

3) Integrate the time evolution equations for the scattering data, i.e. determine the function $s = s(t)$.

4) Construct the "potentials" $\frac{i}{4}w(t,x)\sigma^1$ and $B(t,x)$ of the linear differential operator $L(t)$ from the scattering data $s(t)$ (inverse scattering problem).

5) Extract the Cauchy data at time t: $\{\breve{u}(t,x), \breve{u}_t(t,x) \; -\infty < x < +\infty\}$ from the "potentials" of the operator $L(t)$.

Steps 1) and 5) are immediately carried out. Steps 2), 3) and 4) need some explanation.

The scattering data

In this subsection we shall consider the eigenvalue equations (2.8) or equivalently (2.5) separately for each fixed value of t. Thus, for the time being, we suppress all t-dependence in our notation.

We assume that the Cauchy data satisfy condition (2.14). We set

$$\lambda - \frac{1}{16\lambda} = \nu . \qquad (2.18)$$

Note that $\operatorname{Im}\lambda > 0$ implies $\operatorname{Im}\nu > 0$ and vice versa. For real $\lambda \neq 0$ the eigenvalue equation (2.5) has two linearly independent solutions $f_1(x,\lambda)$ and $f_2(x,\lambda)$ normalized at $x = +\infty$:

$$f_1(x,\lambda) \underset{x\to+\infty}{=} e^{i\nu x}\begin{pmatrix}1\\ i\end{pmatrix} + o(1), \; f_2(x,\lambda) \underset{x\to+\infty}{=} e^{-i\nu x}\begin{pmatrix}1\\ -i\end{pmatrix} + o(1) \qquad (2.19)$$

and two linearly independent solutions $g_1(x,\lambda)$ and $g_2(x,\lambda)$ normalized at $x = -\infty$:

$$g_1(x,\lambda) \underset{x\to-\infty}{=} e^{i\nu x}\binom{1}{i} + o(1)\ , \quad g_2(x,\lambda) \underset{x\to-\infty}{=} e^{-i\nu x}\binom{1}{-i} + o(1)\ . \tag{2.20}$$

The two linearly independent solutions of the free equation

$$-i\sigma^2 \frac{d}{dx}\Psi + \frac{\sigma^4}{16\lambda}\Psi = \lambda\Psi \tag{2.21}$$

are
$$e_1(x,\lambda) = e^{i\nu x}\binom{1}{i}$$
and
$$e_2(x,\lambda) = e^{-i\nu x}\binom{1}{-i} = e_1(x,\lambda)^*\ . \tag{2.22}$$

We comprise the respective pairs of solutions to matrix solutions

$$F(x,\lambda) = (f_1(x,\lambda), f_2(x,\lambda)),\ G(x,\lambda) = (g_1(x,\lambda), g_2(x,\lambda)),\ E(x,\lambda) = (e_1(x,\lambda), e_2(x,\lambda)) \tag{2.23}$$

The Wronski determinants coincide with the usual determinants

$$-2i = \det F = \det G = \det E\ . \tag{2.24}$$

The differential equation (2.5) and the respective normalizations can be converted into integral equations of Volterra type

$$F(x,\lambda) = E(x,\lambda) + i\int_x^\infty dx'\sigma^2 e^{-i\nu\sigma^2(x'-x)}\left\{\frac{i}{4}w(x')\sigma^1 + \frac{B^2(x') - \frac{\sigma^4}{16}}{\lambda}\right\}F(x',\lambda) \tag{2.25}$$

$$G(x,\lambda) = E(x,\lambda) - i\int_{-\infty}^x dx'\sigma^2 e^{-i\nu\sigma^2(x'-x)}\left\{\frac{i}{4}w(x')\sigma^1 + \frac{B^2(x') - \frac{\sigma^4}{16}}{\lambda}\right\}G(x',\lambda)\ . \tag{2.26}$$

These integral equations can be solved by the method of successive approximations which always converges. In particular, it follows that $f_1(x,\lambda)$ and $g_2(x,\lambda)$ are holomorphic in the upper half λ-plane and continuous in the closed upper half λ-plane minus the origin $\lambda=0$. For $x>0$ $(x<0)$, $f_1(x,\lambda)$ $(g_2(x,\lambda))$ is continuous in the entire closed upper half λ-plane.
The following bounds hold

$$f_1(x,\lambda)e^{-i\nu x} \underset{\substack{\mathrm{Im}\,\lambda\geq 0\\ |\lambda|\to\infty}}{=} \binom{1}{i} + o(1) \tag{2.27}$$

$$g_2(x,\lambda)e^{i\nu x} \underset{\substack{\mathrm{Im}\,\lambda\geq 0\\ |\lambda|\to\infty}}{=} \binom{1}{-i} + o(1)\ . \tag{2.28}$$

$f_2(x,\lambda)$ and $g_1(x,\lambda)$ have analogous properties in the lower half λ-plane.

The f-solutions can be expressed in terms of the g-solutions and vice versa:

$$F(x,\lambda) = G(x,\lambda)T(\lambda), \quad G(x,\lambda) = F(x,\lambda)T(\lambda)^{-1} \tag{2.29}$$

$T(\lambda)$ is called the transition matrix. The equality of $\det F$ and $\det G$ implies

$$\det T(\lambda) = 1 \quad . \tag{2.30}$$

The symmetries of the integral equations (2.25) and (2.26)

$$-i\sigma^2 F(x,\lambda)^*\sigma^2 = F(x,\lambda) \; , \; i\sigma^1 F(x,-\lambda)^*\sigma^3 = F(x,\lambda) \tag{2.31}$$

$$-i\sigma^2 G(x,\lambda)^*\sigma^2 = G(x,\lambda), \; i\sigma^1 G(x,-\lambda)^*\sigma^3 = G(x,\lambda) \tag{2.32}$$

ensure the unitarity of the transition matrix. Moreover, together with equation (2.30) the symmetries imply the following form for $T(\lambda)$, λ real:

$$T(\lambda) = \begin{pmatrix} a(\lambda), & -b(\lambda)^* \\ b(\lambda), & a(\lambda)^* \end{pmatrix} \tag{2.33}$$

with

$$a(-\lambda)^* = a(\lambda), -b(-\lambda)^* = b(\lambda), |a(\lambda)|^2 + |b(\lambda)|^2 = 1 . \tag{2.34}$$

In terms of the f- and g-solutions the coefficients $a(\lambda)$ and $b(\lambda)$ are given by

$$\begin{aligned} a(\lambda) &= \tfrac{i}{2} \det(f_1(x,\lambda), g_2(x,\lambda)) \\ b(\lambda) &= -\tfrac{i}{2} \det(f_1(x,\lambda), g_1(x,\lambda)) . \end{aligned} \tag{2.35}$$

It follows that $a(\lambda)$ can be analytically continued to a holomorphic function in the upper half λ-plane which is continuous in the closed upper half λ-plane including the point $\lambda = 0$ (see below). At infinity $a(\lambda)$ behaves like

$$a(\lambda) \underset{\substack{\operatorname{Im}\lambda \geq 0 \\ |\lambda| \to \infty}}{=} 1 + o(1) \quad . \tag{2.36}$$

Actually, with $a(\lambda)$ also $a(-\lambda^*)^*$ is a holomorphic (continuous) function in the (closed) upper half λ-plane. On the real axis both functions coincide. Hence they coincide everywhere in the closed upper half plane. From this we infer that $a(\lambda)$ is real on the imaginary axis and that the zeros of $a(\lambda)$: $\lambda = \zeta_j$ lie symmetrically with respect to the imaginary axis.

We shall assume that $a(\lambda)$ does not have zeros on the real axis. This restriction guarantees that there is only a finite number of zeros of $a(\lambda)$ in the upper half λ-plane, which, moreover, are all of finite order. The zeros of $a(\lambda)$ for $\operatorname{Im}\lambda>0$ can only accumulate to points of the continuous spectrum [12]!

In addition, for the sake of transparency, we shall assume that all zeros of $a(\lambda): \lambda=\zeta_j$ $j=1,\cdots,N$ are simple

$$a(\lambda)=\dot{a}(\zeta_j)(\lambda-\zeta_j)+O(|\lambda-\zeta_j|^2)\,,\quad \dot{a}(\zeta_j)\neq 0 \tag{2.37}$$

For $\lambda=\zeta_j$, the solutions f_1 and g_2 become linearly dependent

$$f_1(x,\zeta_j)=b_j\, g_2(x,\zeta_j) \tag{2.38}$$

The solution $f_1(x,\zeta_j)$ decreases exponentially at both ends of the x-axis. Thus the zeros ζ_j of $a(\lambda)$ in the upper half λ-plane correspond to bound states, ζ_j $j=1,\cdots,N$ are the discrete eigenvalues of the operator L.

We enumerate first the zeros on the imaginary axis

$$\zeta_l = i\varkappa_l \qquad l=1,\cdots,n_1 \tag{2.39}$$

and then the zeros off the imaginary axis which come in pairs

$$\zeta_{n_1+2k-1}=\lambda_k\,,\ \zeta_{n_1+2k}=-\lambda_k^* \qquad k=1,\cdots,n_2 \tag{2.40}$$
$$0<\arg\lambda_k<\frac{\pi}{2}\,,\quad n_1+2n_2=N\,.$$

The scattering data of the operator L consist of the reflection coefficient $r(\lambda)$ $-\infty<\lambda<+\infty$

$$r(\lambda)=\frac{b(\lambda)}{a(\lambda)}\,, \tag{2.41}$$

the discrete eigenvalues ζ_j and the quotients

$$m_j=\frac{b_j}{i\dot{a}(\zeta_j)} \qquad j=1,\cdots,N\ : \tag{2.42}$$

$$S=\{r(\lambda)\ -\infty<\lambda<+\infty\,;\ \zeta_j\,,\ m_j\ \ j=1,\cdots,N\}. \tag{2.43}$$

The coefficients $a(\lambda)$ and $b(\lambda)$ can be completely recovered from the scattering data according to

$$a(\lambda)=\lim_{\varepsilon\downarrow 0}\exp\left\{\frac{1}{2\pi i}\int_{-\infty}^{+\infty}d\mu\,\frac{\ln[1+|r(\mu)|^2]}{(\lambda+i\varepsilon)-\mu}\right\}\prod_{j=1}^{N}\left(\frac{(\lambda+i\varepsilon)-\zeta_j}{(\lambda+i\varepsilon)-\zeta_j^*}\right)$$
$$b(\lambda)=r(\lambda)\cdot a(\lambda)\,. \tag{2.44}$$

Next, let me sketch how to solve the direct scattering problem and the inverse scattering problem[13]. As a by-product we shall obtain the continuity of $a(\lambda)$ from the upper half λ-plane at $\lambda = 0$.

The direct scattering problem

Expand the solutions $f_1(x,\lambda)$, ... , $g_2(x,\lambda)$ in terms of the eigenfunctions $e_1(x,\lambda)$, $e_2(x,\lambda)$ of the free eigenvalue equation (2.21)

$$F(x,\lambda) = E(x,\lambda) + \int_x^{\infty} dy\, \mathfrak{A}_1(x,y)\, E(y,\lambda) + \frac{1}{\lambda}\int_x^{\infty} dy\, \mathfrak{A}_2(x,y)\, E(y,\lambda) \quad (2.45)$$

$$G(x,\lambda) = E(x,\lambda) + \int_{-\infty}^{x} dy\, \mathfrak{B}_1(x,y)\, E(y,\lambda) + \frac{1}{\lambda}\int_{-\infty}^{x} dy\, \mathfrak{B}_2(x,y)\, E(y,\lambda). \quad (2.46)$$

Formally, one arrives at the representation for, say, $f_1(x,\lambda)$ in the following way: One inserts the orthonormality relation

$$\delta(\mu-\lambda) = \left(1+\frac{1}{16\lambda\mu}\right)\frac{1}{4\pi}\int_{-\infty}^{+\infty} dy\, e_2^T(y,\mu)\, e_1(y,\lambda) \quad (2.47)$$

into the identity

$$f_1(x,\lambda) - e_1(x,\lambda) = \int_{-\infty}^{+\infty} d\mu \left(f_1(x,\mu) - e_1(x,\mu)\right)\delta(\mu-\lambda)$$

interchanges the order of integrations and converts the analyticity and growth property of $(f_1(x,\lambda) - e_1(x,\lambda))$ in λ into support properties of $\mathfrak{A}_1(x,y)$ and $\mathfrak{A}_2(x,y)$ (Paley-Wiener theorem). The "transformation operators" $\mathfrak{A}_1(x,y)$ and $\mathfrak{A}_2(x,y)$ satisfy a system of coupled integro-differential equations of "Volterra type", the kernels of which are determined by the Cauchy data. The same is true for the pair of transformation operators $\mathfrak{B}_1(x,y)$ and $\mathfrak{B}_2(x,y)$, e.g.

$$0 = \left[\sigma^2, \mathfrak{B}_1(x,y)\right] - \frac{1}{4} w\left(\frac{x+y}{2}\right)\sigma^1 + \int_{\frac{x+y}{2}}^{x} dz \left\{ \frac{w(z)}{4}\left[\sigma^2, \sigma^1 \cdot \right.\right.$$

$$\left.\left. \cdot \mathfrak{B}_1(z, x+y-z)\sigma^2\right] + 16i\left[\sigma^2, \sigma^2\left(B^2(z) - \frac{\sigma^4}{16}\right)\mathfrak{B}_2(z, x+y-z)\right]\right\}. \quad (2.48)$$

$\mathfrak{B}_1(x,x)$ and $\mathfrak{B}_2(x,x)$ are related to the potentials $\frac{1}{4} w(x)\sigma^1$ and $B^2(x)$ through the equations

$$0 = \left[\sigma^2, \mathfrak{B}_1(x,x)\right] - \frac{1}{4} w(x)\, \sigma^1 \quad (2.49)$$

and

$$-i\sigma^2 \mathfrak{B}_2(x,x) + 16\, i\, B^2(x)\, \mathfrak{B}_2(x,x)\, \sigma^2 + B^2(x) - \frac{\sigma^4}{16} = 0 \, . \quad (2.50)$$

The integro-differential equations in question are derived by inserting the above representations for $F(x,\lambda)$ and $G(x,\lambda)$ into the integral equations (2.25) and (2.26). They can be solved by the method of successive approximations which always converges for potentials satisfying condition (2.14). It can be shown that the behaviour of $F(x,\lambda)$ and $G(x,\lambda)$ in the appropriate half planes near the point $\lambda = 0$ is controlled by the behaviour of $E(x,\lambda)$ (compare the representations (2.45) and (2.46)). Hence $a(\lambda) = \frac{i}{2} \det (f_1(x,\lambda), g_2(x,\lambda))$ varies continuously as the point $\lambda = 0$ is approached from the upper half λ-plane.

The expressions for the coefficients $a(\lambda)$ and $b(\lambda)$ in terms of the transformation operators $\mathcal{L}_1$ and $\mathcal{L}_2$ are:

$$a(\lambda) = 1 + \frac{i}{16\lambda} \int_{-\infty}^{+\infty} dx \left[\cos \check{u}(x) - 1\right] + i \int_{-\infty}^{0} dy\, e^{-2i\lambda y} (1, i) \Big[\int_{-\infty}^{+\infty} dx\, V_\lambda(x) \cdot \\ \cdot \left(\mathcal{L}_1(x, x+2y) + \frac{1}{\lambda} \mathcal{L}_2(x, x+2y) \right) \Big] \binom{1}{-i} \qquad (2.51)$$

$$b(\lambda)^* = -\frac{i}{2} \int_{-\infty}^{+\infty} dy\, e^{-2i\lambda y} (1, -i) \left\{ V_\lambda(y) + 2 \int_{y}^{\infty} dx\, V_\lambda(x) \left(\mathcal{L}_1(x, 2y-x) + \frac{1}{\lambda} \mathcal{L}_2(x, 2y-x) \right) \right\} \binom{1}{-i}$$

where

$$V_\lambda(x) = \frac{i}{4} w(x) \sigma^1 + \frac{B^2(x) - \frac{\sigma^4}{16}}{\lambda} \quad . \qquad (2.52)$$

In order to find the initial scattering data

$$s(0) = \left\{ r(\lambda, 0) \quad -\infty < \lambda < +\infty \; ; \; \zeta_j, \; m_j(0) \quad j = 1, \cdots, N \right\}$$

we compute $\mathcal{L}_1(x,y)$ and $\mathcal{L}_2(x,y)$ from the integro-differential equations corresponding to the Cauchy initial data. We then insert the result of the computation into the expressions (2.51) for $a(\lambda)$ and $b(\lambda)$, whence we determine the reflection coefficients $r(\lambda, 0) \;\; -\infty < \lambda < +\infty$, the zeros ζ_j of $a(\lambda)$ in the upper half λ-plane together with the quotients $m_j(0)$.

Time evolution of the scattering data

The function $s = s(t)$ is easily determined (compare the end of the Introduction). The scattering data evolve in time like

$$s(t): r(\lambda, t) = \exp\left\{-2i\left(\lambda + \frac{1}{16\lambda}\right)t\right\} r(\lambda, 0) \quad -\infty < \lambda < +\infty$$

$$\zeta_j(t)=\zeta_j(0)=\zeta_j,\; m_j(t)=\exp\{-2i(\zeta_j+\frac{1}{16\zeta_j})t\}m_j(0)\quad j=1,\dots,N. \tag{2.53}$$

The inverse scattering problem

How do we recover the potentials $\frac{i}{4}w(x)\sigma^1$ and $B(x)$ from the scattering data s?
We write the equations defining the coefficients $a(\lambda)$ and $b(\lambda)$ in the following modified form

$$\left[\frac{1}{a(\lambda)}-1\right]f_1(x,\lambda)=r(\lambda)[g_2(x,\lambda)-e_2(x,\lambda)]+r(\lambda)e_2(x,\lambda)$$
$$+[g_1(x,\lambda)-e_1(x,\lambda)]-[f_1(x,\lambda)-e_1(x,\lambda)] \tag{2.54}$$
$$\left[\frac{1}{a(\lambda)^*}-1\right]f_2(x,\lambda)=\ \cdots$$

multiply the first equation by $e_2^T(y,\lambda)/4\pi$ $(e_2^T(y,\lambda)/4\pi\lambda)$, the second one by $e_1^T(y,\lambda)/4\pi$ $(e_1^T(y,\lambda)/4\pi\lambda)$, integrate both sides of the two equations for $x>y$ over the real λ-axis and evaluate the l.h.s. by the residue theorem in the upper and lower λ-half plane, respectively. Finally, we take the sum, note that for $x>y$ as a consequence of the analyticity and growth properties of $f_1(x,\lambda)$ and $f_2(x,\lambda)$

$$\int_{-\infty}^{+\infty}d\lambda[f_1(x,\lambda)-e_1(x,\lambda)]e_2^T(y,\lambda)/4\pi(\lambda)=0=\int_{-\infty}^{+\infty}d\lambda[f_2(x,\lambda)-e_2(x,\lambda)]e_1^T(y,\lambda)/4\pi(\lambda)$$

and insert the representation (2.46) for $g_j(x,\lambda)$ $j=1,2$. In this way we arrive at the linear Gelfand-Levitan-Marchenko integral equations

$$x>y:\ 0=K_1(x,y)+\mathcal{F}_1(x,y)+\int_{-\infty}^{x}du\,K_1(x,u)\mathcal{F}_1(u,y)+\int_{-\infty}^{x}du\,K_2(x,u)\mathcal{F}_2(u,y)$$

$$0=16K_2(x,y)+\mathcal{F}_2(x,y)+\int_{-\infty}^{x}du\,K_1(x,u)\mathcal{F}_2(u,y)+\int_{-\infty}^{x}du\,K_2(x,u)\mathcal{F}_3(u,y) \tag{2.55}$$

whose kernels $\mathcal{F}_\ell(x,y)$ $\ell=1,2,3$ are determined by the scattering data s through

$$\mathcal{F}_\ell(x,y)=\frac{1}{4\pi}\int_{-\infty}^{+\infty}\frac{d\lambda}{(\lambda)^{\ell-1}}\{r(\lambda)e_2(x,\lambda)e_2^T(y,\lambda)-r(\lambda)^*e_1(x,\lambda)e_1^T(y,\lambda)\} \tag{2.56}$$

$$+\frac{i}{2}\sum_{j=1}^{N}\left\{\frac{m_j}{(\zeta_j)^{\ell-1}}e_2(x,\zeta_j)e_2^T(y,\zeta_j)-\frac{m_j^*}{(\zeta_j^*)^{\ell-1}}e_1(x,\zeta_j^*)e_1^T(y,\zeta_j^*)\right\}$$

Their solutions $K_1(x,y)$ and $K_2(x,y)$ are essentially identical with the transformation operators $\mathcal{L}_1(x,y)$ and $\mathcal{L}_2(x,y)$ whence

$$[\sigma^2, K_1(x,x)] = \frac{1}{4} w(x)\sigma^1 \tag{2.49}$$

$$-i\sigma^2 K_2(x,x) + 16i\, B^2(x)\, K_2(x,x)\sigma^2 + B^2(x) - \frac{\sigma^4}{16} = 0 \tag{2.50}$$

Thus, in order to recover the potentials $\frac{i}{4} w(x)\sigma^1$ and $B(x)$ from the scattering data s, we compute $K_1(x,y)$ and $K_2(x,y)$ from the GLM equations for s and apply the last two relations.

II.2. Transformation to Action Angle Variables

As a first goal we aim at expressing the symplectic form Ω in terms of the scattering data. For that we need a formula which relates the infinitesimal changes of the Cauchy data to the infinitesimal changes of the corresponding scattering data. We start by comparing two scattering problems (not necessarily infinitesimally close to each other) described by the respective linear differential operators $L^{(1)}$ and $L^{(2)}$ or, equivalently, by the respective scattering data $s^{(1)}$ and $s^{(2)}$.

According to ref.[12] the solutions of problem (2): $f_1^{(2)}(x,\lambda), \cdots\cdots\cdots g_2^{(2)}(x,\lambda)$ can be expanded in terms of the solutions of problem (1): $f_1^{(1)}(x,\lambda), \cdots, g_2^{(1)}(x,\lambda)$. Thus, there exist transformation operators $\hat{\mathcal{O}}_1(x,y), \cdots, \hat{\mathcal{L}}_2(x,y)$ such that the following integral representations hold

$$F^{(2)}(x,\lambda) = F^{(1)}(x,\lambda) + \int_x^\infty dy\, \hat{\mathcal{O}}_1(x,y) F^{(1)}(y,\lambda) + \frac{1}{\lambda}\int_x^\infty dy\, \hat{\mathcal{O}}_2(x,y) F^{(1)}(y,\lambda) \tag{2.57}$$

$$G^{(2)}(x,\lambda) = G^{(1)}(x,\lambda) + \int_{-\infty}^x dy\, \hat{\mathcal{L}}_1(x,y) G^{(1)}(y,\lambda) + \frac{1}{\lambda}\int_{-\infty}^x dy\, \hat{\mathcal{L}}_2(x,y)\, G^{(1)}(y,\lambda). \tag{2.58}$$

Formally, the integral representation for, say, $f_1^{(2)}(x,\lambda)$ is obtained by inserting the orthonormality relation

$$\delta(\mu-\lambda) = \frac{1}{4\pi a^{(1)}(\mu)} \int_{-\infty}^{+\infty} dy\, g_2^{(1)T}(y,\mu)\left[\sigma^4 + \frac{[B^{(1)}(y)]^2}{\lambda\mu}\right] f_1^{(1)}(y,\lambda) \tag{2.59}$$

into the identity

$$f_1^{(2)}(x,\lambda) - f_1^{(1)}(x,\lambda) = \int_{-\infty}^{+\infty} d\mu \left(f_1^{(2)}(x,\mu) - f_1^{(1)}(x,\mu)\right)\delta(\mu-\lambda).$$

The properties of the transformation operators $\hat{\mathcal{A}}_1(x,y), \cdots, \hat{\mathcal{B}}_2(x,y)$ - in particular relation of $\hat{\mathcal{B}}_1(x,x)$ and $\hat{\mathcal{B}}_2(x,x)$ to the potentials $\frac{i}{4}w^{(j)}(x)\sigma^1$ and $B^{(j)}(x)$ $j=1,2$ - can be inferred from the integro-differential equations of which $\hat{\mathcal{A}}_1(x,y)$, $\hat{\mathcal{A}}_2(x,y)$ and $\hat{\mathcal{B}}_1(x,y)$, $\hat{\mathcal{B}}_2(x,y)$ are solutions, respectively. In their turn the integro-differential equations are derived by inserting the above representations into the integral equations of Volterra type

$$F^{(2)}(x,\lambda) = F^{(1)}(x,\lambda) - \frac{i}{2}\int_x^{\infty} dx' \Big\{\big[f_1^{(1)}(x,\lambda)f_1^{(1)\dagger}(x',\lambda)+f_2^{(1)}(x,\lambda)f_2^{(1)\dagger}(x',\lambda)\big] \sigma^2\Big[\frac{i}{4}\big(w^{(2)}(x')-w^{(1)}(x')\big)\sigma^1+\frac{[B^{(2)}(x')]^2-[B^{(1)}(x')]^2}{\lambda}\Big]\Big\} F^{(2)}(x',\lambda) \tag{2.60}$$

$$G^{(2)}(x,\lambda) = \quad \cdot \quad \cdot \quad \cdot$$

Next we derive appropriate generalized Gelfand-Levitan-Marchenko equations. For $x>y$, we integrate the following combinations of the equations defining the coefficients $a^{(1)}(\lambda), \cdots, b^{(2)}(\lambda)$ over the entire real λ-axis:

$$\Big\{\Big[\frac{1}{a^{(2)}(\lambda)}-1\Big]f_1^{(2)}(x,\lambda)-\Big[\frac{1}{a^{(1)}(\lambda)}-1\Big]f_1^{(1)}(x,\lambda)\Big\}g_2^{(1)T}(y,\lambda)/4\pi(\lambda) \tag{2.61}$$
$$+\Big\{\Big[\frac{1}{a^{(2)}(\lambda)^*}-1\Big]f_2^{(2)}(x,\lambda)-\Big[\frac{1}{a^{(1)}(\lambda)^*}-1\Big]f_2^{(1)}(x,\lambda)\Big\}g_1^{(1)T}(y,\lambda)/4\pi(\lambda) = \cdot\cdot$$

We evaluate the l.h.s. with the help of the residue theorem and insert the representations for $g_j^{(2)}(x,\lambda)$ in terms of $g_j^{(1)}(y,\lambda)$ $j=1,2$ (compare eq. (2.58)). In this manner we arrive at linear integral equations of the same form as the GLM equations (2.55). The kernels are different, though:

$$\mathcal{F}_\ell(x,y)=\frac{1}{4\pi}\int_{-\infty}^{+\infty}d\lambda\Big\{\frac{r^{(2)}(\lambda)-r^{(1)}(\lambda)}{(\lambda)^{\ell-1}}g_2^{(1)}(x,\lambda)g_2^{(1)T}(y,\lambda)-\frac{r^{(2)}(\lambda)^*-r^{(1)}(\lambda)^*}{(\lambda)^{\ell-1}}g_1^{(1)}(x,\lambda)g_1^{(1)T}(y,\lambda)\Big\}$$
$$+\frac{1}{2}\sum_{j=1}^{N_2}\Big\{\frac{m_j^{(2)}}{(\zeta_j^{(2)})^{\ell-1}}g_2^{(1)}(x,\zeta_j^{(2)})g_2^{(1)T}(y,\zeta_j^{(2)})-\frac{m_j^{(2)*}}{(\zeta_j^{(2)*})^{\ell-1}}g_1^{(1)}(x,\zeta_j^{(2)*})g_1^{(1)T}(y,\zeta_j^{(2)*})\Big\}$$
$$-\frac{1}{2}\sum_{j=1}^{N_1}\Big\{m_j^{(2)}\to m_j^{(1)},\ \zeta_j^{(2)}\to\zeta_j^{(1)}\Big\} \qquad \ell=1,2,3. \tag{2.62}$$

Again, the solutions $K_1(x,y)$ and $K_2(x,y)$ are essentially identical with the transformation operators $\hat{\mathcal{F}}_1(x,y)$ and $\hat{\mathcal{F}}_2(x,y)$ respectively. Consequently,

$$[\sigma^2, K_1(x,x)] = \frac{1}{4}\left(w^{(2)}(x) - w^{(1)}(x)\right)\sigma^1 \tag{2.63}$$

$$-i\sigma^2 K_2(x,x) + i[B^{(2)}(x)]^2 K_2(x,x)[B^{(1)}(x)]^{-2}\sigma^2 + [B^{(2)}(x)]^2 - [B^{(1)}(x)]^2 = 0. \tag{2.64}$$

Specializing to scattering problems which are inifinitesimally close to each other

$$r^{(1)}(\lambda) = r(\lambda) \quad, N_1 = N \quad, \zeta_j^{(1)} = \zeta_j \,, \quad m_j^{(1)} = m_j$$

$$r^{(2)}(\lambda) = r(\lambda) + dr(\lambda), \; N_2 = N \,, \; \zeta_j^{(2)} = \zeta_j + d\zeta_j \,, \; m_j^{(2)} = m_j + dm_j$$

we note that $K_j(x,y)$ and $\mathcal{F}_l(x,y)$ $j=1,2$ $l=1,2,3$ are of first order in inifinitesimal changes of the scattering data. Thus, by keeping only first order terms in the generalized GLM equations, we find

$$x > y: \qquad K_1(x,y) = -\mathcal{F}_1(x,y), \; K_2(x,y) = -\mathcal{F}_2(x,y) \tag{2.65}$$

and further

$$[\sigma^2, \mathcal{F}_1(x,x)] = -\frac{1}{4} dw(x)\sigma^1 \tag{2.66}$$

$$d\tilde{u}(x)\sigma^3 = -\sigma^2 \mathcal{F}_2(x,x)[B(x)]^{-2} + [16 B(x)]^2 \mathcal{F}_2(x,x)\sigma^2 . \tag{2.67}$$

Thereby, we have established the relation of the infinitesimal changes of the Cauchy data and the infinitesimal changes of the corresponding scattering data:

$$d\tilde{u}(x) = \frac{i}{\pi}\int_{-\infty}^{+\infty} d\lambda\, dr(\lambda)\, g(x,\lambda) + 2i\sum_{j=1}^{N}\left\{g(x,\zeta_j)\,dm_j + m_j \frac{\partial}{\partial\lambda} g(x,\lambda)\Big|_{\lambda=\zeta_j} d\zeta_j\right\} \tag{2.68}$$

$$dw(x) = \frac{2i}{\pi}\int_{-\infty}^{+\infty} d\mu\, dr(\mu)\, f(x,\mu) + 4i\sum_{j=1}^{N}\left\{f(x,\zeta_j)\,dm_j + m_j \frac{\partial}{\partial\mu} f(x,\mu)\Big|_{\mu=\zeta_j} d\zeta_j\right\} \tag{2.69}$$

with

$$g(x,\lambda) = \frac{1}{\lambda}\, g_{2,1}(x,\lambda)\, g_{2,2}(x,\lambda) \tag{2.70}$$

and

$$f(x,\mu) = [g_{2,2}(x,\mu)]^2 - [g_{2,1}(x,\mu)]^2$$

$g_{2,j}$ $j = 1, 2$ being the jth component of the solution vector g_2 .

We insert this result into the expression (2.3) for Ω and evaluate the remaining x-integration with the help of the differential equation (2.5). The integrand turns out to be a total x-derivative of products of $g_2(x,\lambda)$ and $g_2(x,\mu)$ components, of $g_2(x,\lambda)$ and $g_2(x,\zeta_j)$ components and the like. Thus the integral immediately can be expressed in terms of the asymptotics of $g_2(x,\lambda)$, i.e. in terms of the scattering data. However, in terms of the differentials $dr(\lambda)$, $d\zeta_j$ and dm_j, Ω does not have the canonical symplectic form. In particular, in Ω the differentials $dr(\lambda)$ and $dr(-\lambda)$ which are not independent occur side by side.

However, Ω does have the canonical symplectic form in the variables

$$\begin{aligned}
&\rho(\lambda) = -\frac{8}{\pi\gamma\lambda}\ln|a(\lambda)|\,, \quad \varphi(\lambda) = -\arg b(\lambda) \quad \lambda > 0 \\
&p_l = \frac{1}{\gamma}\ln \varkappa_l\,, \quad q_l = 8\ln|c_l| \quad l=1,\cdots,n_1;\; c_l = -ib_l = m_l\dot{a}(i\varkappa_l) \\
&\left.\begin{aligned}&\xi_k = \frac{4}{\gamma}\ln|\lambda_k|\,, \quad \eta_k = 4\ln|d_k| \\ &\Theta_k = \arg\lambda_k\,, \quad \varphi_k = -\frac{16}{\gamma}\arg d_k\end{aligned}\right\} k=1,\cdots,n_2;\; d_k = m_k\dot{a}(\lambda_k)
\end{aligned} \tag{2.71}$$

i.e.

$$\Omega = \int_0^\infty d\lambda\, d\rho(\lambda)\wedge d\varphi(\lambda) + \sum_{l=1}^{n_1} dp_l\wedge dq_l + \sum_{k=1}^{n_2}\{d\xi_k\wedge d\eta_k + d\Theta_k\wedge d\varphi_k\}. \tag{2.72}$$

Hence, the above variables are canonical. They arise from the original variables $\check{u}(x)$ and $\frac{1}{\gamma}\check{u}_t(x)$ by a canonical transformation. As we shall see shortly, they are of the type of action angle variables.

Under spatial translations of the system by a distance a:

$$w(x) \to w(x+a)\,, \quad B(x) \to B(x+a) \tag{2.73}$$

the scattering data change according to

$$\begin{aligned}
&r(\lambda) \to \exp\{2i(\lambda - \tfrac{1}{16\lambda})a\}\, r(\lambda) \qquad -\infty < \lambda < +\infty \\
&\zeta_j \to \zeta_j\,, \quad m_j \to \exp\{2i(\zeta_j - \tfrac{1}{16\zeta_j})a\}\, m_j \quad j=1,\cdots,N.
\end{aligned} \tag{2.74}$$

Combining this result with the time evolution for the scattering data (2.53) and the "equations of motion"

$$\frac{\partial}{\partial t}F = \{F, P_0\}\,, \quad \frac{\partial}{\partial x}F = \{F, P_1\} \tag{2.75}$$

we infer that the Poisson brackets $\{\rho(\lambda),P_0\}, \{\rho(\lambda),P_1\} \ldots\, \{\theta_k,P_0\}, \{\theta_k,P_1\}$ all vanish whereas the remaining Poisson brackets have the values:

$$\{\varphi(\lambda),P_0\} = 2\left(\lambda+\frac{1}{16\lambda}\right), \quad \{\varphi(\lambda),P_1\} = -2\left(\lambda-\frac{1}{16\lambda}\right)$$

$$\cdots\cdots\cdots$$

$$\{\varphi_k,P_0\} = \frac{32}{\gamma}\cos\theta_k\left(|\lambda_k|+\frac{1}{16|\lambda_k|}\right) \tag{2.76}$$

$$\{\varphi_k,P_1\} = -\frac{32}{\gamma}\cos\theta_k\left(|\lambda_k|-\frac{1}{16|\lambda_k|}\right)$$

From these relations we obtain the following expressions for P_0 and P_1 in terms of the above generalized coordinates

$$P_{\substack{0\\1}} = P^{(1)}_{\substack{0\\1}} + P^{(2)}_{\substack{0\\1}} + P^{(3)}_{\substack{0\\1}} \tag{2.77}$$

where

$$P^{(1)}_{\substack{0\\1}} = \int_0^\infty d\lambda\, \rho(\lambda)\left(\frac{1}{8\lambda} \pm 2\lambda\right)$$

$$P^{(2)}_{\substack{0\\1}} = \frac{1}{\gamma}\sum_{l=1}^{n_1}\left(\frac{1}{\varkappa_l} \pm 16\varkappa_l\right) \tag{2.78}$$

$$P^{(3)}_{\substack{0\\1}} = \frac{2}{\gamma}\sum_{k=1}^{n_2}\sin\theta_k\left(\frac{1}{|\lambda_k|} \pm 16|\lambda_k|\right).$$

Clearly, P_0 and P_1 are cyclic in the variables $\varphi(\lambda), \ldots, \varphi_k$. Thus, the generalized coordinates formed from the scattering data are indeed of action-angle type.

The expressions (2.78) are special cases of the so-called trace identities.

II.3. Trace Identities

The values of the action variables completely determine the submanifold of phase space on which the motion of the system takes place globally. On the other hand, the action variables are in a one to one correspondence to the time-independent analytic function $a(\lambda)$. In its turn - just because of its analyticity - the function $a(\lambda)$ is fixed by a countable set of numbers, say, the coefficients of its Taylor series around the point $\lambda=+i$. These coefficients would provide a complete set of constants of motion in involution defined for all Cauchy data satisfying condition (2.14).

If we suppose - as we shall do - that $w(x)$ and

$(e^{iu(x)}-1)$ are test functions from the class $\mathcal{S}$, then in the upper half λ-plane the function $\ln a(\lambda)$ possesses an asymptotic expansion in odd inverse powers of λ around $\lambda=\infty$, and an asymptotic expansion in - apart from a constant - odd powers of λ around $\lambda=0$. The coefficients of these asymptotic expansions are independent in view of the essential singularity of $a(\lambda)$ at $\lambda=\infty$ and at $\lambda=0$. Thus for a restricted class of Cauchy data they can serve as a complete set of constants of motion in involution. This set is essentially identical to the system of charges of those infinitely conserved currents mentioned in the Introduction (compare the discussion around equation (1.25) and remember that $\lambda=i/4s$).

With the help of the eigenvalue equation (2.5), these same coefficients can be expressed in terms of the Cauchy data. The resulting relations are called trace identities [14].

To see that under the above mentioned restriction on the Cauchy data the function $a(\lambda)$ can be expanded around $\lambda=\infty$ and $\lambda=0$ as claimed, we note that now the reflection coefficient $r(\lambda)$ belongs to the testfunction class $\mathcal{S}$ (compare ref. [13]), and that the same is true for $r(1/16(-\lambda))$. The latter statement follows from the fact that $r(1/16(-\lambda))$ ist equal to the reflection coefficient for eigenvalue problem (2.5) with potentials $\frac{i}{4}(w(x)-2\tilde{u}_x(x))\sigma^1$ and $\frac{1}{16}B^{-1}(x)$. Hence $r(\lambda)$ is infinite differentiable and vanishes with all derivatives faster than any inverse power of λ at $\lambda=\pm\infty$, and faster than any power of λ at $\lambda=0$. From eq. (2.44) we obtain

$$\ln\left\{\prod_1^N\left(\frac{\lambda-\zeta_j^*}{\lambda-\zeta_j}\right)a(\lambda)\right\}$$

$$\underset{\substack{|\lambda|\to\infty\\ \operatorname{Im}\lambda>0}}{\sim}\sum_0^\infty\frac{\lambda^{-2n-1}}{2\pi i}\int_{-\infty}^{+\infty}d\mu\,\mu^{2n}\ln\left[1+|r(\mu)|^2\right] \tag{2.79}$$

$$\underset{\substack{|\lambda|\to 0\\ \operatorname{Im}\lambda>0}}{\sim}-\sum_0^\infty\frac{\lambda^{2n+1}}{2\pi i}\int_{-\infty}^{+\infty}\frac{d\mu}{\mu^{2n+2}}\ln\left[1+|r(\mu)|^2\right]$$

or

$$\ln a(\lambda)=\sum_0^\infty I_{2n+1}(2i\lambda)^{-2n-1}\qquad |\lambda|\to\infty,\ \operatorname{Im}\lambda>0 \tag{2.80}$$

with

$$I_{2n+1} = \frac{(2i)^{2n+1}}{2\pi i}\int_{-\infty}^{+\infty} d\mu\, \mu^{2n} \ln[1+|r(\mu)|^2] - \frac{(2i)^{2n+1}}{2n+1}\sum_1^N \{\zeta_j^{2n+1} - \zeta_j^{*2n+1}\}$$

and

$$\ln a(\lambda) = \sum_1^N \ln\left(\frac{\zeta_j}{\zeta_j^*}\right) + \sum_0^\infty I_{-2n-1}\lambda^{2n+1} \tag{2.81}$$

with $\qquad |\lambda|\to 0,\ \mathrm{Im}\,\lambda > 0$

$$I_{-2n-1} = -\frac{1}{2\pi i}\int_{-\infty}^{+\infty} \frac{d\mu}{\mu^{2n+2}} \ln[1+|r(\mu)|^2] - \frac{1}{2n+1}\sum_1^N \{\zeta_j^{-2n-1} - \zeta_j^{*-2n-1}\}.$$

Next we shall sketch how to establish the functional dependence of the coefficients I_{2n+1} $n=0,\pm1,\pm2,\cdots$ on the Cauchy data, i.e. the trace identities.

First, with the help of the eigenvalue equation (2.5) we derive a Riccati equation for

$$\omega(x,\lambda) = \frac{f_{1,2}(x,\lambda)}{f_{1,1}(x,\lambda)} - i \tag{2.82}$$

$f_{1,j}$ $j=1,2$ being the jth component of the solution vector f_1. We solve this equation in an asymptotic sense by inserting successively the ansätze

$$\omega(x,\lambda) = \sum_1^\infty \frac{\omega^{(n)}(x)}{(2i\lambda)^n} \text{ for } |\lambda|\to\infty,\quad \omega(x,\lambda) = \sum_0^\infty \omega_{(n)}\lambda^n \text{ for } |\lambda|\to 0,$$

collecting terms of the same power in λ^{-1} (λ) and equating their respective sums to zero. This yields recursion relations for the $\omega^{(n)}$s ($\omega_{(n)}$s) as well as for the $\sigma^{(n)}$s ($\sigma_{(n)}$s), the coefficient functions of the asymptotic expansions for

$$\sigma(x,\lambda) = \left(\lambda - \frac{e^{-iu(x)}}{16\lambda}\right)\omega(x,\lambda) - \left(\frac{i}{16\lambda}e^{-iu(x)} - 1 + \frac{i}{4}w(x)\right)$$

$$= \frac{d}{dx}\ln f_{1,1}(x,\lambda) - i\left(\lambda - \frac{1}{16\lambda}\right) \tag{2.83}$$

$$\simeq \sum_1^\infty \frac{\sigma^{(n)}(x)}{(2i\lambda)^n} \text{ for } |\lambda|\to\infty \tag{2.84}$$

$$\simeq \sum_0^\infty \sigma_{(n)}(x)\lambda^n \text{ for } |\lambda|\to 0 \tag{2.85}$$

For the last equality, again equation (2.5) has been used.

On the other hand, up to asymptotically vanishing

terms (for $|\lambda| \to \infty$ and $|\lambda| \to 0$ respectively), $\ln a(\lambda)$ and $\sigma(x,\lambda)$ are related by

$$\ln a(\lambda) = \int_{-\infty}^{+\infty} dx\, \sigma(x,\lambda) \,. \tag{2.86}$$

When the asymptotic expansions (2.84) and (2.85) are inserted, the trace identities can be read off.

II.4. Interpretation of the Field in Terms of Particles

Equations (2.77) and (2.78) for the energy and the momentum of the system suggest a simple interpretation of the field in terms of particles.

1) For fixed real value of λ, the action variable $\varrho(\lambda)$ takes non-negative values, whereas the angle variable $\varphi(\lambda)$ takes values in the interval $[0,2\pi]$ with the end-points identified, i.e. on a circle. Thus the variables $\varrho(\lambda)$ and $\varphi(\lambda)$ form a pair of canonical variables of the type "particle number" and "phase". Hence we interpret $\varrho(\lambda)$ as the particle density,
   $p(\lambda) = \frac{1}{8\lambda} - 2\lambda$ as the momentum,
   $h(\lambda) = \frac{1}{8\lambda} + 2\lambda$ as the energy, and
   $\sqrt{h^2(\lambda) - p^2(\lambda)} = 1$ as the mass
   of the particles associated with the ordinary excitations. Thus $P_0^{(1)}$ and $P_1^{(1)}$ is the contribution of these particles of mass 1.

2) $P_0^{(2)}$ and $P_1^{(2)}$ are already written in the form of a sum over particles with
   momentum $\frac{1}{\gamma}\left(\frac{1}{\varkappa_\ell} - 16\varkappa_\ell\right)$
   energy $\frac{1}{\gamma}\left(\frac{1}{\varkappa_\ell} + 16\varkappa_\ell\right)$
   mass $M = \sqrt{\left[\frac{1}{\gamma}\left(\frac{1}{\varkappa_\ell}+16\varkappa_\ell\right)\right]^2 - \left[\frac{1}{\gamma}\left(\frac{1}{\varkappa_\ell}-16\varkappa_\ell\right)\right]^2} = \frac{8}{\gamma}$:
   the solitons $(C_\ell > 0)$ and antisolitons $(C_\ell < 0)$.

3) $P_0^{(3)}$ and $P_1^{(3)}$ are sums of contributions from particles with an internal degree of freedom: the corresponding phase spaces are four-dimensional. Their
   momentum is $\frac{2}{\gamma}\sin\Theta_k\left(\frac{1}{|\lambda_k|} - 16|\lambda_k|\right)$,
   energy is $\frac{2}{\gamma}\sin\Theta_k\left(\frac{1}{|\lambda_k|} + 16|\lambda_k|\right)$ and
   mass is $M_k = \frac{16}{\gamma}\sin\Theta_k < 2M, 0 < \Theta_k = \arg\lambda_k < \frac{\pi}{2}$.

   We identify these third particles with the soliton-antisoliton bound states, the breathers.

In this way, we have determined the elementary excitations of the classical sine Gordon system in terms of which the possible trajectories of the system are most conveniently described.

REFERENCES

[1] J.K. Perring and T.H.R. Skyrme: Nucl.Phys.31,550 (1962)

[2] A. Kochendörfer, A. Seeger and H. Donth: Zeits.Phys.134,173 (1964)

[3] U. Enz: Helv.Phys.Acta 37,245 (1964)

[4] B.D. Josephson: Adv.Phys.14,419 (1965)

[5] G.L. Lamb, Jr.: Rev.Mod.Phys.43,99 (1971)

[6] L. Bianchi: Lezioni di Geometria Differentiale, 3rd ed.I, 658

[7] G. Darboux: Lecons Sur La Theorie Générale Des Surfaces III, 432-470

L.P. Eisenhart: Differential Geometry of Curves and Surfaces (Dover, New York, 1960)

[8] M.J. Ablowitz, D.J. Kaup, A.C. Newell and H. Segur: Phys.Rev. Lett.31,125 (1973)

[9] L.A. Takhtadzhyan and L.D. Faddev: Theor. and Math.Phys.21, 1046 (1974)

M.J. Ablowitz, D.J. Kaup, A.C. Newell and H. Segur: Phys.Rev. Lett.30, 1262 (1973)

[10] C.S. Gardner, J. Green, M. Kruskal and R. Miura: Phys.Rev.Lett. 19,1095 (1967)

[11] W.E. Sacharov and A.B. Shabat: JETP 61,118 (1971)

L.A. Takhtadzhyan: JETP 66,476 (1974)

[12] M.A. Naimark: Dokl.Akad.Nauk SSSR,85,41 (1952), 89,213 (1953) and Trudy Mosk.Matem.Obshsh. 3,181 (1954)

B.Ya. Levin: Dokl.Akad.Nauk SSSR 106,187 (1956)

B.S. Pavlov: Topics in Math.Phys. 1, 87 (1967) ed. by.M.Sh. Birman

[13] L.D. Fadeev: Trudy Matem. Inst. im. Steklova 73, 314 (1964)

[14] V.S. Buslaev and L.D. Fadeev: Dokl.Akad. Nauk SSSR 132 (1960)

INTERACTIONS VIA QUADRATIC CONSTRAINTS

K. POHLMEYER

INSTITUT FÜR THEORETISCHE PHYSIK

DER UNIVERSITÄT HEIDELBERG

I. INTRODUCTION

Completely integrable Hamiltonian systems quite naturally lend themselves to semi-classical quantization methods. This is one of the reasons why the pioneers of the semi-classical approach to local quantum field theory have chosen the completely integrable sine Gordon model as the testing ground for their respective formalisms [1] . Certainly, the importance of this model does not end with that. The sine Gordon equation makes its appearance in many different branches of physics. As a Boson quantum field theory in one-time and one-space dimension the sine Gordon model received great attention because of its close relation to the massive Thirring model, which is a Fermion theory [2] .

In this seminar we shall hear about another remarkable relationship, this time on the classical level: The "reduction" of the chiral O_3 (chiral O_n) model involving only scalar fields to the sine Gordon system (its generalizations). The interaction of the chiral model arises solely from a quadratic constraint, namely that the values of the field functions vary on the surface of a fixed sphere. The sine Gordon theory is seen to be but the first member of an infinite sequence of inequivalent relativistically invariant integrable field theories. Of particular interest is the second member since it is the one-space-dimensional version of the non-linear σ-model corresponding to the group $O_4 \approx SU(2) \times SU(2)$ [3] . For the generalizations of the sine Gordon theory corresponding to groups O_n, $n \leq 6$ we are able to set up the isospectral linear eigenvalue problems, which are the key to the inverse scattering method [4]. We determine the time evolution of the scattering data and thereby "solve" the Cauchy initial value problem for the respective equations of motion. In order to "solve" the Cauchy initial value problem for the equations of motion of the corresponding chiral O_n models, we only need to master one more linear differential equation.

For the O_n invariant chiral models we construct a one-parameter family of Bäcklund transformations and an infinite number of integrals of motion. The latter ones are associated with covariant local currents for which the family of Bäcklund transformations serves as a generating functional.

The O_n-invariant classical chiral theories do not possess soliton solutions on account of their scale invariance. This invariance gets broken as we pass to the sine Gordon theory and its generalizations. There exist static solutions of finite energy. In the general case they are nontopological solitons. Their stability rests on dynamical symmetries e.g. shifts of fields by constants

II. NORMALIZATION OF COORDINATES

We start from a classical relativistic theory of a real n-component scalar or pseudoscalar field $\underline{q}(t,x)$

$$\underline{q}(t,x) = (q_1(t,x), \cdots, q_n(t,x))^T$$

subject to the constraint (dimensionless units!)

$$\underline{q}^2(t,x) = \| \underline{q}(t,x) \|^2 = \sum_1^n q_i^2(t,x) = 1 \ .$$

With the help of a Lagrangian multiplier $\lambda(t,x)$ its dynamics is described by the Lagrangian density

$$\mathcal{L}(t,x) = \frac{1}{2} \{ \underline{q}_t(t,x)^2 - \underline{q}_x(t,x)^2 + \lambda(t,x)(\underline{q}^2(t,x) - 1)$$

Indices t and/or x denote differentiation with respect to t and/or x. The surface of constraint, i.e. the surface S_1^{n-1} of the unit sphere in n-dimensional space provides a homogeneous space for the rotation group O_n. In fact, the above Lagrangian is invariant under the action of the internal symmetry group O_n:

$$\underline{q}(t,x) \rightarrow R \, \underline{q}(t,x) \quad , R \in O_n$$

The equations of motion are

$$\underline{q}_{tt} - \underline{q}_{xx} + (\underline{q}_t^2 - \underline{q}_x^2)\underline{q} = 0 \quad , \quad \underline{q}^2 = 1$$

$$\left(\lambda = -(\underline{q}_t^2 - \underline{q}_x^2) \right)$$

We introduce the characteristic coordinates

$$\xi = \frac{t+x}{2} \quad , \quad \eta = \frac{t-x}{2}$$

in which the d'Alembertian $\Box = \frac{\partial^2}{\partial t^2} - \frac{\partial^2}{\partial x^2}$ factorizes:

$$\Box = \frac{\partial^2}{\partial \xi \partial \eta} \quad .$$

Employing the short-hand notation $q = q(\xi,\eta) = \underline{q}(t,x)$,

$$q_\xi = \frac{\partial}{\partial \xi} q(\xi,\eta) \, , \quad q_\eta = \frac{\partial}{\partial \eta} q(\xi,\eta) \, ,$$

$$q_{\xi\xi} = \frac{\partial^2}{\partial \xi^2} q(\xi,\eta) \, , \ldots$$

$$p \cdot q = \sum_1^n p_i q_i$$

the equations of motion read

$$q_{\xi\eta} + (q_\xi \cdot q_\eta) q = 0 \, , \quad q^2 = 1 \, .$$

These equations are forminvariant under general coordinate transformations which map the light cone onto itself, i.e. under the local scale transformations

$$(\xi,\eta) \longrightarrow (\xi',\eta')$$

$$d\xi' = |H(\xi)| d\xi \, , \quad d\eta' = |K(\eta)| d\eta$$

with

$$H(\xi) \neq 0 \neq K(\eta) \, .$$

Note that q_ξ and q_η are orthogonal to q in virtue of $q^2 = 1$, and that $q_{\xi\eta}$ is parallel to q. The sum and difference of the energy and momentum densities of the fields are given by $(1/2) q_\eta^2$ and $(1/2) q_\xi^2$, respectively, and the energy momentum conservation is expressed by the equations

$$\{\tfrac{1}{2} q_\eta^2\}_\xi = 0 \quad , \quad \{\tfrac{1}{2} q_\xi^2\}_\eta = 0 \, .$$

Hence q_η^2 and q_ξ^2 are functions of η and ξ, respectively

$$q_\xi^2 = h^2(\xi) \quad , \quad q_\eta^2 = k^2(\eta) \ .$$

For a Cauchy initial value problem these functions are already determined by the initial data.

If $h(\xi)$ and $k(\eta)$ are different from zero everywhere, we may choose

$$|H(\xi)| = |h(\xi)| \ , \ |H(\eta)| = |h(\eta)| \ .$$

This amounts to rescaling the coordinates such that in the new system of "normalized" coordinates the momentum density vanishes identically and the energy density is equal to 1/2 everywhere. Cauchy initial data with $h(\xi)=0$ for some $\xi=\xi_0$ and/or $k(\eta)=0$ for some $\eta=\eta_0$ are to be approximated by initial data for which $h(\xi)$ and $k(\eta)$ are different from zero everywhere. Without loss of generality we may assume the coordinates as being normalized and omit the identifying primes. Then

$$q^2 \equiv 1 \ , \ q_\xi^2 \equiv 1 \ , \ q_\eta^2 \equiv 1 \ , \ q_\xi \cdot q \equiv 0 \ , \ q_\eta \cdot q \equiv 0$$

$$-1 \leq q_\xi \cdot q_\eta \leq +1 \ .$$

We set

$$(q_\xi \cdot q_\eta) = \cos\alpha \ .$$

III. THE CHIRAL O_3 MODEL AND THE SINE GORDON THEORY

Together with its first ξ- and η-derivatives, the solution vector $q = q(\xi,\eta)$ of the equations of motion

$$q_{\xi\eta} + (q_\xi \cdot q_\eta)\, q = 0 \ , \ q^2 \equiv 1$$

for n = 3 in general spans the entire space $\mathbb{R}^3$. Thus it must be possible for n = 3 to express the second derivatives $q_{\xi\xi}$ and $q_{\eta\eta}$ as linear combinations of q, q_ξ and q_η:

$$q_{\xi\xi} = -q + \alpha_\xi \operatorname{ctg}\alpha \, q_\xi - \alpha_\xi (\sin\alpha)^{-1} q_\eta$$

$$q_{\eta\eta} = -q - \alpha_\eta (\sin\alpha)^{-1} q_\xi + \alpha_\eta \operatorname{ctg}\alpha \, q_\eta \ .$$

Now we can compute the mixed second derivative of

$$\alpha = \arccos(q_\xi \cdot q_\eta) :$$

$$\alpha_{\xi\eta} = -(\sin\alpha)^{-1} \cdot [\alpha_\xi \alpha_\eta \cos\alpha - \cos^2\alpha + (q_{\xi\xi} \cdot q_{\eta\eta})]$$

$$= -\sin\alpha$$

i.e. the angle $\alpha = \sphericalangle(q_\xi, q_\eta)$ satisfies the sine Gordon equation which can be derived from the Lagrangian density

$$L(t,x) = \frac{1}{2\gamma}\{\alpha_t(t,x)^2 - \alpha_x(t,x)^2 + 2(\cos\alpha(t,x) - 1)\}.$$

Conversely, to every solution α of the sine Gordon equation there exists a solution of the equations of motion

$$q_{\xi\eta} + (q_\xi \cdot q_\eta) q = 0 , \quad q^2 \equiv 1$$

with

$$q_\xi^2 \equiv 1, \quad q_\eta^2 \equiv 1 , \quad (q_\xi \cdot q_\eta) = \cos\alpha .$$

It is just the solution of the Cauchy initial value problem for the linear differential equation

$$q_{tt} - q_{xx} + \cos\alpha \, q = 0$$

with initial data

$$q(0,x) = q(x) , \quad q_t(0,x) = \dot{q}(x)$$

satisfying

$$q^2(x) \equiv 1,$$

$$q(x) \cdot \dot{q}(x) \equiv 0 , \quad \dot{q}^2(x) + q_x(x)^2 = 1, \quad \dot{q}(x) \cdot q_x(x) = 0, \quad \dot{q}^2(x) - q_x(x)^2 =$$

$$= \cos\alpha(0,x) , \quad \dot{q}(x) \cdot q_{xx}(x) - q_x(x) \dot{q}_x(x) = -\frac{1}{2}\alpha_t(0,x)\sin\alpha(0,x).$$

The above linear differential equation reproduces the quadratic constraints $q^2 \equiv 1, q_\eta^2 \equiv 1, q_\xi^2 \equiv 1$ once those initial conditions hold for a particular time.

The sine Gordon equation is known to possess a one-parameter family of non-linear Bäcklund transformations

$$T_\zeta , \ \zeta \in \mathbb{R}^1 \setminus \{0\} : \quad \alpha \longrightarrow \alpha(\cdot\,;\zeta)$$

where $\alpha(\cdot\,;\zeta)$ is again a solution of the sine Gordon equation. Up to boundary conditions T_ζ is given by the two ordinary differential equations

$$T_\zeta : \begin{cases} \left\{\frac{\alpha(\cdot;\zeta)+\alpha}{2}\right\}_\xi = \zeta^{-1}\sin\left(\frac{\alpha(\cdot;\zeta)-\alpha}{2}\right) \\ \left\{\frac{\alpha(\cdot;\zeta)-\alpha}{2}\right\}_\eta = -\zeta\sin\left(\frac{\alpha(\cdot;\zeta)+\alpha}{2}\right) \end{cases}$$

ζ being a constant independent of ξ and η [5]. As was explained in the tutorials, this family of Bäcklund transformations provides a generating functional for an infinite number of conservation laws [6].

It is easy to prove that along with a solution

$q = q(\xi,\eta)$ of $q_{\xi\eta} + (q_\xi \cdot q_\eta)q = 0$, $q^2 \equiv 1$ with $q_\xi^2 \equiv 1 \equiv q_\eta^2$, $(q_\xi \cdot q_\eta) = \cos\alpha$, the vector

$$q(\cdot;1) = \cos\left(\frac{\alpha(\cdot;1)}{2}\right)\left(\frac{q_\xi+q_\eta}{2}\right) + \sin\left(\frac{\alpha(\cdot;1)}{2}\right)\left(\frac{q_\xi-q_\eta}{2}\right)$$

is a solution with $q_\xi(\cdot;1)^2 \equiv 1$, $q_\eta(\cdot;1)^2 \equiv 1$ and $(q_\xi(\cdot;1)\cdot q_\eta(\cdot;1)) = \cos\alpha(\cdot;1)$, $\alpha(\cdot;1)$ being the image of α under the Bäcklund transformation T_1. The vectors q and $q(\cdot;1)$ are orthogonal to each other. The requirement that the component $q(\cdot;1)_{\xi\eta}$ in the direction of q vanishes, is equivalent to the conservation law

$$\left\{\cos\left(\frac{\alpha(\cdot;1)+\alpha}{2}\right)\right\}_\xi + \left\{\cos\left(\frac{\alpha(\cdot;1)-\alpha}{2}\right)\right\}_\eta = 0.$$

For general values of the parameter ζ there exists along with the solution vector q also a solution $q(\cdot;\zeta)$ of the equation of motion with

$$q_\xi(\cdot;\zeta)^2 \equiv 1,\ q_\eta(\cdot;\zeta)^2 \equiv 1,\ (q_\xi(\cdot;\zeta)\cdot q_\eta(\cdot;\zeta)) = \cos\alpha(\cdot;\zeta)$$

$\alpha(\cdot;\zeta)$ being the image of α under the Bäcklund transformation T_ζ. However, the geometrical relation between q and $q(\cdot;\zeta)$ is not so simple any more. After the geometrical meaning of the parameter ζ has been clarified in section V, we shall present the resolution of this point in section VI.

IV. GOALS AND STRATEGY

We aim at associating a one-parameter family of isospectral linear eigenvalue equations (first with the η-coordinate, later with the time coordinate as the deformation parameter) to the equations of motion for the invariants of the general chiral group O_n in normalized coordinates. So far, this task is accomplished for $n \leq 6$ only. A less ambitious but attainable goal is the construction of a generating functional for an infinite number of covariant local conserved currents. We shall proceed as follows: We

1) introduce ζ as group parameter for "outer" Lorentz transformations, i.e. transformations which leave the form $p^2 - q^2$ invariant,

2) adjoin one discrete Bäcklund transformation for the chiral fields.

After these first two steps we shall dispose of a one-parameter continuum of Bäcklund transformations T_ζ for the chiral fields which serves as a generating functional for the infinite number of conservation laws. We then

3) pass from the chiral fields to the O_n-invariants,

4) derive a one-parameter family of systems of (n-2) coupled non-linear (at most quadratic) first-order ordinary differential equations,

5) formulate it as a one-parameter family of genuine Riccati equations (if possible)

6) linearize these Riccati equations to obtain the desired isospectral linear eigenvalue equations for characteristic coordinates

7) derive the corresponding isospectral linear eigenvalue equations for time and space coordinates.

V. THE GEOMETRICAL MEANING OF THE PARAMETER

We claim that for every solution q of the equations of motion (in a general system of coordinates)

$$q_{tt} - q_{xx} + (q_t^2 - q_x^2)\, q = 0, \quad q^2 \equiv 1$$

there exists a one-parameter family of solutions $q^{(\zeta)}$, $\zeta \in \mathbb{R}^1 - \{0\}$ with

$$q_t^{(\zeta)\,2} - q_x^{(\zeta)\,2} = q_t^2 - q_x^2 .$$

To prove this, we note that $q_t^{(\varsigma)}$ and $q_x^{(\varsigma)}$ are related to q_t and q_x by an outer Lorentz transformation.

We make the ansatz that they are related by the following special transformation R_ς

$$q_\xi^{(\varsigma)} = \varsigma^{-1} \mathcal{R}^{(\varsigma)} q_\xi, \quad q_\eta^{(\varsigma)} = \varsigma \mathcal{R}^{(\varsigma)} q_\eta$$

where $\mathcal{R}^{(\varsigma)} = \mathcal{R}^{(\varsigma)}(\xi,\eta;q)$ is an orthogonal-matrix valued function, and that in addition $q^{(\varsigma)}$ and q are related by

$$q^{(\varsigma)} = \mathcal{R}^{(\varsigma)} q$$

If this ansatz is consistent, the parameter ς may be interpreted as describing expansions or contractions of the respective sum and difference of the energy and momentum densities of the chiral field vector q :

$$\tfrac{1}{2} q_\eta^2 \longrightarrow \tfrac{1}{2} q_\eta^{(\varsigma)2} = \varsigma^2 \left(\tfrac{1}{2} q_\eta^2\right)$$

$$\tfrac{1}{2} q_\xi^2 \longrightarrow \tfrac{1}{2} q_\xi^{(\varsigma)2} = \varsigma^{-2} \left(\tfrac{1}{2} q_\xi^2\right)$$

without at the same time changing the angle between q_ξ and q_η .

The existence of such coordinate-dependent rotation matrices $\mathcal{R}^{(\varsigma)}$ follows from the consistency of the compatibility equations

$$\mathcal{R}^{(\varsigma)}_\xi = (1-\varsigma^{-1})\, \mathcal{R}^{(\varsigma)} M_{(+)}$$

$$\mathcal{R}^{(\varsigma)}_\eta = (1-\varsigma)\, \mathcal{R}^{(\varsigma)} M_{(-)}$$

$$\mathcal{R}^{(\varsigma)} \cdot \mathcal{R}^{(\varsigma)T} = \mathcal{R}^{(\varsigma)T} \cdot \mathcal{R}^{(\varsigma)} = \mathbb{1}$$

where

$$M_{(\pm)} = \left(M_{(\pm)}(q, q_\xi(, q_\eta))_{k\ell}\right) = q \otimes q^T_{\xi(\eta)} - q_{\xi(\eta)} \otimes q^T$$

and where the symbol T denotes transposition.

$M_{(\pm)k\ell}$ are the sum and difference, respectively, of the zero and one components of the current densities for the chiral field vector q corresponding to a rotation in the (k,ℓ)-plane.

The rotation matrices $\mathcal{R}^{(\xi)}$ are given in the form of "time"-ordered exponentials by

$$T\exp\left\{\int_0^1 d\tau\left[\frac{d\xi}{d\tau}(1-\xi^{-1})M_{(+)}+\frac{d\eta}{d\tau}(1-\xi)M_{(-)}\right]_{\substack{\xi=\xi(\tau)\\ \eta=\eta(\tau)}}\right\}\mathcal{R}^{(\xi)}(\xi_0,\eta_0)$$

which for fixed end points do not depend on the particular differentiable curve $\xi=\xi(\tau)$, $\eta=\eta(\tau)$: $\xi(0)=\xi_0$, $\eta(0)=\eta_0$, $\xi(1)=\xi$, $\eta(1)=\eta$. $\mathcal{R}^{(\xi)}(\xi_0,\eta_0)$ is a constant orthogonal matrix. Moreover, the following transitivity equation holds

$$\mathcal{R}^{(\xi_2\xi_1)}(\cdot\;;q)=\mathcal{R}^{(\xi_2)}(\cdot\;;\mathcal{R}^{(\xi_1)}(\cdot\;;q)q)\cdot$$
$$\cdot\,\mathcal{R}^{(\xi_1)}(\cdot\;;q)\;.$$

VI. BÄCKLUND TRANSFORMATIONS AND CONSERVATION LAWS

For a general system of coordinates there exists a non-linear transformation B_+ which maps the solutions q of the equations of motion for the chiral fields to new solutions q' of the same equations. As opposed to the transformations R_ξ previously considered, B_+ changes the angle between the vectors q_ξ and q_η and leaves their lengths unchanged:

$$(q'_\xi\cdot q'_\eta)\neq(q_\xi\cdot q_\eta)$$
$$q'^2_\xi=q^2_\xi\;,\quad q'^2_\eta=q^2_\eta\;.$$

B_+ is defined up to some coordinate-independent rotations by the four compatible relations

$$(q'+q)_\xi\;\propto\;(q'-q)$$
$$(q'-q)_\eta\;\propto\;(q'+q)$$
$$q'^2\equiv 1\quad,\quad(q'\cdot q)\equiv 0\;.$$

This transformation corresponds to the Bäcklund transformation $\alpha\to\alpha(\cdot;1)$ of the sine Gordon theory. Along with B_+ goes the conservation law:

$$(q'\cdot q_\xi)_\eta+(q'\cdot q_\eta)_\xi=0\;.$$

We obtain a one-parameter family of Bäcklund transformations T_ξ for the chiral fields – to be compared to the maps $\alpha \to \alpha(\cdot;\zeta)$ of the sine Gordon theory - by noting that the transformations B_+ and R_ζ do not commute, and by combining B_+ and R_ζ to form

$$T_{\zeta+} = R_\zeta^{-1} B_+ R_\zeta \qquad \zeta \in \mathbb{R}^1 \setminus \{0\} :$$

$$q \longrightarrow q(\cdot\,;\zeta+) = ((q^{(\zeta)})')^{(1/\zeta)}$$

or schematically

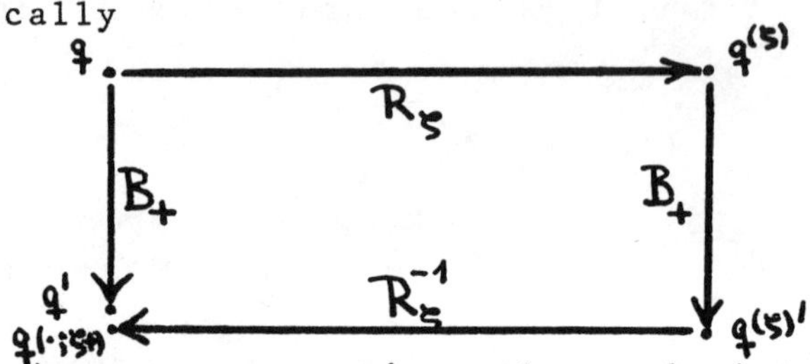

The diagram is not commutative, the angle between $q_\xi(\cdot;\zeta+)$ and $q_\eta(\cdot;\zeta+)$ depends on ζ.
Along with $T_{\zeta+}$ goes the conservation law

$$\left(q^{(\zeta)'} \cdot q_\xi^{(\zeta)}\right)_\eta + \left(q^{(\zeta)'} \cdot q_\eta^{(\zeta)}\right)_\xi =$$

$$= \zeta^{-1}\left(q(\cdot;\zeta+)\cdot \gamma q_\xi\right)_\eta + \zeta\left(q(\cdot;\zeta+)\cdot \gamma q_\eta\right)_\xi = 0$$

with

$$\gamma = \mathcal{R}^{(\zeta)}(\cdot\,; q(\cdot;\zeta+))^T \cdot \mathcal{R}^{(\zeta)}(\cdot\,;q) .$$

By expanding $q^{(\zeta)'}$ near the asymptote of $\frac{\mathcal{R}^{(\zeta)}(\cdot;q)q_\xi}{\|q_\xi\|}$ for $\zeta \sim 0$ into an asymptotic series in ζ, inserting this expansion into the above conservation law, collecting all terms of the same power in ζ and setting their respectiv sums separately equal to zero, we obtain an infinite number of covariant local non-polynomial conservation laws leading to independent integrals of motion which are in involution [4].

Obviously, instead of the discrete non-linear transformation B_+ we could have taken the transformation

$$B_- = B_+ P$$

where the symbol P stands for the symmetry operation: $q \to -q$, the reflection at the origin of $\mathbb{R}^n$: $P \in O_n$.
To B_- we associate a one-parameter family of Bäcklund transformations

$$T_{\zeta-} = R_\zeta^{-1} B_- R_\zeta .$$

Going through a similar routine as before, we obtain an infinite number of new covariant local non-polynomial conservation laws arising from the previous ones by the replacement $\frac{\partial}{\partial \xi} \leftrightarrow \frac{\partial}{\partial \eta}$.

VII. DIFFERENTIAL EQUATIONS FOR THE O_n-INVARIANTS

We choose an arbitrary solution $q = q(\xi,\eta)$ of the equations of motion for the chiral fields. Along with we consider all its Bäcklund-transformed solutions

$$q(\cdot\,;\zeta\pm) = T_{\zeta\pm} \circ q$$

q and $q(\cdot\,;\zeta\pm)$ have common normalized coordinates. In the sequel we shall always work with these special coordinates.

From the solution vector q we construct a basis in $\mathbb{R}^n$ with the first three basis vectors specified

$$\underline{b_{-1}} = q \;,\; \underline{b_0} = q_\xi \;,\; \underline{b_1} = \frac{q_\eta - \cos\alpha\, q_\xi}{\sin\alpha} \;,\; \underline{b_i} \qquad i = 1,\cdots,n-2$$

such as to give $M_{(+)}$ a simple matrix form with constant entries

$$M_{(+)} = \left(\begin{array}{cc|c} 0 & 1 & 0 \\ 1 & 0 & \\ \hline 0 & & 0 \end{array}\right) .$$

In this basis also $M_{(-)}$ takes a simple form

$$M_{(-)} = \left(\begin{array}{ccc|c} 0 & \cos\alpha & \sin\alpha & \\ -\cos\alpha & 0 & 0 & 0 \\ -\sin\alpha & 0 & 0 & \\ \hline & 0 & & 0 \end{array}\right) .$$

$\mathcal{R}^{(\zeta)}(\cdot\,;q)$ transforms this basis into a new one $\underline{b_k}$ $k = -1,0,+1,\cdots,n-2$. With respect to this basis the solution vector $q^{(\zeta)\prime}$ has components X_k

$$X_k = (q^{(\zeta)\prime} \cdot \underline{b_k^{(\zeta)}})$$

which are O_n-invariant. X_{-1} vanishes identically because of the orthogonality of $q^{(\zeta)}$ and $q^{(\zeta)\prime}$. Moreover, the components X_k satisfy a closed system of bilinear ordinary differential equations involving derivatives with respect to ξ :

$$X_{0\xi} = -\zeta^{-1} + \zeta^{-1} X_0^2 - \sum_1^{n-2} (\underline{b_0} \cdot \underline{b_{j\xi}}) X_j$$
$$X_{j\xi} = \zeta^{-1} X_0 X_j + \sum_0^{n-2} (\underline{b_{j\xi}} \cdot \underline{b_\ell}) X_\ell \qquad j = 1, \cdot\cdot, n-2$$

and a similar system involving derivatives with respect to η.

The components X_k are still subject to the constraint

$$\sum_0^{n-2} X_k^2 = 1$$

deriving from the fact that $q^{(5)'}$ is a unit vector. We eliminate the constraint with the help of the stereographic projection

$$Y_j = \frac{2X_j}{1 + X_0} \qquad j = 1, \cdot\cdot, n-2$$

and arrive at two systems of (n−2) coupled bilinear first-order ordinary differential equations

$$Y_{j\xi} = \zeta^{-1} Y_j + \sum_1^{n-2} (\underline{b_{j\xi}} \cdot \underline{b_m}) Y_m + \Big(\sum_1^{n-2} (\underline{b_0} \cdot \underline{b_{m\xi}}) Y_m\Big) Y_j + \frac{1}{2} (\underline{b_0} \cdot \underline{b_{j\xi}}) \Big(1 - \sum_1^{n-2} Y_m^2\Big) \qquad j = 1, \cdot\cdot, n-2,$$

$$Y_{j\eta} = \ldots \qquad \text{"}$$

to be compared to the Riccati equations of the sine Gordon theory [6].

VIII. LINEARIZATION

Having in mind the application of the inverse scattering method to our models, we set out to derive a one-parameter family of isospectral linear eigenvalue problems. The equations obtained so far need to be linearized in order to serve for that purpose. We know how to go about the linearization only if each system can be cast into a single equation for a single unknown function with values in some finite dimensional algebra. This is the case for $n \leq 6$ if we comprise the four unknown functions $Y_1, \cdots, Y_4$ to one quaternion-valued function $\chi = \chi(\xi, \eta)$

$$\chi = Y_1 + iY_2 + jY_3 + kY_4$$

where 1, i, j, k is the conventional basis for the quaternions:

$$\chi_\xi = \zeta^{-1}\chi - \frac{1}{2}\sigma_{(1)}^{(+)}\chi - \frac{1}{2}\chi\sigma_2^{(+)} + \frac{1}{2}\chi\bar{\tau}\chi + \frac{1}{2}\tau$$

$$\chi_\eta = -\zeta \cos\alpha\, \chi - \tfrac{1}{2} \sigma_{(1)}^{(-)} \chi - \tfrac{1}{2} \chi \sigma_{2}^{(-)} + \tfrac{\zeta}{2} \sin\alpha \,(1-\chi^2).$$

The symbols $\sigma_{(r)}^{(\pm)}$ and $\tau, \bar{\tau}$ denote certain quaternions involving the scalar products

$(\underline{b_j} \cdot \underline{b_{m\xi}})$ $j \neq m, j,m = 1,..,4$ $\qquad$ $(\underline{b_0} \cdot \underline{b_{m\xi}})$ $m = 1,..,4$

respectively. Linearization is achieved by the ansatz

$$\chi = (a\varphi)(b\psi)^{-1}$$

with suitably chosen quaternions a and b. In this way, we arrive at the desired family of isospectral linear eigenvalue problems.

We shall continue to discuss only the special case n = 4 which is also the most interesting one because of its relation to the non-linear σ-model.

The SU(2)xSU(2)-invariants in normalized coordinates are

$$\alpha = \arccos(q_\xi \cdot q_\eta)$$

$$u = q_{\xi\xi} \cdot \frac{[q, q_\xi, q_\eta]}{\sin\alpha} \quad , \quad v = q_{\eta\eta} \cdot \frac{[q, q_\xi, q_\eta]}{\sin\alpha} \; .$$

The equations of motion for these invariants are

$$\alpha_{\xi\eta} + \sin\alpha + \frac{uv}{\sin\alpha} = 0$$

$$u_\eta = \frac{\alpha_\xi}{\sin\alpha} v \; , \; v_\xi = \frac{\alpha_\eta}{\sin\alpha} u \; .$$

The last two equations possess $\mathrm{tg}\,\alpha/2$ and $\mathrm{ctg}\,\frac{\alpha}{2}$ as integrating factors. Thus we set

i) $\qquad u = \beta_\xi \,\mathrm{tg}\frac{\alpha}{2} \quad , \; v = -\beta_\eta \,\mathrm{tg}\frac{\alpha}{2}$

ii) $\qquad u = \beta_\xi \,\mathrm{ctg}\frac{\alpha}{2} \quad , \; v = \beta_\eta \,\mathrm{ctg}\frac{\alpha}{2}$

and obtain two hyperbolic equations for the scalar fields α and β

1) $$\alpha_{\xi\eta} + \sin\alpha - \frac{\mathrm{tg}^2\frac{\alpha}{2}}{\sin\alpha} \beta_\xi \beta_\eta = 0$$

$$\beta_{\xi\eta} + \frac{\alpha_\xi \beta_\eta + \alpha_\eta \beta_\xi}{\sin\alpha} = 0$$

derivable from the Lagrangian density

$$L_1(\xi,\eta) = \tfrac{1}{2}\alpha_\xi \alpha_\eta + \tfrac{1}{2}\,\mathrm{tg}^2\tfrac{\alpha}{2}\,\beta_\xi \beta_\eta + \cos\alpha - 1 \quad .$$

2)

$$\alpha_{\xi\eta} + \sin\alpha + \frac{\mathrm{ctg}^2\frac{\alpha}{2}}{\sin\alpha}\beta_\xi \beta_\eta = 0$$

$$\beta_{\xi\eta} = \frac{\beta_\xi \alpha_\eta + \alpha_\xi \beta_\eta}{\sin\alpha}$$

derivable from the Lagrangian density

$$L_2(\xi,\eta) = \tfrac{1}{2}\alpha_\xi \alpha_\eta + \tfrac{1}{2}\,\mathrm{ctg}^2\tfrac{\alpha}{2}\,\beta_\xi \beta_\eta + \cos\alpha - 1 \, .$$

Whereas the dynamics associated with the Lagrangian density L_1 does not admit new static solutions of finite energy (besides the soliton and antisoliton solutions of the sine Gordon theory), the dynamics associated with the Lagrangian density L_2 does.

$$\alpha = 2\arcsin\left[\sqrt{1-A^2}\,\sin\left(2\,\mathrm{arctg}\left(e^{x\sqrt{1-A^2}}\right)\right)\right]$$

$$\beta_x = 2A\,\mathrm{tg}^2\tfrac{\alpha}{2} \quad , \; A = \mathrm{const} : \; |A| \leq 1 \, .$$

The above differential equations can be interpreted as embedding equations for a two-dimensional surface in a three-dimensional sphere which is itself embedded in a four-dimensional Euclidean space [7].

For the formulation of the family of linear isospectral eigenvalue problems associated with the first pair of differential equations we go back to the equations for Y_1 and Y_2 which now read

$$Y_{1\xi} = \zeta^{-1} Y_1 + \check{u}\,\mathrm{ctg}\alpha\, Y_2 - \frac{\alpha_\xi}{2}(Y_1^2 - Y_2^2) - u Y_1 Y_2 - \frac{\alpha_\xi}{2}$$

$$Y_{2\xi} = \zeta^{-1} Y_2 - \check{u}\,\mathrm{ctg}\alpha\, Y_1 - \frac{\alpha_\xi}{2} Y_1 Y_2 + \frac{\check{u}}{2}(Y_1^2 - Y_2^2) - \frac{\check{u}}{2}$$

We can comprise the real functions Y_1 and Y_2 to a complex function $W = Y_1 + iY_2$ and the two differential equations reduce to a single one:

$$W_\xi = \zeta^{-1} W - i\check{u}\,\mathrm{ctg}\alpha\, W - \frac{\alpha_\xi - i\check{u}}{2} W^2 + \frac{\alpha_\xi + i\check{u}}{2} \quad .$$

Define $\omega = \omega(\xi, \eta)$ by the compatible equations

$$\omega_\xi = u \operatorname{ctg}\alpha \ , \quad \omega_\eta = -\frac{v}{\sin\alpha}$$

and set

$$W = \frac{\psi_1 e^{-\frac{i}{2}\omega}}{\psi_2 e^{+\frac{i}{2}\omega}} \ .$$

The differential equation for W is satisfied if

$$\psi_{1\xi} + \frac{\alpha_\xi + i\tilde{u}}{2} e^{i\omega}\psi_2 = \frac{1}{2}\zeta^{-1}\psi_1$$

$$-\psi_{2\xi} + \frac{\alpha_\xi - i\tilde{u}}{2} e^{-i\omega}\psi_1 = \frac{1}{2}\zeta^{-1}\psi_2$$

i.e. if

$$L\psi = k\psi$$

where

$$\psi = \begin{pmatrix}\psi_1 \\ \psi_2\end{pmatrix} , \quad k = \frac{i}{2\zeta}$$

$$L = i\begin{pmatrix}1 & 0 \\ 0 & -1\end{pmatrix}\frac{d}{d\xi} + \frac{i}{2}\begin{pmatrix}0 \ , & (\alpha_\xi + i\tilde{u})e^{i\omega} \\ (\alpha_\xi - i\tilde{u})e^{-i\omega} , & 0\end{pmatrix}$$

The η-evolution of the wave function ψ is given by

$$\frac{\partial\psi}{\partial\eta} = \frac{1}{4ik}\begin{pmatrix}\cos\alpha , & \sin\alpha\, e^{i\omega} \\ \sin\alpha\, e^{-i\omega} , & -\cos\alpha\end{pmatrix}\psi = -M\psi .$$

It is easy to verify the relation

$$\frac{\partial L}{\partial\eta} = [L, M]$$

which guarantees the isospectrality of the family $L(\eta)$.

We take appropriate linear combinations of the eigenvalue and η-evolution equations and change the basis according to

$$\Psi = \frac{1}{\sqrt{2}}\begin{pmatrix}e^{-\frac{i\omega}{2}} , & -i e^{\frac{i\omega}{2}} \\ -i e^{-\frac{i\omega}{2}} , & e^{\frac{i\omega}{2}}\end{pmatrix}\psi$$

Thus we arrive at the linear eigenvalue problem

$$-i\sigma^2\frac{d}{dx}\Psi + A(x)\Psi + \frac{B^2(x)}{\lambda}\Psi = \lambda\Psi$$

and the corresponding time-evolution equation

$$\frac{\partial}{\partial t}\Psi = \frac{d}{dx}\Psi + 2i\sigma^2\left\{\frac{B^2(x)}{\lambda} + \frac{\omega_t-\omega_x}{4}\right\}\Psi .$$

Here we have used the following notation

$$A(x) = i\frac{\alpha_t+\alpha_x}{4}\sigma^1 - i\frac{\breve{u}}{4}\sigma^3 - \frac{\omega_x}{2}\sigma^4$$

$$B(x) = \frac{1}{4}\begin{pmatrix} e^{i\frac{\alpha}{2}}, & 0 \\ 0, & e^{-i\frac{\alpha}{2}} \end{pmatrix}$$

$$\lambda = \frac{i}{4\zeta}$$

σ^i, i = 1, 2, 3 (4) are the three Pauli (2x2 unit) matrices.

The time evolution of the scattering data is

$$r(\lambda,t) = \exp\left\{\frac{\eta}{2i\lambda}t\right\} r(\lambda,0) \qquad -\infty < \lambda < +\infty$$

$$\zeta_j(t) = \zeta_j\,, \quad m_j(t) = \exp\left\{\frac{\eta}{2i\zeta_j}t\right\} m_j(0) \qquad j=1,\cdot\cdot,N \quad [6]$$

Thereby we have "solved" the Cauchy initial value problem for the equations

$$\Box\alpha + \sin\alpha - \frac{tg^2\frac{\alpha}{2}}{\sin\alpha}(\partial_\mu\beta)(\partial^\mu\beta) = 0$$

$$\Box\beta + 2\frac{(\partial_\mu\alpha)(\partial^\mu\beta)}{\sin\alpha} = 0$$

$$\left(\Box\alpha + \sin\alpha + \frac{ctg^2\frac{\alpha}{2}}{\sin\alpha}(\partial_\mu\beta)(\partial^\mu\beta) = 0,\ \Box\beta - 2\frac{(\partial_\mu\alpha)(\partial^\mu\beta)}{\sin\alpha} = 0\right)$$

which arise in the context of the one-space-dimensional version of the non-linear σ-model.

REFERENCES

[1] R. Dashen, B. Hasslacher and A. Neveu: Phys. Rev. D10, 4114, 4130, 4138 (1974)

W.E. Korepin and L.D. Faddev: Teor. i Mat. Fiz. 25, 147 (1975)

A. Klein and F. Krejs: Phys. Rev. D13, 3295 (1976)

N.H. Christ and T.D. Lee: Phys. Rev. D12, 1606 (1975)

[2] S. Coleman: Phys. Rev. D11, 2088 (1975)

B. Schroer and T.T. Truong: "Equivalence of Sine-Gordon and Massive Thirring Model and Cumulative Mass Effects", FUB HEP 6/76

[3] S. Weinberg: Phys. Rev. 166, 1568 (1968)

[4] K. Pohlmeyer: Commun. Math. Phys. 46, 207 (1976)

The inverse scattering method originates with

C.S. Gardner, J.M. Green, M.D. Kruskal and R.M. Miura: Phys. Rev. Lett. 19, 1095 (1967)

[5] E. Goursat: Memorial Sci. Math. Fasc. 6 (Gauthiers-Villars, Paris, 1925)

[6] K. Pohlmeyer: "The Classical Sine Gordon Theory", Proceedings of this Summer Institute

[7] K. Meetz: Seminar Talk, Hamburg, March 1976 (unpublished)

F. Lund: "Note on the Geometry of the Nonlinear σ-Model in Two Dimensions", IAS preprint, October 1976

SEMICLASSICAL METHODS IN FIELD THEORY

Andre Neveu

The Institute for Advanced Study

Princeton, New Jersey 08540

This is an introduction to some of the methods used in the last two years to exploit at the quantum level part of the knowledge one may have on a classical field theory.

The interest of classical equations of motion in quantum mechanics is already manifest for one degree of freedom: the WKB approximation for bound states is in general excellent, even for the ground state in the extreme strong coupling limit (for numerical results in the anharmonic oscillator, see [1]). When there are many degrees of freedom, one tries to separate them. If one can, the WKB method is applied to each of them separately. If one cannot, whether for fundamental or practical reasons, the problem is more complicated, and, the knowledge of the classical system being more restricted, so will be the validity of semiclassical quantization. In practice, for such non-separable systems, one knows only a very special set of simple classical motions (rather than all the motions). It is still possible to extract useful quantum mechanical information. Examination of simple systems leads to the general heuristic feeling that the results of semiclassical quantization will be good at least when the quantum fluctuations (around the classical motion which is quantized) remain small enough that the effect of nonlinearities is small. A general quantitative statement is difficult to formulate. Each system has to be examined separately. For field theory, it turns out that the region of validity includes the weak coupling region. This has the advantage that one can compare with the results of ordinary perturbation theory. Of course, in field theory, one has to deal with divergences and renormalization. This is rather straightforward, and the result is that all divergences are handled by the ordinary one-loop counterterms. All this fits with the widespread and rather vague belief that "WKB = trees (classical) + one loop". Actually, semiclassical results cut across the whole perturbations expansion, picking in each order diagrams or pieces of diagrams that cannot be identified in any simple fashion.

The approach to semiclassical quantization used in ref [2] involves the trace of the resolvent operator:

$$G(E) = \mathrm{tr}\ \frac{1}{H-E} = \sum_n \frac{1}{E_n - E} \tag{1}$$

$G(E)$ has poles for $E = E_n$, $n\underline{th}$ eigenvalue of the quantum-mechanical hamiltonian H. We do not worry here about the convergence of the series in (1); subtractions could be necessary, but would not affect the location of the poles. The next step involves writing $G(E)$ as

$$G(E) = i\ \mathrm{tr} \int_0^\infty \frac{dT}{\hbar} \exp\ \left[i\ (E-H) \frac{T}{\hbar} \right] \tag{2}$$

Here again, we do not worry about possible divergences in the T integration. In principle, an $i\varepsilon$ provides convergence. Let us only remark that one is not restricted to an integration along the real T axis, but that complex values of T can be considered, as long as they are compatible with the $i\varepsilon$ prescription. Allowing T to be complex would correspond to including possible tunnelling phenomena, for which we do not yet know the general formalism. We now outline the general method for the calculation of the right-hand side of eq.(2), referring the reader to the original papers for the details. The strategy by which classical solutions appear naturally involves using Feynman path integrals:

$$\mathrm{tr}\ e^{-iHT/\hbar} = \int \mathscr{D}x(\tau) e^{iS/\hbar} \tag{3}$$

$$S = \int_0^T \left[\tfrac{1}{2} \dot{x}^2 - V(x)\right] d\tau \tag{4}$$

where S is the action computed along the path $x(\tau)$. Here $x(\tau)$ is only a generic name for all the degrees of freedom of the theory: in the simplest case, it is just the position of a particle in a one-dimensional potential $V(x)$, but it can have many components, for a motion in a multidimensional potential; and for field theory it has an infinite number of components, namely the values of the fields at each point in space. In this case, $V(x)$ contains both space derivatives of the field and interaction terms. Finally, the functional integration in (3) is to be done on periodic paths only: $x(0) = x(T)$. This condition is just the translation in position space language of the trace operation of the left-hand-side.

The connection with classical mechanics is now evident: the semiclassical (= small h) approximation consists in computing the functional integral of eq. (3) by stationary phase around classical periodic orbits of period T.

By examining separable systems, one can see that there are in general many periodic orbits with the same period. In such a case, it can be shown [2] that including all of them in the stationary phase calculation of (3) leads to the same result as first separating the variables and then quantizing each of them a la WKB. In field theory, one cannot hope to retain all possible periodic classical motions, but only a limited set. The larger the set, the more information one will get on the quantum system. Here, we will restrict the discussion to one set of period orbits with the period being the only varying parameter; a typical example of such a set of motions is the doublet of the sine-Gordon theory: the classical energy varies continuously with the period, and quantization will restrict it to discrete values.

The stationary phase calculation of eq. (3) is no different in principle from ordinary stationary phase calculation of a simple definite integral: shifting the integration variable to the stationary phase point and expanding the exponent to second order makes the integration gaussian:

$$\mathrm{tr}\, e^{-iHT/\hbar} \simeq e^{iS_{c\ell}/\hbar} \int \mathcal{D}x(\tau) e^{i\tilde{S}/\hbar} \tag{5}$$

where $S_{c\ell}$ is the classical action around one orbit, and

$$\tilde{S} = \int_0^T [\tfrac{1}{2}\dot{x}^2 - \tfrac{1}{2} x V''(x_{c\ell}(\tau))x]dt \tag{6}$$

$x_{c\ell}$ being the classical trajectory.

The x integration is now gaussian. Performing this integration gives the inverse of the square root of the product of the eigenvalues of the differential operator $\partial_\tau^2 + V''(x_{c\ell}(\tau))$. This eigen product is computed in ref.[4] in terms of the stability angles of the classical trajectory $x_{c\ell}(\tau)$. The stability angles ν_α are defined by the solutions of

$$[\partial_\tau^2 + V''(x_{c\ell}(\tau))]\, y_\alpha(\tau) = 0 \tag{7}$$

such that

$$y_\alpha(\tau+T) = e^{-i\nu_\alpha}\, y_\alpha(\tau) \tag{8}$$

The classical trajectory is stable if all the ν's are real. We will assume that this is the case: quantum fluctuations remain small, and do not take the system to regions of phase space far from the classical trajectory.

Zero stability angles require a special treatment: any continuous symmetry of the classical system generates a corresponding zero stability angle. In practice, for field theory, it will be space-time translation and internal symmetries. The integration over the corresponding modes is not gaussian, and formula (5) has to be modified to take this into account. We refer the reader to the literature on this delicate subject (ref.[5]).

Finally, one should take into account the fact that if a trajectory with period T is known, traversing it n times trivially defines a trajectory with period n T: one has to sum over n. After which, the approximate form of G(E) turns out to be

$$G(E) = \int_0^\infty dT \sum_{n=1}^\infty \sum_{\{q_\alpha\}} e^{i\frac{n}{\hbar}\{(S_{c\ell}+ET) - \sum_\alpha (q_\alpha + \frac{1}{2})\nu_\alpha \hbar\}} \tag{9}$$

$\{q_\alpha\}$ being any set of positive integers (or zero).

The remaining integration over T is also done by stationary phase. The stationary phase point is

$$E = E_{c\ell} + \hbar \sum_\alpha (q_\alpha + \tfrac{1}{2}) \frac{d\nu_\alpha}{dT} \tag{10}$$

and the summation over n in eq. (9) gives poles at

$$\oint p\, dq + \hbar \sum_\alpha (q_\alpha + \tfrac{1}{2}) \left(T \frac{d\nu_\alpha}{dT} - \nu_\alpha\right) = 2n\pi\hbar \tag{11}$$

This is the generalization of the ordinary WKB formula

to many degrees of freedom. The "main" quantum number n quantizes motions along the classical trajectory. The integers q_α (in number equal to the number of degrees of freedom minus one, in general) quantize the oscillations around that classical orbit. Since these have been treated in the linear approximation, the validity range of (11) is large n but small q_α. In practice, in field theory, one will be looking for bounds states, and most of the ν_α's correspond to travelling waves, for which q_α will be taken equal to zero. In field theory, because of the infinite number of degrees of freedom, the sums in (10) and (11) diverge and have to be renormalized: it turns out that for a renormalizable theory, (10) and (11) are made finite by simply subtracting from E the vacuum energy in the one loop approximation and using in the functional integral (3) the Lagrangian with the ordinary one-loop counter-terms included.

A time independent solution is a particular case to which formula (10) can apply. In that case, $\nu_\alpha = \omega_\alpha T$, where ω_α's are the frequencies of the small oscillations around that solution, treated in the harmonic approximation.

Examples of these methods have been worked out in some detail for two-dimensional model field theories where calculations could be done analytically, and compared with exact results. Both static and time-dependent solutions have been considered.

The first and most remarkable example is the sine-Gordon theory, defined by the Lagrangian

$$\mathcal{L} = - \frac{1}{2}(\partial_\mu\phi)^2 + \frac{m^4}{\lambda}\left[\cos\frac{\sqrt{\lambda}}{m}\phi - 1\right] \tag{12}$$

In (12), the variables have been defined so that the small ϕ expansion

$$\mathcal{L} = - \frac{1}{2}(\partial_\mu\phi)^2 - \frac{1}{2}m^2\phi^2 + \frac{\lambda}{4!}\phi^4 + \ldots \tag{13}$$

corresponds to a field theory of a boson of mass m with a weakly attractive contact interaction. On the other hand, the rescaling

$$\phi \to \frac{m}{\sqrt{\lambda}}\phi \quad , \; x \to \frac{x}{m} \tag{14}$$

brings the Lagrangian (12) under the form

$$\mathcal{L} = \frac{m^4}{\lambda}\left[-\frac{1}{2}(\partial_\mu\phi)^2 + \cos\phi - 1\right] \tag{15}$$

The classical behavior of the sine-Gordon theory is completely known. See ref. [6]. From (15), we see that the dimensionless coupling constant λ/m^2 will play in the functional integral (3) the role of $\hbar$. Hence, the results of semiclassical calculations overlap with the validity range of perturbation theory. This is a completely general feature, independant of the space-time dimension and of the specific lagrangian.

From the work of ref. [6] on the classical sine-Gordon theory, we learn that there are two types of solutions that, when quantized, will give particles and bound states. In their rest frame, these are the soliton:

$$\phi = 4 \text{ Arctan } e^x \tag{16}$$

and the doublet (or breather), which is a soliton-anti-soliton bound state:

$$\phi = 4 \text{ Arctan } \frac{\varepsilon \sin[t(1+\varepsilon^2)^{-\frac{1}{2}}]}{\cosh[\varepsilon x(1+\varepsilon^2)^{-\frac{1}{2}}]} \tag{17}$$

(ε is any real positive number).

The theory being Lorentz invariant, solutions (16) and (17) can be boosted to arbitrary velocities. By putting the system in a periodic box, there results further periodic motions. Quantization of these motions can be done [4]: this is actually a case of separation of variables, with the center of mass position being one of the variables. One then just gets the expected quantization of momentum in the box.

Application of eq. (10-11) to the solutions (16-17) is described in ref. [4]. All calculations can be done analytically, thanks to the fact that the classical sine-Gordon system is integrable: in particular, one can find analytically the solutions of eq. (7-8). The final results are: the mass of the soliton is

$$M(\text{soliton}) = \frac{8m^3}{\lambda} - \frac{m}{\pi} \tag{18}$$

The first term on the right hand side of eq. (18) is the classical mass. The second is the contribution of small oscillations around the static solution. It turns out in the course of the calculation of the first quantum correction that the only non-zero stability angles are part of the continuum: there is no soliton-meson bound state. This is in contrast with $\lambda\phi^4$ (see below). For weak coupling ($\lambda/m^2 << 1$), the soliton is a very heavy particle. It cannot decay into ordinary particles because of its topological properties. The topological conservation laws of static field theory classical solutions is discussed in ref. [7].

The doublet (17) produces the remaining series of states at masses

$$M_n = \frac{16m}{\gamma'} \sin \frac{n\gamma'}{16} \qquad n = 1,2,3,--- \quad <8\pi/\gamma' \tag{19}$$

with $\gamma' = (\lambda/m^2)(1 - \lambda/8\pi m^2)^{-1}$

The original "elementary particle" of the theory is the $n = 1$ state in eq. (19), as can be seen in the small coupling limit $\lambda << m^2$. The other $n>1$ states can be considered as bound states (at least in weak coupling) of the elementary particle. It is interesting that one can either consider ϕ as the fundamental field, of which the soliton is a complicated collective excitation or the soliton as the fundamental object, other states, including the elementary particle, being soliton-antisoliton bound states. Indeed, Coleman has shown that the sine-Gordon theory is equivalent to the massive Thirring model, the fermion field being identified as the field of the soliton. Thanks to that equivalence, one can show that the mass ratios as given by eq. (19) are exact to all orders in λ/m^2: see A. Luther's lectures at this school. This is of course an accident, analogous to the non-relativistic hydrogen atom.

The other model field theory considered in ref. [4] is $\lambda\phi^4$ in the two-phase region:

$$\mathcal{L} = -\tfrac{1}{2}(\partial_\mu\phi)^2 + \tfrac{1}{2}m^2\phi^2 - \frac{1}{4}\lambda\phi^4 \tag{20}$$

which, after the rescaling (14) becomes

$$\mathcal{L} = \frac{m^4}{\lambda}\left[-\tfrac{1}{2}(\partial_\mu\phi)^2 + \tfrac{1}{2}\phi^2 - \frac{1}{4}\phi^4\right] \tag{21}$$

Ordinary perturbation theory of (21) involves first shifting the field to its vacuum value + 1 (it could also be -1), thus spontaneously breaking the $\phi \to -\phi$ discrete symmetry. The first state above the vacuum is a particle state, with mass $m\sqrt{2}$ in lowest order of perturbation theory. Here again, perturbation theory is made in powers of λ/m^2.

There is a static space dependent solution of (21), the

kink, analogous to the sine-Gordon soliton:

$$\phi = \tanh \frac{1}{\sqrt{2}} x \tag{22}$$

This solution connects the two vacua ± 1. It is stable, both classically and quantum-mechanically [7]. The computation of the first quantum mechanical correction to the mass of the solution (22) is performed in ref. [8]. Contrary to the case of sine-Gordon, there is one isolated non-zero stability angle, which means that there can be bound states of a kink and an ordinary particle, labeled by an integer $q \geq 0$.

The mass of such states is

$$M_q = \frac{2}{3}\sqrt{2}\,\frac{m^3}{\lambda} + m\left(-\frac{3}{\pi\sqrt{2}} + \frac{1}{2\sqrt{6}}\right) + m\sqrt{\frac{3}{2}}\,q \tag{23}$$

where the first term on the right hand side is the classical mass of the kink, the second the first quantum mechanical correction. The state $q = 0$ is the unexcited kink. The state $q = 1$ is a kink-meson bound state, and is stable because of energy conservation. States with $q > 1$ can decay into an unexcited kink and a meson; they would gain a width in higher order of perturbation theory.

Non-trivial time-dependant solutions of the ϕ^4 theory are not known analytically. However, one can find them in perturbation by analogy with eq. (17), in which ε is considered a small parameter. The strategy is to expand simultaneously in harmonics of the fundamental frequency and in powers of ε. This is explained in detail in ref.[4]. The result is a classical motion qualitatively analogous to the doublet [17], which, when quantized, gives a set of bound states built out of $n(n \geq 1)$ elementary particles. This solution also seems to be a kink-antikink bound state [4]. States with $n > 2$ are not expected to retain their stability in higher orders: ϕ^4 does not have all the higher conservation laws of sine-Gordon which stabilize all the states of eq. (19).

The non-linear Schroedinger equation [9], which is the non-relativistic limit of the sine-Gordon theory can also be quantized semiclassically with this method. There too, the semi-classical results are exact. See ref.[10] for details.

The introduction of fermions in semi-classical methods is delicate: the only way one can reach a classical limit (= large quantum numbers) is by the introduction of a large number of fermion species, so that there can be many fermions in the same state. A two-dimensional model of this type has been considered in ref. [11] and gives a remarkably rich spectrum of bound states; it could also be a new system with an infinite number of conservation laws; it also has the advantage of being renormalizable and asymptotically free, rather than superrenormalizable; it also exhibits dynamical symmetry breaking.

In higher space-time dimensions, all known particle-like solutions of classical field theory are static or have a rather trivial time-dependence; thus, so far, there has been few applications of the semiclassical method to higher dimensions.

Finally, we mention the new developments on semi-classical tunnelling in field theory. They make use of imaginary time (euclidean) finite action solutions which connect topologically distinct classical vacua. This gives rise to a new class of phenomena, which are being investigated, and I can only refer the reader to the recent literature [12].

REFERENCES

1. J. Kilpatrick and M. Kilpatrick, J. Chem. Phys. 16, 781 (1948).

2. R. Dashen, B. Hasslacher and A. Neveu, Phys. Rev. D10, 4114 (1974). See also ref. 3.

3. M. Gutzwiller, J. Math. Phys. 12, 343 (1971); 11, 1791 (1970); 10, 1004 (1969); 8, 1979 (1967). J. Keller, Ann. Phys. 4 180 (1958).
V. Maslov, Theor. Math. Phys. 2, 21 (1970).

4. R. Dashen, B. Hasslacher and A. Neveu, Phys. Rev. D 11, 3424 (1975).

5. J. L. Gervais, A. Jevicki and B. Sakita, in proceedings of the conference on extended systems in field theory, Paris, June, 1975, Physics Reports 23, 281 (1976).

6. L. A. Takhtadzhyan and L.D. Fadeev, Theor. Math. Phys. 21, 160 (1974).

7. S. Coleman, Lectures delivered at the 1975 International School of Subnuclear Physics "Ettore Majorana", Erice, Italy.

8. R. Dashen, B. Hasslacher and A. Neveu, Phys. Rev. D10, 4130 (1974).

9. A. Scott, F. Chu and D. McLaughlin, Proc. IEEE, 61, 1443 (1973).

10. C. Nohl, Princeton University preprint (1975).

11. R. Dashen, B. Hasslacher and A. Neveu, Phys. Rev. D12, 2443 (1975).

12. G. 't Hooft, Harvard University preprint (1975), and references therein. See also P. Korepin, Pizma JETP 23, 224 (1976).

QUASI-CLASSICAL USE OF TRACE IDENTITIES IN ONE AND HIGHER DIMENSIONAL FIELD THEORIES

B. Hasslacher

California Institute of Technology

Pasadena, California 91125

Suppose one suspected that a non-linear system in 1+1 dimensions was perfect (completely integrable)[1], but did not know all the classical static or time-dependent solutions. If one wished to consider such a system as a field theory and apply functional WKB methods[2] to compute quasi-classical quantities, it is generally assumed that one would have to know all these solutions to the original non-linear system. The reason is that they determine the saddle point of the functional integral defining the energy spectrum and the accuracy of the WKB is in direct proportion to control over the saddle points. If one had the complete set of classical solutions, then a strong coupling approximation is possible, since one is no longer doing perturbation theory in some small parameter, but rather a functional expansion in solution space. However, the complete spectrum of classical solutions to non-linear systems is notorious for the analytic effort required to find it.

There is a way around this difficulty which is very powerful, but requires the use of notions unfamiliar to most field theorists. There exists a method which fixes the quasi-classical energy spectrum directly and simultaneously generates all of the classical solutions to the original non-linear system. It hinges on the existence of certain sum rules connected to the non-linear system, called trace identities. Of course it works in 1+1 dimensions only if the system is perfect, but has a natural extension to higher dimensions which is the definition of the higher dimensional analog to perfection in 1+1.

For example, certain non-abelian gauge models in 3+1 dimensions, that support monopole solitons[3] and others which in the euclidean sector support pseudoparticles[4], or instantons look suspiciously like perfect systems. By that one means that a one-loop WKB calculation, done over all the saddle points of the system is exact[5]. There would be no further quantum corrections. Such a situation could occur in systems with dynamically generated, conserved topological quantum number, connected to the solitons.

This is of course a conjecture, but if it is true, the method of trace identities should in principle be powerful enough to show it. If it is false, they would give no statement. The next sections will be a general introduction to trace identity sum rules and the inverse scattering method. Since this is supposed to be tutorial in nature, I will go into a minimum of analytic detail, which can be found in the references, but rather motivate concepts as we go along, and sketch their use.

THE INVERSE SCATTERING METHOD[1]

The inverse scattering method arose in the study of certain classes of one-dimensional non-linear fields. Its essence consists of associating with the classical field under study, a certain linear differential operator, with coefficients from this field, whose spectral characteristics change with time in a known way. The Cauchy problem for the original non-linear system is reduced to the known problem for the direct and inverse spectral problem for a linear operator. The main linear operators used are the Schrödinger and modified Dirac. This description is somewhat compact. In outline the steps are as follows.

We want to study a general non-linear wave equation $\phi_t = k(\phi)$ where $k(\phi)$ denotes a suitably defined non-linear operator. If we can find linear operators L and B which are functions of ϕ so that $i L_t = [B,L]$, then if B is self-adjoint, the eigenvalues E of L i.e., solutions of $L\psi = E\psi$, are time independent, even though L depends on time through ϕ. Further, the eigenfunctions ψ of L evolve in time according to $i\psi_t = B\psi$. In such a case it is possible to associate a scattering problem with the linear operator L, and given the initial data $\phi(x,0)$, we can construct $\phi(x,t)$. The steps are:

a) Calculate asymptotic data for ψ (reflection coefficients and bound state eigenvalues for L) at $|x| = \infty$ and t=0, from a knowledge of $\phi(x,0)$ or by other means.
b) Using the evolution equation $i\psi_t = B\psi$, compute the time evolution of the scattering data.
c) The inverse problem: knowing the scattering data as a function of time, reconstruct the potential $\phi(x,t)$.

The reconstruction is done with the Gel'fand-Levitan-Marchenko equation. For the Schrödinger kernel L, this procedure is as follows:

The integral equation

$$g(x,y) + K(x,y) + \int_x^\infty K(y+y')\, g(x,y')\, dy' = 0 \quad ; \quad y > x \tag{1.1}$$

is solved for g(x,y), where the kernel K is given by

$$K(x+y) \equiv R(x+y) + \sum_{n=1}^{M} m_N \exp(-k_n(x+y)) \tag{1.2}$$

$$R(x+y) \equiv \frac{1}{2\pi} \int_{-\infty}^{+\infty} R(k) \exp(ik(x+y))\, dk \tag{1.3}$$

where R(k) is the reflection coefficient, M the number of bound states with eigenvalues K_n^2. The factors m_N are certain functions fixed by the evolution equation. The potential $\phi(x,t)$ is then reconstructed by

$$\phi(x) = -2\, d/dx\; g(x,x) \tag{1.4}$$

(we have suppressed the t dependence for simplicity)

So, a non-linear wave equation has been replaced by the linear GLM integral equation, which in many cases are easier to solve. These equations have been generalized to 3+1 dimensions by Faddeev and Newton[6]. Schematically we have:

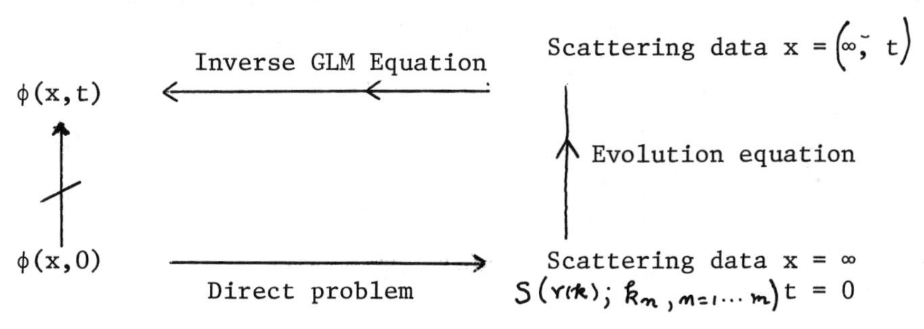

For our purposes we do not know $\phi(x,0)$, but in static cases want to find it and the evolution equation is unnecessary. Our problem will be schematically

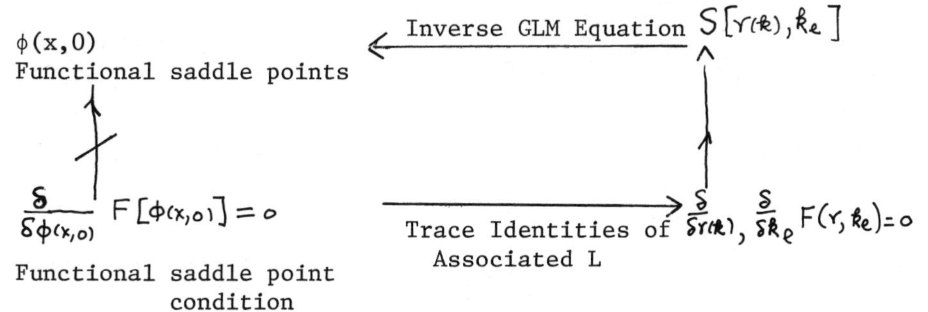

Again $\{r(k), k_\ell\}$ are reflection coefficients and bound state data.

TRACE IDENTITIES IN ONE AND HIGHER DIMENSIONS

Let me show you a trace identity in 1+1 for the Schrödinger equation[7]

$-\psi_{xx} + U(x)\psi = k^2\psi$ which has the asymptotic scattering data:

$S\{r(k),\ k_\ell,\ \ell=1,\ldots,m\}$. There are an infinite family of them. The first is

$$-\frac{1}{2i}\int_{-\infty}^{+\infty} U(x)\,dx = \frac{1}{2\pi i}\int_{-\infty}^{+\infty} \ln\left[1 - |r(k)|^2\right]dk - 2\sum_{\ell=1}^{m}(ik_\ell) \tag{2.1}$$

This is a remarkable formula, since it expresses an unknown potential $U(x)$ directly as functions over separate pieces of the asymptotic data. In other words, it is a function of completely factorized S matrix data. From the previous section we know that if we can fix the scattering data, we can reconstruct the potential by GLM. If the original equation that $U(x)$ obeyed is perfect, it will be possible to fix the $r(k)$ and k_ℓ completely. If not, one will simply not get enough constraints on these parameters to reconstruct solutions. This is a very efficient method for finding the saddle points of functional integrals.

Before proceeding to sketch how such formulas are used in quasi-classical calculations, I will give a loose derivation, first due to Faddeev and Buslaev[8], on how such sum rules come about in arbitrary dimensions and what assumptions on L go into them.

Generalized Trace Identities

This is all for the Schrödinger kernel in up to three space dimensions. We will need some theorems.

(1) $$\mathrm{tr}(R_\lambda - R_0) = -\frac{d}{d\lambda}\ln M(\sqrt{\lambda}) \tag{2.2}$$

where $M(\sqrt{\lambda})$ is essentially the S matrix, R is the resolvent kernel for the Schrödinger operator and R_0 is the resolvent in the absence of a potential.

(2) Consider the properties of the auxiliary function

$$g(s) = s^{2z}\frac{d}{ds}\ln M(s),\ z \text{ complex, in the complex } \lambda \text{ plane.}$$

We will assume only very loose restrictions on locations of poles and cuts, so as to fit the argument to the widest class of kernels. By definition, $\ln M(s) = i\,\eta(s) + \ln\Lambda(s)$, which by partial integration

$$\Rightarrow \int g(s)ds = \int s^{2z}\frac{d}{ds}\ln M(s) = -2z\int\left(i s^{2z-1}\eta(s)ds + s^{2z-1}\ln\Lambda(s)ds\right) \tag{2.3}$$

Do the LHS by contour integration on the lower half plane of λ. Since $s = \sqrt{\lambda}$ (s = the energy), $d/ds \ln m(s) = 2\sqrt{\lambda}\, d/d\lambda \ln m(\sqrt{\lambda}) = -2\sqrt{\lambda}\,\mathrm{tr}\, 1/H-\lambda$ (by (1)) $= -2\sqrt{\lambda}\ \sum_n 1/E_n-\lambda$. Now in the spectrum of H, assume only poles along neg λ axis, excluding the point $\lambda = 0$, and that all integrals vanish over contours at ∞ and 0

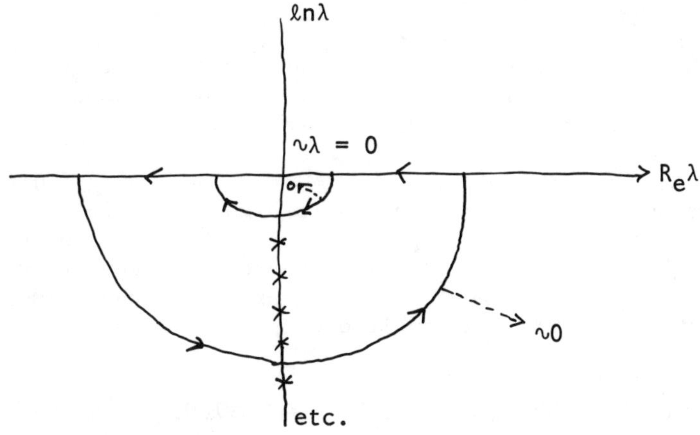

POLES AT $\sqrt{\lambda} = s = ik_e$, BOUND STATE ENERGIES

Then:

$$\oint s^{2z}\frac{d}{ds}\ln M(s)ds = -\oint s^{2z}\,\mathrm{tr}\frac{1}{H-\lambda}d\lambda = -2\pi i\sum_e (ik_e)^{2z}$$

$$\Rightarrow 2\pi i\sum_e k_e^{2z} = -2z\left(\cos\pi z - i\sin\pi z\right)\int_{-\infty}^{+\infty}\left(i s^{2z-1}\eta(s) + s^{2z-1}\ln\Lambda(s)\right)ds \tag{2.4}$$

Equating imaginary parts in eq. (2.4)

$$\frac{\pi}{2}\sum_e k_e^{2z} = -z\cos\pi z\int_0^\infty s^{2z-1}\eta(s)ds - z\sin\pi z\int_0^\infty s^{2z-1}\ln\Lambda(s)ds \tag{2.5}$$

To go further we need an asymptotic expansion for $\eta(s)$

$$\eta(s) \underset{s\to\infty}{=} \frac{\alpha_0}{s} + \eta_1(s)\ ;\ \eta_1(s) \sim O\left(\frac{1}{s^2}\right) \tag{2.6}$$

Then analytically continue eq. (2.5) by observing the action of the poles.

$$-z\cos\pi z \int_0^\infty s^{2z-1}\eta(s)\,ds \longrightarrow \cos\frac{\pi}{2}\,\frac{\alpha_0}{2z-1}\,s^0\Big|_{z=1/2}$$

so for z = 1/2 we get a sum rule

$$\alpha_0 = \pi\sum_\ell k_\ell + \int_0^\infty \ln\Lambda(s)\,ds \tag{2.7}$$

α_0 is the first trace coefficient, or first term in the asymptotic expansion for the phase shift, k_ℓ are the bound state data and $\ell n\Lambda(s)$ are S matrix elements, which in 1+1 dimension become the reflection coefficients.

The sum rule can be extended to higher coefficients in the asymptotic expansion. The point is that the α_i's are just moments of the potential and its derivatives and one needs efficient ways of generating such expansions in higher dimensions. Also one clearly needs some factorization conditions on Re M(s)

APPLICATION TO A QUASI-CLASSICAL CALCULATION, THE GROSS-NEVEU MODEL

The Gross-Neveu model[9] was the first case in which trace identities were used to evaluate the quasi-classical energy spectrum. We don't need to know very much about it, since it is used here only as an example of what we've been discussing.

The Lagrangian for this system is

$$\mathcal{L} = i\bar\psi\partial\!\!\!/\psi + \tfrac{1}{2}g^2(\bar\psi\psi)^2 \tag{3.1}$$

where $i\bar\psi\partial\!\!\!/\psi \equiv \sum_{k=1}^{N} i\bar\psi_k\partial\!\!\!/\psi_k$; $\bar\psi\psi \equiv \sum_{k=1}^{N}\bar\psi_k\psi_k$, g is a dimensionless parameter

and $\partial\!\!\!/ = \gamma_0\partial_0 - \gamma_1\partial_1$ γ's being the usual Dirac matrices.

Examine the ordinary vacuum functional for this system.

$$\langle 0|0\rangle = \int \mathcal{D}(\bar\psi)\,\mathcal{D}(\psi)\,\exp\left[i\int\mathcal{L}\,dx\right] \tag{3.2}$$

By introducing an auxiliary scalar field $\sigma(x)$, this may be rewritten as

$$\langle 0|0\rangle = \int \mathcal{D}(\bar{\psi})\,\mathcal{D}(\psi)\,\mathcal{D}(\sigma)\,\exp\left[i\int\left(-\tfrac{1}{2}\sigma^2 + \bar{\psi}(i\partial\!\!\!/ - g\sigma)\psi\right)\right] \tag{3.3}$$

One could integrate the fermions out of the system at this point, since everything is quadratic to get

$$\langle 0|0\rangle = \int \mathcal{D}(\sigma)\,\exp i\left[-\frac{\sigma^2}{2} + \mathrm{tr}\,\ln(i\partial\!\!\!/ - g\sigma)\right] \tag{3.4}$$

The integrand of this equation is a perfect system of a type not studied in the literature. To see how trace identities enter we are going to do the functional fermion integration in another way for the representation of tr exp [iHT], the fundamental object of the WKB method.

For this model, it can be shown that

$$\mathrm{Exp}[-iHT] = \sum_{\{n\}} C(N,\{n\})\int \mathcal{D}(\sigma)\exp\left[i\int_0^T dt\int_{-\infty}^{+\infty} dx\left(-\frac{\sigma^2}{2} + iN\varphi(\sigma) - i\sum_k n_k\,a_k(\sigma)\right)\right] \tag{3.5}$$

where $\varphi = \sum \alpha_i$, $0 \leq n_i \leq 2N$ N = number of fermion flavors, over all

$\sigma(x,t)$ such that $\sigma(x,t+T) = \sigma(x,t)$ and $|\sigma(x,t)| \to |\sigma_0|$ as $|x| \to \infty$, $C\{N,\{n\}\}$ are some binomial coefficients and the α_i are the Floquet indices for the Dirac problem $(i\partial - g\sigma)\psi_j = 0$ where $\psi_j(x,t+T) = \exp(-i\alpha_j)\psi_j(x,t)$. The problem is to find the stationary

phase points of eq. (3.5). To simplify matters we will look only for time independent ones, which means we have to consider the simpler problem

$$\frac{\delta}{\delta\varphi(x)}\left\{-\int_{-\infty}^{+\infty}\frac{\mathcal{T}}{2}\left(\sigma^2(x') - \sigma_0^2\right)dx' + N\left[-\int\frac{\delta\,d\omega}{\pi} + \omega_0(\sigma) + g\sigma_0 - n_0\,\omega_0(\sigma)\right]\right\} = 0 \tag{3.6}$$

where $\int \frac{\delta\,d\omega}{\pi}$ expresses the phase shift experienced on putting the

system in a box with periodic b.c. and carefully counting modes.

For our purposes we can write this even more schematically as

$$\frac{\delta}{\delta\varphi(x)}\left\{-\int_{-\infty}^{+\infty}\frac{\mathcal{T}}{2}\left(\sigma^2(x') - \sigma_0^2\right)dx' + \bar{f}\left[\sigma(x),\sigma_0\right]\right\} = 0 \tag{3.7}$$

with the same assumed b.c. as above. To solve an equation like eq. (3.7), consider $\sigma(x)$ as a time independent potential of a Dirac equation, whose solution will be a functional of $\sigma(x)$. So $(i\partial - g\sigma)\psi = 0$ and setting $\partial_t^2 = \omega^2$, premultiplying by $(i\partial + g\sigma)$ gives the Schrödinger form

$$\psi'' - [g^2(\sigma^2 - \sigma_0^2) \pm g\sigma']\psi = -(\omega^2 - g^2\sigma_0^2)\psi \tag{3.8}$$

where we have just added and subtracted a constant field σ_0. Make the obvious identification $k^2 \equiv (\omega^2 - g^2\sigma_0^2)$ and $u(x) \equiv [g^2(\sigma^2 - \sigma_0^2) \pm g\sigma']$ with b.c. $\sigma(x) \to \sigma_0$ $|x| \to \infty$.

Using the trace identity eq. (2.1), we can immediately write $\int u(x)dx$ as a function of the reflection coefficients and bound state eigenvalues of this associated scattering problem, in a factorized form. This allows the variation $\delta/\delta\sigma(x)$ to become independent variations on $(\delta/\delta r(k),\ \delta/\delta k_\ell,\ \ell = 1, \ldots m)$.

If one writes out everything completely and carefully performs the variation[9], the saddle point conditions are: (1) $r(k) = 0$, (2) algebraic relations from the k_ℓ variations, which imply the energy spectrum $E_{n_0} = 2/\pi\, g\sigma_0 N \sin(\theta)$, $\theta = \pi/2\, n_0/N$. The result (1) says that $u(x)$ is a reflectionless potential, which is characteristic of 1+1 dimensional perfect systems, the reason for which emerges in its clearest form here. By using GLM equations and the value for θ and $r(k) = 0$, one reconstructs $\sigma(x)$ as

$$\sigma(x) = \sigma_0 + \sigma_0 y \tanh\left(g\sigma_0 y x - \frac{1}{4}\ln\frac{(1+y)}{(1-y)}\right) - \sigma_0 y \tanh\left(g\sigma_0 y x + \frac{1}{4}\ln\frac{(1+y)}{(1-y)}\right);\quad y = \sin(\theta) \tag{3.9}$$

which is a soliton-antisoliton pair.

RESULTS AND PROBLEMS IN HIGHER DIMENSIONS

In summary, let us list some general features of the trace identity method. In 1+1 dimensions it is:

(1) a method that is deductive, for rather than guessing all the classical saddle points, it involves only the choice of a clever scattering problem. The classical solutions are generated as output.

(2) It avoids having to separate out the various zero symmetry modes in the functional measure.

(3) Generates analytic conditions on the energy spectrum, in many cases a simple algebraic equation.

How much of this generalizes to higher dimensions? The solid results are

(a) There is a generalization of the inverse scattering reconstruction procedure in 3+1 dimensions, for many kernels, due to Faddeev and Newton[6].

(b) The existence of trace identities and their explicit form, for a wide variety of kernels, is known in 3+1, mainly due to Buslaev[10].

This was part of the reason for going through the rather arcane demonstration of sec. (2). It makes no reference to dimension and a variant of the argument can be made as tight as one pleases. The problems with higher dimensional versions are the following:

(a) The potentials generated by 3+1 inverse scattering methods are not unique. This is not fatal: Quantum considerations may put on enough constraints to pick out one of the set. This is already a salient feature of 1+1 dimensional problems.

(b) While there are strong factorization theorems for the S matrix in higher dimensions (e.g., triangularzation), the explicit implementation of them has not been worked out and is strongly dependent on the particular scattering problem one picks.

(c) Methods for efficiently generating asymptotic expansions for $\eta(k)$, for any kernel, simply don't exist in the literature. There exist methods[10,11], but they are very cumbersome to operate.

This last point has been overcome in the last year by Neveu and myself[12]. We have managed to develop very fast ways of computing the coefficients in $\eta(k) = \sum_{n=1}^{\infty} \frac{a_n}{k^n}$ by using either functional integral representations for the partition function and doing a high temperature expansion, giving the a(k) directly, or using a contour representation of the partition function coupled with a Feynman diagram technique, that is good even for gauge theories and in many dimensions. We do not know how to handle the factorization of the S matrix (i.e., what are the independent scattering data) for models with gauge fields, but feel that this problem can eventually be overcome. The development of such methods would open up the practical possibility of doing strong coupling approximations in non-abelian gauge models.

REFERENCES

1. See review article by A. Scott, F. Chu and D. McLaughlin in Proc. IEEE 61 1443 (1973) for a comprehensive discussion of known 1+1 dimension perfect systems and inverse scattering methods applied to them.

2. R. Dashen, B. Hasslacher, A. Neveu, Phys. Rev. D10 4114 (1974); 10, 4130 (1974); D11, 3424 (1975); "Some Non-Perturbative Semi-Classical Methods in QFT." D. Rajaraman, Phys. Reports 21, 227 (1975).

3. G. 't Hooft, Nucl. Phys. B79, 276 (1974).

4. Belavin, Polyakov, Schwartz, Tyupkin, Phys. Letters 59B, 85(1975); "Computation of Quantum Effects due to a 4-Dimensional Pseudo-particle," G. 't Hooft, Harvard Univ. preprint (1976).

5. "Quark Confinement and the Topology of Gauge Groups," A. M. Polyakov, Nordita preprint Oct. 1976 wherein he makes a similar conjecture, based on different reasoning.

6. L. D. Faddeev, 3-Dimensional Inverse Problems in Quantum Theory of Scattering, preprint ITP-71-106E, Kiev, 1971; R. G. Newton, Lectures at 1974 Summer Seminar on Inverse Problems, AMS, Aug. 1974.

7. V. E. Zakharov and L. D. Faddeev, Transl. in Func. Anal. Appl. 5 280, (1972).

8. V. S. Buslaev and L. D. Faddeev, Dokl. Akad. Nauk SSR, 132 (1) 1960.

9. R. Dashen, B. Hasslacher and A. Neveu, Phys. Rev. D12, 2443 (1975

10. V. S. Buslaev, "Trace Formulas and some Asymptotic Estimates of the Resolvent Kernel of the 3-Dimensional Schrödinger Equation." Topics in Math. Phys. Vol. 1, Spectral Theory and Wave Processes, M. Sh. Birman, ed. Leningrad University Press (1966).

11. I. Percival, Proc. Phys. Soc. 80(6) (1962); I. Percival and M. Roberts, Proc. Phys. Soc. 82(4) (1963).

12. B. Hasslacher and A. Neveu (Forthcoming Caltech preprint).

DYNAMICAL CONSERVATION LAWS IN THE QUANTIZED MASSIVE THIRRING MODEL +

M. Lüscher

II. Institut für Theoretische Physik

der Universität Hamburg, Hamburg, Germany

It has been shown recently by A. Luther [1] that the quantized, renormalized massive Thirring model can be considered as the continuum limit of a tractable lattice quantum field theory. This latter model lives on a one dimensional lattice with spacing a (time remains continuous). Its Hamiltonian is given by

$$H = \sum_n \Big\{ \frac{i}{2a} v(G)(\phi_n^+ \phi_{n+1} - \phi_{n+1}^+ \phi_n) + (-1)^n \frac{m_o}{2}(\phi_n^+ \phi_{n+1}^+ + \phi_{n+1} \phi_n) - \\ - \frac{G}{2a}(\phi_n^+ \phi_n - \tfrac{1}{2})(\phi_{n+1}^+ \phi_{n+1} - \tfrac{1}{2}) \Big\} - E_o$$

Here, ϕ_n denotes a (one component) fermi operator at site n, G is a renormalized coupling constant, m_o is the bare mass, v (G) is a finite renormalization constant keeping the speed of light equal to one and E_o makes the ground state energy vanish.

The low-lying spectrum of H can be calculated explicitly. This is due to the fact that H is essentially a logarithmic derivative (with respect to a certain coupling constant V) of the transfer matrix T(V) of the Baxter model. The largest resp. second largest eigenvalues of T(V) have been evaluated by Baxter [3] resp. Johnson, Krinsky and McCoy [4]. As expected from the Coleman equivalence theorem [5] one finds soliton, antisoliton and breather states.

+ This is a brief summary of results that emerged from an analysis of the lattice massive Thirring model as proposed by A. Luther [1] . Full details are given in reference [2] .

The relationship between the lattice massive Thirring model and the Baxter model also makes it possible to explicitly construct an infinite set of local, conserved charges Q_n for the fermion model. They correspond to higher logarithmic derivatives of the transfer matrix T(V). Since the transfer matrices commute for different values of the parameter V [3] so do all the charges Q_n (because H is equal to the lowest charge Q_o this implies especially that the Q_n's are conserved).

The one particle states with momentum q and energy E(q) are simultaneous eigenstates of all charges. The corresponding eigenvalues are:

$$Q_n(q) = \left(E(q)\frac{\partial}{\partial q}\right)^n E(q)$$

Consider now a general scattering process involving solitons, antisolitons and breathers. Assume that there are m incoming particles with momenta $q_1, \ldots, q_m$ and m' outgoing particles with momenta $q'_1, \ldots, q'_{m'}$. Since the charges Q_n are local they are asymptotically additive, i.e.

$$Q_n(q_1) + \ldots + Q_n(q_m) = Q_n(q'_1) + \ldots + Q_n(q'_{m'})$$

This infinite set of conservation laws restricts the possible scattering processes severely. In fact, one finds that

a) the total number of fermions (solitons, antisolitons) and the number of breather modes with internal quantum number n are conserved separately,

b) the sets of momenta q_i (q'_i) of incoming and outgoing fermions are equal,

c) the same as b) holds for each type of breather mode separately.

These statements are independent of the cutoff (i.e. the lattice constant a) and are therefore true for the continuum, renormalized massive Thirring model as well.

References:

1 A. Luther: Phys. Rev. B14, 2153 (1976)

2 M. Lüscher: DESY 76/31, to appear in Nucl. Phys. B

3 R. J. Baxter: Ann. Phys. (N.Y.) 70 193, 323 (1970)

4 J.D. Johnson, S. Krinsky, B.M. McCoy: Phys. Rev. A8 2526 (1973)

5 S. Coleman: Phys.Rev. D11, 2088 (1975).

HIGHER CONSERVATION LAWS IN THE MASSIVE THIRRING MODEL

R. Flume

Theoretical Physics Division - CERN

1211 Geneva 23

A.

Among field theories in two space-time dimensions the massive Thirring model (MTM) and the Sine-Gordon theory (SGT) play a prominent rôle in the recent discussion about the so-called soliton phenomenon.

The SGT considered as a classical field theory is an integrable system [1]. This fact makes this model an ideal test ground for the application of semi-classical approximation methods [2]. A by-product of the integrability of the SGT is the existence of an infinite number of non-trivial local conservation laws.

The MTM on the other hand has been discussed extensively under the aspect of a correspondence with the SGT [3]. We will deal here with an additional characteristic feature of the MTM displaying some analogies to the SGT, namely the existence of an infinite number of non-trivial conservation laws. We do not know at present whether these conservation laws (and their analogues in the SGT) can be brought into connection with the correspondence mentioned above. The material presented in this contribution is compared to the oral version of the seminar given at the Cargèse summer school supplemented by some results not yet known at that time.

B.

The MTM is defined through the formal equation of motion

$$(i\partial_\mu \gamma^\mu - m)\psi = g\,\bar{\psi}\gamma^\mu\psi\gamma_\mu\psi \tag{1}$$

for an anticommuting spinor field ψ in two space-time dimensions. To start with we derive from Eq. (1) the simplest higher conservation law where all terms carry canonical dimensions not higher than five. By use of the equation of motion one easily verifies the identity

$$\begin{aligned} \partial_\mu (i\bar{\psi}\gamma^\mu \psi_{,3u} + h.c.) &= \partial_u (\ldots.) \\ &- 6g\,\bar{\psi}_{,u}\gamma_\mu \psi_{,u}\,\partial_u(\bar{\psi}\gamma^\mu\psi), \end{aligned} \tag{2}$$

$$u = \frac{t+x}{2}, \quad v = \frac{t-x}{2}, \quad (\psi_{,nu} = \partial_u^n \psi).$$

Rewriting the second term on the right-hand side of (2) in spinor components and applying the equations of motion in characteristic co-ordinates

$$\begin{aligned} i\,\psi_{2,u} - m\psi_1 &= 2g\,\psi_1^+\psi_1\psi_2 \\ i\,\psi_{1,v} - m\psi_2 &= 2g\,\psi_2^+\psi_2\psi_1 \end{aligned}$$

one obtains

$$\bar{\psi}_{,u}\gamma_\mu\psi_{,u}\,\partial_u(\bar{\psi}\gamma^\mu\psi) = -2\psi_{1,u}^+\psi_{1,u}\partial_v(\psi_1^+\psi_1) \tag{3}$$

In the last equality use has been made of the Grassman algebra property of the ψ field : terms with two or more equal spinor components vanish identically. By the same recipes (equation of motion and Fermi statistics) one realizes that the right-hand side of Eq. (3) can be rewritten as a total derivative

$$\psi_{1,u}^+\psi_{1,u}\partial_v(\psi_1^+\psi_1) = \partial_v(\psi_{1,u}^+\psi_{1,u}\psi_1^+\psi_1) \tag{4}$$

The information of Eqs. (2)-(4) can now be summarized in the conservation law

$$\partial_\mu (i\bar{\psi}\gamma^\mu \psi_{,3u} + h.c.) + \partial_u (\ldots)$$
$$- 12 g \partial_v (\psi^+_{1,u} \psi_{1,u} \psi_1^+ \psi_1) = 0 . \tag{5}$$

By use of the equation of motion and exploitation of the Grassman property of the spinor fields in the spirit of the procedure demonstrated above several conservation laws were found in Refs. 4) and 5)

$$\partial_\mu (i\bar{\psi}\gamma^\mu \psi_{,5u} + h.c.) + \ldots = 0$$
$$\partial_\mu (i\bar{\psi}\gamma^\mu \psi_{,7u} + h.c.) + \ldots = 0$$

The dots denote those terms in the conservation laws which are more than bilinear in the spinor fields. In Ref. 4) also the conservation law with nine u derivatives has been calculated. A compact recursion formula generating all conservation laws

$$\partial_\mu (i\bar{\psi}\gamma^\mu \psi_{,(2n+1)u} + h.c.) + \ldots = 0$$
$$1 \leq n < \infty \tag{6}$$

has been given by Kulish and Nissimov [6]. This recursion formula can be most conveniently quoted in terms of a generating relation [7]

$$\partial_u \chi = \frac{m}{\gamma}\chi - \frac{2m}{\gamma}\psi_1 - i g \psi_1^+ \psi_1 \chi - - i g \chi^+ \chi \psi_1 , \tag{7a}$$

$$\partial_v \chi = -m\gamma\chi - 2mi\psi_2 - i g \psi_2^+ \psi_2 \chi - - \gamma g \chi^+ \chi \psi_2 , \tag{7b}$$

where χ denotes a generic function to be developed in powers of the real parameter γ. The integrability condition of Eqs. (7a) and (7b) being satisfied in virtue of the equation of motion (1) for ψ renders if developed in powers of γ the conservation laws (6). Using the generating relation (7) one can derive a Bäcklund transformation [8] for the MTM [7]. For the SGT a neat connection has been established between its Bäcklund transformation and an associated linear eigenvalue problem, which supplies via the inverse scattering method a proof for integrability [8]. Having this in mind one might speculate that the MTM considered as a classical Grassman number system is integrable.

We turn now to the meaning of the conservation laws (6) for the quantized MTM. It has been observed by Polyakov [9] that, provided conservation laws of the form of Eq. (6) are respected by the quantum corrections to the classical (tree graph) approximation, the S matrix will be a pure phase, i.e., there will be no particle production and no non-trivial multiparticle scattering. The renormalization of the first higher conservation law [Eq. (5)] has been carried out in Ref. 10). It was found that the possible anomalies in the integrated Ward identities can be absorbed multiplicatively. It seems rather plausible that all conservation laws (6) can be renormalized with the same result found for the renormalized version of Eq. (5). From the conservation law, Eq. (5), alone one can infer the following consequences for the S matrix :

(a) particle production $2 \rightarrow 2n$, $n \geq 2$ is absent [11] ;

(b) the connected three-particle scattering amplitude factorizes into two-body amplitudes [12].

(a) can be easily verified if one restricts the integrated Ward identity to the mass shell which gives

$$\sum P_{iu,in}^{3} + \sum P_{ju,out}^{3} = 0 \tag{8}$$

$P_{ju,(in,out)}$ denotes the momentum of an incoming resp. outgoing particle conjugate to the variable u. A simple calculation [11] shows that for no kinematical situation of particle production $2 \rightarrow 2n$, $n \geq 2$ the polynomial on the left-hand side of (8) vanishes. (b) can be derived using results of analytic S matrix theory taking into account the vanishing of production amplitudes $2 \rightarrow 2n$ [12].

REFERENCES

1) M.J. Ablowitz, D.J. Kaup, A.C. Newell and H. Segur - Phys. Rev.Letters 30, 1262 (1973) ;

L.D. Faddeev and L.A. Takhtajan - JINR Preprint E27998 (1974).

2) "Extended Systems in Field Theory" - Proceedings of the Meeting held at the Ecole Normale Supérieure (June 1975), organized by J.L. Gervais and A. Neveu, Physics Reports 23C, 3 (1976), and references therein.

3) See the contributions of S. Coleman and S. Mandelstam in Ref. 2) ;

B. Schroer and T.T. Truong - Preprint Berlin FUB, HEP6 (1976).

4) B. Berg, M. Karowski and H.J. Thun - "Conserved Currents in the Massive Thirring Model", FUB Preprint, May 76/5.

5) R. Flume, P.K. Mitter and N. Papanicolaou - Phys.Letters B, in press.

6) P.P. Kulish and E.R. Nissimov - Pis'ma v. JETP (in Russian) 24, 247 (1976).

7) J. Stehr - Private communication ; to appear as DESY preprint.

8) K. Polhmeyer - Contribution to these proceedings.

9) Quotation in : L.D. Faddeev, "Quantization of Solitons", Preprint Institute for Advanced Study, Princeton (1975).

10) R. Flume and S. Meyer - CERN Preprint TH. 2243 (1976).

11) This observation is due to V. Glaser ; private communication.

12) D. Iagolnitzer - Private communication.

QUANTUM FIELD THEORY OF KINKS IN TWO-DIMENSIONAL SPACE-TIME

B. Schroer

CERN - Geneve +)

ABSTRACT

We discuss the construction of interpolating quantum fields for topological kinks in two dimensional quantum field theories.

+) Permanent address: Institut für Theoretische Physik
Freie Universität Berlin, Berlin, Germany

I. BRIEF OUTLINE OF CONTENT

After a short glossary of classical kink-aspects in the second section we will discuss the relation between the Thirring-model and the sine-Gordon-model (section 3). Our presentation is based on short distance properties of the massive Thirring model [1]. We obtain the Coleman equivalence [2] only for $\beta^2 < 4\pi$ (using Coleman's terminology). For the free massive Dirac equation $\beta^2 = 4\pi$ and for $4\pi \leq \beta^2 < 8\pi$ the equivalence relations are modified due to the occurrence of nonleading short-distance singularities [3]. These short distance problems also slightly modify the "bosonization" picture of the massive Thirring-Schwinger model as well as generalizations to models involving several fermions. In the fourth section we will comment briefly on quantum field theoretical aspects of "soliton" properties (i. e. an infinite number of non-topological conservation laws). These properties seem to be most specific for two-dimensional field theories and will probably lead to the explicit solubility of such models. In the subsequent section 5 we will demonstrate that the sine-Gordon kinks are statistical "schizons". If one asks for local fields which carry the sine-Gordon topological charge one finds two local fields which generate the same big Hilbert-space: The massive Thirring fermion field and a noncanonical (with dimension $\frac{1}{4}$ if the associated Thirring field has canonical dimension) bosonic field. This ambiguity in statistics persists to other nonlocal two-dimensional kinks as we explain in section 6 for the A^4-kink. This section also contains some remarks on the still incomplete picture of $D > 2$ quantum kinks (i. e. monopoles and vortices).

II. GLOSSARY OF CLASSICAL ASPECTS

The physical interest and relevance concerning topological concepts in relativistic field theory has been discussed in several recent publications [4]. Usually the investigation of these phenomena on the quantum level starts from the reinterpretation of the classical "kink"-solutions as the lowest step of some quasiclassical approximation scheme [5, 6, 7, 8]. In the following we will not make use of quasiclassical methods because the existence and properties of quantum kinks is a basic issue of general quantum field theory which should not be burdened with approximation methods which in most cases have only an asymptotic validity for small (suitably chosen) coupling parameter [5, 6, 7, 8]. So the main reason for starting with classical field equations is (as in reference 4) historical and pedagogical. Consider first the A_2^4 theory in the broken mode:

$$\mathcal{L} = \frac{1}{2}\partial_\mu A \partial^\mu A - \mathcal{U}(A) \tag{1}$$

with $\mathcal{U}(A) = \frac{\lambda}{2}(A^2 - a^2)^2$

$a^2 = \frac{\mu^2}{\lambda}$ i. e. μ^2 = coefficient of mass term

This Lagrangian leads to spontaneous symmetry-breaking of $A(x) \longrightarrow -A(x)$.

The requirement of finite energies

$$E = \int \mathcal{H}(x)\, dx \quad ; \quad \mathcal{H}(x) = \frac{1}{2}[\dot{A}^2 + (\partial_x A)^2] + \mathcal{U}(A) \tag{2}$$

and the continuity of temporal development leads immediately to the following asymptotic statement [1]:

$$A(t,x) \xrightarrow[x \to \mp\infty]{} \{A(t,\mp\infty)\} = \{a, -a\} \tag{3}$$

The asymptotes are independent of t and hence define four inequivalent classes of solution; an element of one class for example $A(t, \mp\infty) = a$ cannot be continuously deformed into another class for instance $A(t, \mp\infty) = \mp a$ without penetrating an "infinite energy barrier" [5]. Keeping an eye on the corresponding quantum field theory which is built on one vacuum state, it is customary to work only with two classes:

$$K_0 = \{ A ; A(t, \mp\infty) = a \}$$
$$K_1 = \{ A ; A(t, \mp\infty) = \pm a \} \qquad (4)$$

Using the invariance: $A(x) \to A(-x)$ we then identify the class $A(t, \mp\infty) = \mp a$ with K_1 ; in constructing asymptotic "two kink" solutions $A_{1,2}$ (necessarily time dependent solutions) we thus follow Coleman [4] and compose the $A_1(x)$ function which we assume to be localized to the far left of the origin with the $A_2(-x)$ function so that the resulting function looks as $A_1(x)$ far to the left and as $A_2(-x)$ to the far right. This situation of a vacuum class and a one-kink class is typical of a broken finite symmetry group which leads always to a kink number modulo the order of the group [4].

The generalization of this topological idea to classical field theories with more space-dimensions leads to vortices and monopoles [9,10,11]. The common feature of all generalizations is that relativistic classical fields A at spatial infinity define a mapping [4] of manifolds:

$$A: \quad M \text{ (space at infinity)} \longrightarrow N \text{ (internal symmetry manifold)}$$
which has an integer degree: deg A $\qquad (5)$

These mappings may be divided into homotopy classes of fields which cannot be continuously deformed into each other without going through infinite energy barriers.

A physically slightly different but mathematically similar situation is provided by the classical sine-Gordon equation:

$$D = 2 \quad , \mathcal{L} = \tfrac{1}{2}\partial_\mu \varphi \, \partial^\mu \varphi \; - (1 - \cos\varphi) \tag{6}$$

Here the classical observables e.g. the energy density

$$\mathcal{H}(x) = \frac{1}{2}(\pi^2 + \mathrm{grad}^2\varphi) + (1 - \cos\varphi) \tag{7}$$

have the periodicity: $\varphi \rightarrow \varphi + 2\pi$

Therefore we may consider as our field manifold

$$S^1 = \{\varphi \mod 2\pi\}$$

which then leads to the mapping:

$$\varphi : \quad R^1 \text{ (space)} \longrightarrow S^1 \text{(field values)} \tag{8}$$

Without loss of generality (because of the structure of the observable field manifold) we may restrict ourselves to solutions with $\varphi(-\infty) = 0$. Since the finite energy condition requires $\varphi(+\infty) = 2n\pi$ we therefore have:

$$\varphi : \quad S^1 \text{(compactified space)} \longrightarrow S^1 \text{(field manifold)} \tag{9}$$

Such mappings are characterized by winding numbers n :

$$K_n = \{\varphi ; \; \varphi(\infty) - \varphi(-\infty) = 2n\pi\} \tag{10}$$

The generalization of this idea to higher dimensions leads to field theories with an intrinsically nonlinear field manifold. A prototype of these theories is the nonlinear σ-model:

$$D = 4 \quad , \mathcal{L} = \tfrac{1}{2}\partial_\mu \vec{\varphi} \, \partial^\mu \vec{\varphi} \quad , \vec{\varphi}^2 \equiv \sum_1^4 \varphi_i^2 = 1 \tag{11}$$

The vacuum remains unbroken and without loss of generality the $\vec{\varphi}$ at spatial infinity may be assumed to approach a universal value, for example: $\lim_{\vec{x}\to\infty} \varphi_i(\vec{x}) = \delta_{i1}$. One then may interpret $\vec{\varphi}$ as a mapping:

$$\vec{\varphi} \quad : \quad S^3 \text{(compactified space)} \longrightarrow S^3 \text{("internal sphere")} \tag{12}$$

For reasons of brevity we will say that the field "winds" and the degree function will be referred to as the "winding number". We are aware of the fact that in the case of A_2^4 this constitutes a slight misuse of terminology. Classical solutions with nontrivial winding numbers will be called "kinks". So our kinks are in Coleman's language [4] topological "lumps". The name soliton we reserve for solutions of Lagrangians which have additional (i. e. in addition to pure topological considerations) infinitely many conserved currents [12, 13], as the sine-Gordon Lagrangians. The discussion of nonrelativistic models show that a soliton situation can arise independently of "winding".

The first type of kinks for which the symmetry breaking of vacuum states is important and which in higher dimension leads to Landau-Ginsberg-Abrikosov-Nielsen-Olsen [8] vortices and 't Hooft-Polyakov monopoles [9, 10] will be called a "vacuum-kink", whereas the sine-Gordon and the σ-model kink are examples of a "nonlinear field kink". The study of the latter has been advocated by Fadejev [14] and the main reason why (except the sine-Gordon model) they do not enjoy the same amount of popularity as the vacuum kinks is that the corresponding quantum Lagrangians are perturbatively nonrenormalizable.

A quantum field theory of kinks has to answer the following questions:

1. Is there a local conserved current whose charges count winding numbers?

2. Are their interpolating kink fields whose repeated applications to the vacuum generate all the higher "winding"-charge sectors?

3. What are the locality properties of interpolating kink operators, in particular is it possible to find a local representative for an interpolating field? What are their transformation properties and the commutation relations for large space-like separations?

4. What particle statistics does the LSZ scattering theory (in the case of local interpolating fields) or in the general case the Haag-Ruelle scattering theory lead to?

For all questions one is very far away from being able to give satisfactory and general answers. The first question only makes sense if the winding numbers consist of the set of all integers. A topological class number mod 2 as for the A_2^4 theory should be interpreted as a multiplicative quantum number, as for example the multiplicative univalence rule (or C-parity) of a Majorana spinor field. In the sine-Gordon case, the winding number current is

$$j_\mu(x) = \varepsilon_{\mu\nu}\partial^\nu\varphi(x) \tag{13}$$

and for the 't Hooft-Polyakov monopoles such a current is given by the monopole density:

$$k^\mu = \varepsilon^{\mu\nu\lambda\sigma}\partial_\nu F_{\lambda\sigma} \tag{14}$$

with $F_{\lambda\sigma}$ being the inner product of the nonabelian field strength $G_{\lambda\sigma}$ with the Higgs field ϕ :

$$F_{\lambda\sigma} = \phi_{vac}^{-1} \vec{\phi} \vec{G}_{\lambda\sigma} \quad , \quad \phi_{vac} = |\langle 0|\vec{\phi}|0\rangle| \tag{15}$$

The renormalization aspects of these currents are a part of the renormalization of the composite fields of the Lagrangian theory.

For the nonlinear field kinks in $D > 2$ Fadejev [14] has given examples of topological conserved currents, however, their quantum field theoretical aspects are still obscure due to the lack of renormalizability.

The second question concerning interpolating kink operators is of a more subtle nature. One may have for example classical kinks arising from topological considerations which do not have quantum counter-parts. This happens if the theory lacks symmetry. An illustration of this is provided by a D=2 scalar field Lagrangian which maintains the classical vacuum degeneracy of the (Higgs mode) A_2^4 theory but lacks the $A \to -A$ symmetry. There are no quantum kinks because of infinite energy fluctuations [15] even though the classical kinks still exist [section 6].

Even if the requirements of topology and symmetry are met, the interpolating field must still be chosen with great care in order to avoid infinite energy fluctuations. For example interpolating fields of the type used for coherent states will (with the exception of the sine-Gordon theory for dim $\cos \beta\phi < 1$) produce infinite energy fluctuations [16]. Experience with two-dimensional models suggest that interpolating kink fields in that case involve at least bilinear exponential functions of the basic fields [17].

A discussion of the problem of commutation relations for asymptotically space-like separated interpolating kink fields clearly depends on ones ability to solve the previous problem of their explicit construction. Up to now this has not been possible for $D > 2$, but in $D = 2$ models the picture is fairly clear and will be explained in sections 5 and 6. The last problem, namely the spin and statistics of kink particles via a Haag-Ruelle scattering theory [18] of the interpolating field is simple if these fields act in a Hilbert-space of positive metric and without zero mass states. For the interesting case of the 't Hooft-Polyakov monopoles the scattering theory, i. e. the structure of multiparticle states is troubled by precisely these problems.

III. SHORT DISTANCE BEHAVIOR IN THE MASSIVE THIRRING-MODEL AND CORRESPONDENCE WITH THE SINE-GORDON-MODEL

The quantum field theory of the sine-Gordon kinks can be studied in a remarkable explicit fashion thanks to S. Coleman's observation that this field theory may be embedded into the massive Thirring model.

We will briefly analyse this correspondence by using the short-distance behavior of the massive Thirring-model [1]. This model belongs to the Lagrangian:

$$\mathcal{L} = i\bar{\psi}\gamma_\mu\partial^\mu\psi - g\, j_\mu j^\mu - m\bar{\psi}\psi \tag{16}$$

Our starting point is the statement that a pseudo-current which has an affiliated conserved current always leads to a local pseudo-potential [1]:

$$j_{\mu 5} = N[\bar{\psi}\gamma_\mu\gamma^5\psi] \quad , \quad j_\mu = N[\bar{\psi}\gamma_\mu\psi] = \varepsilon_{\mu\nu} j_5^\nu \tag{17}$$

$$\partial^\mu \varepsilon_{\mu\nu} j_5^\nu = 0 \quad \rightarrow \quad j_{\mu 5} = -\frac{1}{\sqrt{\pi}} \partial_\mu \phi \tag{18}$$

which satisfies the equation

$$\frac{1}{\sqrt{\pi}} \partial^2 \phi = 2im\, N[\bar{\psi}\gamma^5\psi] \tag{19}$$

This field has the following Källén-Lehmann representation:

$$\langle \phi(x)\, \phi(0) \rangle = \int \rho(s^2)\, i\Delta^{(+)}_{ms}(x)\, ds^2 \tag{20}$$

with

$$i\Delta^{(+)}_{ms}(x) = \frac{\pi}{4} H_0^1\left(ms\sqrt{-x^2+i\varepsilon x_0}\right) \tag{21}$$

$$\rho(s^2) = c\,(s^2)^{a-3} \qquad \text{for} \qquad s \to \infty \tag{22}$$

$$a = \dim \bar{\psi}\psi$$

Only in this asymptotic statement for the Lehmann-Källén spectral function we used for the first time a specific property of the model: the massive Thirring-model is for $a < 2$ asymptotically scale-invariant for short distances. In the framework of Callan-Symanzik equations [19, 20] this means that $\beta \equiv 0$ [21] and the validity of the homogeneous differential equations for short distances (or nonexceptional scaled up euclidean momenta). These properties of the theory are also sufficient [1] in order to prove:

$$\langle :\phi^n: \rangle = c_n < \infty \tag{23}$$

where the double dot denotes Wick-ordering in the sense of two-point function subtractions. In other words, the short-distance behaviour of ϕ is that of a free field as a result of asymptotic scale invariance. Note, however, that the canonical conjugate $\pi = \dot{\phi}$ develops a more singular short-distance behaviour than that of a free field as soon as the coupling $g \gtrsim 0$ so that dim $\bar{\psi}\psi \gtrsim 1$ (in Coleman's terminology $\beta^2 \gtrsim 4\pi$).

$$\langle \pi(0,x)\pi(0,y)\rangle = \int ds^2 s^2 \rho(s^2) \, i\Delta^{(+)}_{m_s}(x-y) \quad + \text{grad. terms} \tag{24}$$

Clearly the short distance behaviour of the first term is:

$$-\frac{1}{4\pi}\int ds^2 \rho(s^2) s^2 \cdot \ln(-\xi^2 + i\varepsilon\xi_0)m^2 \tag{25}$$

The logarithmically ultraviolet-divergent integral in front of the log which one obtains for $a = 1$ indicates potential short-distance trouble although the canonical formalism (which requires that the space-smeared π is a well-defined operator) only breaks down for $a \gtrsim 3/2$.

We will return to the regime dim $\bar{\psi}\psi \gtrsim 1$ of the model after we obtained a good insight into its dim $\bar{\psi}\psi < 1$ regime.

The local pseudopotential ϕ is only determined by (18) up to a multiple of the charge operator. We make the choice:

$$\phi = \sqrt{\pi}\int_{-\infty}^{x} j_0(x')\,dx' \tag{26}$$

which leads to the space-like commutation relations

$$\frac{1}{\sqrt{\pi}}[\phi(x), \psi(y)] = -\theta(x^1-y^1)\psi(y) \tag{27}$$

$$[:e^{-2i\sqrt{\pi}\lambda\phi(x)}:, \psi(y)] = \theta(x^1-y^1)(e^{2\pi i\lambda}-1)\cdot$$
$$\cdot\psi(y):e^{-2i\sqrt{\pi}\lambda\phi(x)}: \tag{28}$$

Formula (28) is of a formal nature; although the Wick-products for polynomials are defined (23) we still need the finiteness of the exponentials.

Let us now study the linear combinations of two local operators:

$$B(x) = N[\psi_1^+\psi_2](x) - cm :e^{-2i\sqrt{\pi}\phi(x)}: \tag{29}$$

where the constant c will be chosen appropriately. The operator B is not only local with respect to itself but also relatively to the ψ's and $\bar{\psi}$'s. The scale dimension of the exponential of ϕ is determined by the Schwinger term b :

$$\langle\phi(x)\,\phi(0)\rangle = \frac{-b}{4\pi}\ln(-x^2)m^2 \tag{30}$$

$$b = \int\rho(s^2)\,ds^2$$

namely

$$\dim :e^{-2i\sqrt{\pi}\phi}: = b$$

In order to understand the equality of b with a let us look at the "bosonization" of zero mass fermions [22, 23]. A zero mass

Thirring spinor field can be written as [1]:

$$\psi_1(u,v) = \sqrt{\frac{m}{2\pi}} \; :e^{i\alpha_1\varphi_1(u) + i\alpha_2\varphi_2(v)}:$$
$$\psi_2(u,v) = \sqrt{\frac{m}{2\pi}} \; :e^{-i\alpha_2\varphi_1(u) - i\alpha_1\varphi_2(v)}: \tag{31}$$

$$u = t+x \quad , \quad v = t-x$$

in terms of "light cone" Bose-fields

$$\langle 0|\varphi_1(u)\varphi_1(u')|0\rangle = -\frac{1}{4\pi}\ln i(u-u')m \tag{32}$$

$$\langle 0|\varphi_2(v)\varphi_2(v')|0\rangle = -\frac{1}{4\pi}\ln i(v-v')m \tag{33}$$

Exponentials of Bose-fields have a built-in charge selection rule (in our case $Q \pm Q_5$). The only aspect in which (31) differs from the massless Thirring field is in the <u>relative</u> commutation property of the two components. If one wants to convert this into an anticommutation relation by a Klein-transformation, one just adds charges in the exponent (i. e. φ's at infinity) and thus obtains the Mandelstam representation [24]

$$\psi = \sqrt{\frac{m}{2\pi}} \; :\exp\left\{-i\frac{\alpha_1+\alpha_2}{2}\gamma^5\varphi - i\frac{\alpha_1-\alpha_2}{2}\int_x^\infty \dot{\varphi}(x')\,dx'\right\}: \tag{34}$$

with $\varphi = \varphi_1(u) + \varphi_2(v)$

The axial current is obtained from a split point limit [25] as:

$$N[\bar{\psi}\gamma_\mu\gamma^5\psi] = -\frac{\alpha_1+\alpha_2}{2\pi}\partial_\mu\varphi \tag{35}$$

which leads to the identification

$$\phi = \frac{\alpha_1+\alpha_2}{2\sqrt{\pi}}\varphi \tag{36}$$

for the zero mass limit.

(31) or (34) gives directly the identity

$$N[\psi_1^+\psi_2] = \frac{m}{2\pi} : e^{-2i\sqrt{\pi}\phi}: \tag{37}$$

in the massless theory (here m is just a normalization mass from (32), (33), (34)).

On the basis of the inhomogeneous Callan-Symanzik differential equations the zero mass bosonic description (37) hold for the leading short-distance part of the massive correlation function up to a multiplicative normalization constant c (g) which accounts for how we renormalized our massive theory [26]. Therefore the two-point function of (29) for $a < 2$ has cancelling leading parts:

$$\langle B(x) B(y) \rangle = \text{non leading singularity} \tag{38}$$

The Callan-Symanzik analysis also allows to make precise statements about nonleading singularities [27, 28, 29]. Applied to the case at hand one obtains a nonleading singularity which is by a factor $(\xi^2 m^2)^{2-a}$ less singular than the leading one [30]. So for dim there will be no nonleading short distance singularity: Hence we can apply the following theorem which replaces Schur's Lemma for canonical fields:

Theorem: If local field B which is local relative to some basis fields (generating the Hilbert-space from the vacuum) has a finite two-point function at short distances, it must be a multiple of the identity operator.

Proof: from the Lehmann-Källèn spectral representation one concludes $B|0\rangle = \lambda|0\rangle$ and from spacelike locality with respect to the complete set of basis fields (in our case ψ and $\bar{\psi}$) follows that $B = \lambda \mathbb{1}$.

Therefore by making an additive adjustment we finally obtain the desired identity:

$$N[\psi_1^+\psi_2](x) = c\, m\left(:e^{-2i\sqrt{\pi}\phi(x)}: -1\right) \tag{39}$$

valid for dim $\bar{\psi}\psi < 1$

For dim $\bar{\psi}\psi \geq 1$ not only does our proof breaks down, but also the exponential series of Wick-ordered polynomnial ceases to make sense. The origin is a very subtle "cumulative mass effect" which manifests itself as a insufficient decrease of the numbers (23) for large n. Note that in case of a zero mass theory all terms with n 2 in (23) would be zero. For the pseudo-potential of a free massive Dirac field they turn out to lead to a unit convergence radius [1] in λ for $:e^{-2i\lambda\sqrt{\pi}\phi}:$ This cumulative mass effect precisely sets in at dim $\bar{\psi}\psi = 1$ where according to the Callan-Symanzik picture the spinor correlation functions exhibit for the first time non-leading singularities which are well known for the massive free field:

$$\langle \psi_1(x)\, \psi_2^+(0) \rangle = m\, i\Delta_m^{(+)}(x) \tag{40}$$

What really happens in the regime dim $\bar{\psi}\psi \geq 1$? Is the relation (39) between normal ordered $\psi_1^+\psi_2$ and "appropriately" ordered exponentials of the pseudo-potential completely lost? For the case of the free field theory Lehmann and Stehr [31] have shown by an explicit computation based on "triple ordered" exponentials in which <u>all</u> vacuum expectation values (not just the two point function) have been removed, that a relation of the form:

$$N[\psi_1^+\psi_2] = :\phi\left(e^{-2i\sqrt{\pi}\phi} - 1\right): \tag{41}$$

still holds. (41) would lead to a modified sine-Gordon equation which lacks the classical periodicity property. It turns out that the triple ordering has some unusual properties which restore the intrinsic periodicity on the operator level [31]. In order to recover the classical relations it is helpful to introduce yet another normal product [1] via an operator short distance expansion of exponentials with half the λ -value (inside the convergence radius of the cumulative mass effect) :

$$:e^{-i\sqrt{\pi}\phi(x)}::e^{-i\sqrt{\pi}\phi(y)}: = \langle \dots \rangle 1 + c(\xi) N[e^{-2i\sqrt{\pi}\phi}] + \dots \quad (42)$$

For this normal product N the correspondence (39) continues to hold for $a = 1$ [1] and we expect its validity in the whole range $1 \leq a < 2$.

The same short-distance problem as in the massive Thirring model also occurs in the massive Schwinger-Thirring model for dim $\bar{\psi}\psi \geq 1$ i.e. in particular for the massive Schwinger model dim $\bar{\psi}\psi = 1$. One has to pay particular attention to the breakdown of the simple Wick-ordering (used by Coleman) in writing down spinor bilocals in terms of Bose fields. Cumulative mass effects are also relevant for certain generalizations with internal symmetries for example the QED_2-interaction on a massive SU_2 fermion field as recently studied by Coleman [32]. As an even simpler example consider the hamiltonian of a free SU_2 massive Dirac field:

$$\mathcal{H} = \sum_{i=1,2} \mathcal{H}_{0i} + m\bar{\psi}\psi \quad ; \mathcal{H}_{0i} = \text{kinetic part} \quad (43)$$

According to the general theorem on vector-potentials the commuting pseudocurrents have a local potential

$$:\bar{\psi}_i \gamma_\mu \gamma^5 \psi_i: = -\frac{1}{\sqrt{\pi}} \partial_\mu \phi_i \tag{44}$$

If one lets electromagnetism act on such a field, it is convenient to work with

$$\phi_\pm = \frac{1}{\sqrt{2}} (\phi_1 \pm \phi_2)$$

In terms of these fields the kinetic part is still diagonal, however, the mass term will be:

$$m \bar{\psi} \psi \longrightarrow m^2 \cos \sqrt{2\pi}\, \phi_+ \cos \sqrt{2\pi}\, \phi_- \tag{45}$$

Each single factor can be just ordered with double dots (the coefficient inside the cos has only half the "critical" value), however, for the ordering of the total product the Wick-ordering will be insufficient due to cumulative mass effects. If one wants to keep the periodic form one has to use our normal products defined by a split point limiting procedure similar to (42).

IV. SOLITON-CONSERVATION LAWS AND S-MATRIX

Two-dimensional models of relativistic field theories with non-vanishing masses often show a remarkable, often unexpected simplicity of their S-matrix. Consider for example two massive free Dirac-fields [1, 2] which are coupled by an interaction [33]:

$$\mathcal{L}_{int} = \lambda\, j_\mu^{(1)} j^\mu_5{}^{(2)} \tag{46}$$

This model is simpler than the massive Thirring model. It leads to a "semitrivial" S-matrix:

$$|p^{(1)}p^{(2)}\rangle = e^{\mp 2i\pi\lambda}\, |p^{(1)}p^{(2)}\rangle \tag{47}$$

The higher particle sector S-matrices being products of this two particle S. The sign of the energy-independent phase depends on whether species (1) enters of the right or left of species (2). This S-matrix cannot be transformed to $\mathbb{1}$ by multiplying it with functions of the charges $Q_{(1,2)}$ and it is for this reason that we do not call it "trivial"[34].

The local operator solution for (46) is:

$$\psi^{(1)} = :\exp -2i\sqrt{\pi}\lambda\,\phi^{(2)}:\,\psi_0^{(1)} \tag{48a}$$

$$\psi^{(2)} = :\exp 2i\sqrt{\pi}\lambda\,\phi^{(1)}:\,\psi_0^{(2)} \tag{48b}$$

Generalising the algorithm of fermion ordered exponentials of Lehmann and Stehr[31], one can evaluate the correlation functions to a remarkable explicit degree[34]. The semi-trivality of the S-matrix is a direct result of (48). For the massive Thirring model the correlation functions are unknown and one expects them to be much more complicated. How then can one see anything of the structure of the S-matrix of this model? The answer is hidden in a new set of higher conservation laws. Additional conservation laws to the convential energy-momentum conservations have been extensively studied for the classical sine-Gordon equation and related equation[35]. These "soliton" conservation laws, if valid in the quantum theory, lead to an S-matrix which describes purely elastic processes[36]. Perturbative S-matrix computations for the sine-Gordon[37] as well as the massive Thirring equation[38] have demonstrated the absence of creation of particles up to the level of one-loop diagrams. On the classical level it was subsequently shown that the massive Thirring Lagrangian supplemented by a Grassmann structure for the classical ψ also has an infinite set of conservation laws [39, 40]. The absence of anomalies in the Ward identities of these conserved currents has been recently shown for the lowest nontrivial current. The pure elastic nature of $S^{(3)}$ is already a result of the lowest proven quantum conservation law [41]. Macrocausality restricts the elastic n-particle S-matrix to be products of just the two particle matrix. Let us look for simplicity at the n-particle S-matrix of the sine-Gordon model. From the product structure

$$S^{(n)} = \prod_{i<k} S(\nu_{ik}) \tag{49}$$

with $\quad \nu_{ik} = \frac{p_i p_k}{m^2} = \cosh \theta_{ik}$

one immediately concludes [42, 43].

<u>Statement</u>: A bound state in S (ν) induces a bound state in the n-particle S-matrix (49). This family of bound states lie on the quasiclassical sine-Gordon trajectory

$$m_n = 2\mu \sin \tfrac{n}{2}\theta \tag{50}$$

Here μ and θ are constants which have been adjusted such that m_1 and m_2 are the elementary respectively first bound state. This result is somewhat puzzling since it gives a spectrum which never stops. The quasiclassical result, however [5, 36], has an n_{max} corresponding to the last state below the lowest two particle continuum. The relevant point here is, that the above statement does not take into account that higher poles in S may be compensated by CDD zeros [44] accompanying CCD poles in $S^{(2)}$.

So the existence of an n_{max} should actually be interpreted as a consistency requirement on the function which appears in the crossing symmetric unitary $S^{(2)}$:

$$S(\nu) = \frac{8\sqrt{\nu^2-1}\,g(\nu)-i}{8\sqrt{\nu^2-1}\,g(\nu)+i} \tag{51}$$

There is the justified hope that the sine-Gordon field theory respectively the Thirring model via their simple S-matrix structure lead to the explicit construction of their Heisenberg fields [42]. In this context it is interesting that A. Luther [45] recently succeeded to relate a suitably defined massive Thirring lattice model with the XYZ model. On the latter a considerable amount of rigorous work concerning its spectrum has been carried out [46]. The quasi-classical sine-Gordon spectrum emerges from the known part of the two point function spectrum in a suitable limit.

It is also very encouraging to see that the soliton conservation laws have a lattice counterpart in that model [47]. It is our hope that the lattice point of view as well as the direct continuous approach complement each other in such a way as to make an explicit solution possible.

V. SPIN AND STATISTICS FOR THE SINE-GORDON KINKS

Particle states in a D=2 dimensional world transform according to

$$U(\Lambda)\,|p\rangle = |\Lambda p\rangle \tag{52}$$

under Lorentz transformations. There is no physical spin in the sense of Wigner [48]. However, one may want to introduce states

$$|p\rangle_s = \left(\frac{p^0+p^1}{m}\right)^s |p\rangle \tag{53}$$

and corresponding free fields: which transform with the "Lorentz-spin":

$$\mathcal{U}(\Lambda)\,|p\rangle_s = e^{s\chi}\,|\Lambda p\rangle_s \tag{54}$$

It is evident that the choice of this s is a matter of convention. It is somewhat less trivial to see that particle statistics is also a matter of convention [49]. Assume for example that $a(p)$ and $a^+(p)$ are canonical boson annihilation and creation operators in a Fock-space and consider the line-integrals

$$b^+(p) = a^+(p)\exp -i\pi\int_p^\infty n(p')\,dp' \tag{55}$$

$$\text{with} \qquad n(p) = \frac{1}{p_0}\,a^+(p)\,a(p)$$

the particle number momentum density. One immediately checks that b and b^+ form a covariant canonical fermion system:

$$\{b^+(p),\, b(p')\} = \delta(p-p')\,p^0 \tag{56}$$

Although in higher dimension such line integrals can still be used to change the statistics, the lack of Lorentz-covariance (and causality for the suitably Fourier transformed operators) renders such a construction uninteresting.

In the case of particles which arise from topological kinks, one can go further and change the statistics of interpolating fields. If local interpolating fields as for the sine-Gordon equation can be found, then the field with the changed statistics can also be chosen local. Let us explain this using the language developped in section 3. We already have a fermionic interpolating field ψ for the sine-Gordon kink. Let us now construct a bosonic field [49] with the help of the short-distance expansion:

$$\psi_{\frac{1}{2}}(x)\, N[\exp \pm i\sqrt{\pi}\phi(y)] \sim c_{\frac{1}{2}}(\xi_0,\xi_1) B_{\frac{1}{2}}(x) + \dots \qquad (57)$$

Here c is a directional dependent most singular coefficient function and B are zero Lorentz-spin Bose-fields of dimension:

$$\dim B_{\frac{1}{2}} = \frac{1}{4} \dim \bar{\psi}\psi \qquad (58)$$

Although this can be demonstrated on the basis of the short-distance behaviour similar to the discussions in section 3, let us here be satisfied with an explicit understanding of the case $\dim \bar{\psi}\psi = 1$, where we can replace the general normal product by the "triple" ordered expression:

$$\vdots\, e^{\pm i\sqrt{\pi}\phi} \,\vdots = {}^{*}_{*}\, e^{L_\pm} \,{}^{*}_{*} \qquad (59)$$

The "star-ordering" on the right hand side is the ordering in terms of free fermion which one may always use [31] when the underlying fermions are free.

$$L_\pm(x) = \mp \frac{1}{2\pi}\Big[\int \frac{e^{-\hat{\theta}}}{\cosh\hat{\theta}} \left(e^{i(p+q)x} a_p^+ b_q^+ + e^{-i(p+q)x} a_p b_q\right) d\theta_p d\theta_q \qquad (60)$$

$$+ \int \frac{e^{-\hat{\theta}}}{-\sinh(\hat{\theta}+i\varepsilon)} \left(e^{i(p-q)x} e^{i\pi\lambda} a_p^+ a_q + e^{-i(p-q)x} e^{-i\pi\lambda} b_q^+ b_p\right) d\theta_p d\theta_q\Big]$$

where $\hat{\theta} = \frac{1}{2}(\theta_p - \theta_q)$

It is now very easy to determine the singular coefficient function and the operator $B_{\frac{1}{2}}$ just by Fermion-contractions [42]. The result is:

$$C_{\frac{1}{2}} = \text{const.} \left(\frac{\xi^0 + \xi^1 + i\varepsilon}{\xi^0 - \xi^1 + i\varepsilon} \right)^{\pm 1/4} K_{1/2}(m\sqrt{-\xi^2}) \tag{61}$$

and $$B_{\frac{1}{2}} = {*\atop *} \int \left(b^+(q) + e^{i\pi/2} a(q) \right) d\theta_q \, e^{L_\pm} {*\atop *} \tag{62}$$

The dimension $\frac{1}{4}$ and the vanishing spin of B is consistent with the dimensional balance and the Lorentz-transformation properties of both sides in (57). The space-like local commutativity of B follows from the locality properties of the left hand side of (57) with the help of (28).

As a curious side aspect we mention that the two operators $B_{\frac{1}{2}}$ should actually be equal if the Mandelstam-representation[24)] (34) would hold for ψ. However, this representation seems to break down for dim $\bar{\psi}\psi \geq 1$.

VI. THE A^4-KINK, SPIN AND STATISTICS

All relativistic topological kinks for D = 2 are generalizations (coupled systems) of either the sine-Gordon kink or the A^4-kink. The understanding of the former was greatly facilitated by its relation to the massive Thirring model. The quantum field theory of the A^4 kink is much more difficult.

How can we assert that there is any interpolating kink field with its affiliated new particle for a finite coupling constant, i.e. outside the quasiclassical considerations which only have asymptotic validity for $g \to 0$?

In order to develop some intuition for the difference between quasiclassical kinks and genuine quantum kinks let us elaborate the remarks made at the end of section 2. Imagine a Higgs potential of the A_2^4 type with a small unsymmetric perturbation which is suitably chosen in order not to destroy the classical vacuum degeneracy. For the quasi classical kink it is only this degeneracy of the vacuum which counts. However, in the quantum field theoretical treatment one also has to pay attention to the energy fluctuation in "would be" kink states. Let $|0_i\rangle$, $i = 1, 2$ be the two vacua which have an energy degeneracy. Due to the asymmetry of the theory these vacua lead to different correlation functions (phase-transitions without symmetry!). The energy density correlation functions in a "would be" kink state $|s\rangle$ has the following asymptotic behaviour:

$$\langle s|\mathcal{H}(x)\mathcal{H}(y)|s\rangle = \langle 0_1|\mathcal{H}(x)\mathcal{H}(y)|0_1\rangle \qquad (63)$$
$$+ \theta(x)\theta(y)\left(\langle 0_2|\mathcal{H}(x)\mathcal{H}(y)|0_2\rangle - \langle 0_1|\mathcal{H}(x)\mathcal{H}(y)|0_1\rangle\right)$$
$$+ F(x,y)$$

where the θ can be taken as some smothered step function in the spatial coordinates and $F(x, y)$ goes to zero if either x' or y' approach infinity. The second term in (63) will cause infinite energy fluctuations if one integrates over space and uses the representation of the two point function for the energy momentum tensor. Even if topological <u>and</u> symmetry prerequisites are met, one still must avoid choosing unsuitable interpolating kink fields with infinite energy fluctuations. As was pointed out recently,[16)] interpolating fields of the form used to create coherent states (with $f \neq 0$ on one side of space):

$$\exp.\ i \int \pi(x,0) f(x) dx \tag{64}$$

are not suited to describe A^4-kinks because they generate infinite energy fluctuations. So even when the short-distance behaviour is favourable and Coleman's criticism [2] on the use of coherent states for kinks does not apply, these states may still be rendered useless by their pathological large distance fluctuations.

A reasonable way to think about an interpolating kink field for this model is the following [49]: imagine a continuous rotation of A into -A as x sweeps over space. With a one component field as ours such a rotation must necessarily be thought of as a phase space rotation

$$\begin{pmatrix} A(x) \\ \pi(y) \end{pmatrix}_r = \begin{pmatrix} \cos\theta(x), & -\sin\theta(x) \\ \sin\theta(x), & \cos\theta(x) \end{pmatrix} \begin{pmatrix} A(x) \\ \pi(x) \end{pmatrix} \tag{65}$$

where θ is some function with

$$\lim \theta(x) = \begin{cases} 0 & x \to -\infty \\ \pi & x \to +\infty \end{cases}$$

Formally:

$$\exp\ i/_2 \int_{-\infty}^{\infty} \theta(x) :\pi^2(x) + A^2(x): dx \tag{66}$$

would implement such a rotation, however, this expression has an uncurable short distance disease in that (even for free fields!) the quadratic exponential has an infinite norm. In sophisticated language this failure of (66) means that a <u>local</u> rotation in phase space does not define an automorphism [50] of the field algebra. One may, however, obtain a non-local (but still quasi-local) automorphism which at spatial infinity does the required job [49) 51)]:

$$V = \exp \int_{-\infty}^{\infty} \theta(x)\, a^{+}(x)\, a(x)\, dx \tag{67}$$

$$a(x) = \frac{1}{2}\left[(-\Delta+m^2)^{1/4} A(x) - \frac{i}{(-\Delta+m^2)^{1/4}}\, \pi(x)\right] \tag{68}$$

i. e.

$$\begin{pmatrix} A \\ \frac{1}{\sqrt{-\Delta+m^2}}\pi \end{pmatrix} \xrightarrow{V} \begin{pmatrix} \cos\theta, & -\sin\theta \\ \sin\theta, & \cos\theta \end{pmatrix} \begin{pmatrix} A \\ \frac{1}{\sqrt{-\Delta+m^2}}\pi \end{pmatrix} \tag{69}$$

Since I cannot envisage any local automorphism which continuously links A with - A, I believe that the A_2^4-kink is intrinsically non-local and that the best kink operator is a quasi-local kink operator of the type (67). This operator can be thought of as a quasi-local object [18)] centered around x = 0. Defining:

$$V(a) = U(a)\, V\, U^{+}(a)$$

$$U(a) = \text{translation}$$

we obtain

$$\lim_{a\to\infty} [V(0), V(a)] = 0 \tag{70}$$

Note that (67) has the correct multiplicative composition property mentioned in the 2nd section: the product of two kink operators is equivalent to zero kink. Again one may construct an anticommuting operator:

$$\psi(0) = V(0) \int A(x) g(x) dx \tag{71}$$

with g say a test function of compact support.

Since V acts on the far right localized A as a rotation into - A whereas a far left localized A is left uneffected, we obtain

$$\lim_{a\to\infty} \{\psi(0)\,\psi(a)\} = 0 \tag{72}$$

Both fields V and ψ are expected to have nonvanishing matrix-elements with a kink state :

$$\langle s | \left\{ \begin{matrix} V \\ \psi \end{matrix} \right\} | 0 \rangle \neq 0 \tag{73}$$

and therefore either field may be used for the construction of Haag-Ruelle scattering states [18]. Hence for the A_2^4-kink one again obtains a statistics ambiguity.

VII. SOME REMARKS ON KINKS FOR D > 2

From the previous experience with the A^4-kinks one expects that quantum vortices for D = 3 and 't Hooft-Polyakov quantum monopoles for D = 4 should be described in terms of interpolating fields involving bilinear operators in the exponentials, the latter representing space-dependent internal rotations. We will make no attempt to write down such expressions since we have been unable to understand their short-distance properties. Perhaps their relation to conserved currents (in contrast to the phase-space rotations in the previous section) will facilitate the understanding of the short distance problem even if the fields in the exponential are non-canonical.

Encountering thus the road via interpolating kink fields blocked by nasty technical problems (including the infrared problems mentioned at the end of the 2nd section), we may want to look at other indications and hints for the kink-statistics. The present wisdom about this problem based on the semiclassical form of the conserved angular momentum operator [52, 53] and on quantum mechanical consistency considerations [54] is the following: 't Hooft-Polyakov monopoles or dyons (monopoles with non-vanishing electric charge) in a theory involving only fields with integer spin and iso-spin lead to a conserved angular momentum with integer eigenvalues. Therefore one would hope on the basis of the spin and statistics theorem in local quantum field theory (i.e. a situation which is not quite applicable to kinks) that the affiliated particles turn out to have Bose-statistics. For dyons in a theory involving fields of half-integer isospin and integer space-spin there is the possibility of conversion of half-integer iso-spin into half-integer ordinary spin [52, 53]. In that case one would hope for Fermi-statistics. Indeed, by making suitable technical assumptions on the coordinate permutation properties of wave functions in a particular gauge, one may obtain consistency with Fermi-statistics [54, 55].

The very subtle nature of the kink statistics problem becomes evident if one looks at statistical mechanics systems where the difference between bosons and fermions whould show up most clearly. If we take the usual computation procedure for canonical or grand canonical ensembles at face value, then we are in a somewhat awkward position: enclosing vacuum-type kinks (see section 2) into a box takes away their topological "raison d'ètre".

ACKNOWLEDGEMENT

I thank H. Lehmann, M. Lüscher and P. Weisz for discussions. I am also indebted to P. Weisz for reading the manuscript.

REFERENCES AND FOOTNOTES

1. B. SCHROER and T. TRUONG; "Equivalence of sine-Gordon- and Thirring-Model and Cumulative Mass Effects", FUB HEP 6. To be published in Phys. Rev. D.

2. S. COLEMAN; Phys. Rev. D11, 2088 (1973)

3. Nonleading short distance singularities originate from superrenormalizable interactions (generalized mass terms) on massless renormalizable theories. See also reference 27) 28) and 29)

4. S. COLEMAN "Classical Lumps and their Quantum Descendants", 1975 Erice Lecture notes

5. R. DASHEN, B. HASSLACHER and A. NEVEU; Phys. Rev. D10, 4114, 4130 (1974)

6. J. GOLDSTONE and R. JACKIW; Phys. Rev. D11, 1486 (1975)

7. J.L. GERVAIS and B. SAKITA; Phys. Rev. D12, 1038 (1975)

8. N. CHRIST and T.D. LEE; Phys. Rev. D12, 1606 (1973)

9. H. NIELSEN and P. OLESEN; Nucl. Phys. B61, 45 (1973)

10. G. 't HOOFT; Nucl. Phys. B79, 276 (1974)

11. A. M. POLYAKOV; JETP Lett. 20, 194 (1974)

12. M. J. ABLOWITZ, O.J. KAUP, A.C. NEWELL and H. SEGUR; Phys. Rev. Lett. 31, 125 (1973)

13. L. A. TAKHTADZHYAN and L. D. FADEEV; Theor. Math. Phys. 21, 1046 (1973)

14. L. D. FADEEV; "Quantization of Solitons" Princeton preprint (1975), see also the reference to Skyrme contained therein

15. The relevance of energy fluctuations was first pointed out to me by J. A. Swieca, private communication; see also reference 16)

16) J. A. SWIECA; "Solitons and confinement", lectures presented at the Latin-American School of Physics Caracas, Venezuela, PUC Rio de Janeiro preprent 1976

17. R. HAAG; Phys. Rev. 112, 669 (1958)
D. RUELLE; Hebr. Phys. Acta 35, 177 (1962)

18. For an application to kink physics see J. FRÖHLICH "New Superselection Sectors (Soliton States) in Two Dimensional Bose Quantum Field Models"; University of Princeton preprint, 1976

19. C. G. CALLAN; Phys. Rev. D2, 227 (1970)

20. K. SYMANZIK; Commun. Math. Phys. 18, 227 (1970)

21. M. GOMES and J. H. LOWENSTEIN; Nucl. Phys. B45, 252 (1972)

22. B. SCHROER and J. A. SWIECA; Phys. Rev. D10, 480 (1974) here the boson form of fermion fields was used for D=2 model discussions of conformal invariant theories

23. In two unpublished notes B. SCHROER "Notes on Bosonization of Fermions in Two Dimensional Field Theories" Jan. 1975 and FUB HEP 5, April 1975, we discussed such problems. We apologize for the bad word "Bosonization" which we will not use for the rest of this paper.

24. S. MANDELSTAM; Phys. Rev. D11, 3026 (1975)

25. K. JOHNSON; Nuovo Cim. 20, 773 (1961)

26. In case we work with "soft mass" quantization the zero mass limit and the high energy asymptotic agree without re-parametrization, see
B. SCHROER in "Renormalization Theory", Erice lecture notes 1975 edited by G. Velo and A. S. Wightman, D. Reidel Publishing Company 1976

27. K. WILSON; Phys. Rev. 179, 1499 (1969)

28. K. SYMANZIK in "Particles, Quantum Fields and Statistical Mechanics", Lecture Notes in Physics 32, Springer Verlag 1973. In this notes and in reference 29) the "Wilson Spurion" analysis of superrenormalizable interactions has been worked out and made rigorous in the framework of the Callan Symanzik differential equations

29. N. K. NIELSEN; "Non-Analytic Mass Behaviour and the Renormalization Group", to appear

30. Observe that because of chirality conservation in the scale invariant limit the next term in the two point-function of comes from two insertions of the mass operator

31. H. LEHMANN and J. STEHR; DESY preprint 1976. See also the lecture at this school of H. Lehmann

32. S. COLEMAN; "More about the Massive Schwinger Model" Harvard University preprint 1976

33. P. FEDERBUSH; Phys. Rev. 121, 1247 (1960)

34. B. SCHROER, T. T. TRUONG and P. WEISZ; "Model Study of Nonleading Mass Singularities I", FUB HEP 7 (1976) and references to earlier papers

35. See the lectures of K. POHLMEYER at this school and references quoted therein

36. See reference 14) and other references quoted therein

37. I. Ya. ARAF'EVA and V. E. KOREPIN; JETP Lett. 20, 312 (1974)

38. B. BERG, M. KAROWSKI and H. -J. THUN; FUB HEP 76/2 to be published in Physics Letters

39. B. BERG, M. KAROWSKI and H. -J. THUN; FUB HEP 76/5

40. P. P. KULISH and E. R. NUSSIMOV-PIS'MA; JETP 24, 247 (1976)

41. See lectures of R. FLUME at this school

42. B. SCHROER, T. T. TRUONG and P. WEISZ; Phys. Lett. 63, 422 (1976)

43. The work of S. NUSSINOV, Princeton Institute for Advanced Studies preprint, seems to be closely related to our kinematical observation

44. L. CASTILLEJO, R. H. DALITZ and F. J. DYSON; Phys. Rev. 101, 453 (1956)

45. A. LUTHER; "Eigenvalue Spectrum of Interacting Massive Fermions in One Dimension", Nordita 76/8

46. J. D. JOHNSON, S. KRINSKY and B. M. McCOY; Phys. Rev. A8, 2526 (1973)

47. M. LÜSCHER; "Dynamical Charges in the quantized, renormalized massive Thirring model", DESY preprint 1976

48. E. P. WIGNER; Ann. Math. 40, 149 (1939)

49. B. SCHROER and J. A. SWIECA; Ref. TH-2218 CERN 1976

50. S. DOPLICHER, R. HAAG and J.E. ROBERTS; Commun. Math. Phys. 15, 199 (1969)

51. V. KURAK; to be appear as a PUC Rio de Janeiro preprint

52. P. HASENFRATZ and G. 't HOOFT; Phys. Rev. Lett. 36, 1119 (1976)

53. R. JACKIW and C. REBBI; Phys. Rev. Lett. 36, 1116 (1976)

54. A. S. GOLDHABER, Phys. Rev. Lett. 36, 1122 (1976)

55. R. A. BRANDT and J. R. PRIMACK; Ref. TH 2212-CERN

PROBABILISTIC APPROACH TO CRITICAL BEHAVIOR

G.Jona-Lasinio

Istituto di Fisica and Gruppo GNSM

Università - Roma

1. - Motivation

The aim of these lectures is to give an introduction to some recent developments in probability theory which, besides having an independent interest, seem to provide a very effective tool in view of a mathematically rigorous description of critical phenomena. What is involved is the construction of a systematic theory of limit distributions for sums of "strongly dependent" random variables. The notion of strong dependence will be made precise later. For the moment, as a working definition, we take it to mean that the central limit theorem is not valid.

On the side of physics, the problem may be described as the construction of non trivial large scale (or small scale. See Sec.5) observables for systems with an infinite number of degrees of freedom such as thermodynamic systems undergoing a second order phase transition. Let us try to be more explicit. Let X_i be zero expectation random variables referring to the subsystems of a homogeneous macroscopic object; then any sum of the form

$$Z_n = \frac{\sum_1^n X_i}{B_n} \tag{1}$$

if the normalization factor B_n is properly chosen, may be expected to have for large n a smooth distribution i.e. something which does not involve δ functions. If this is the case we shall say that Z is a good "large scale" or "collective" variable. However in many physical situations the dynamical content of such a variable will be rather trivial. In fact, if the forces acting at the microscopic level are short range the X_i in the sum (1) as $n \to \infty$ will tend to become independent of each other and Z will fluctuate as if the X_i were actually independent variables. Mathematically this usually amounts to the validity of the central limit theorem, i.e. to the statement that

$$F_n(a) = P(Z_n < a) \underset{n \to \infty}{\longrightarrow} \frac{1}{\sqrt{2\pi}} \int_{-\infty}^{a} e^{-\frac{x^2}{2}} dx \tag{2}$$

together with

$$B_n \sim n^{1/2} \tag{3}$$

which represents the well known square root law of fluctuations.(*)

There are situations in which the above argument does not apply. The critical point of a second order phase transitions so far represents in physics the most important instance where the central limit theorem breaks down. This may happen in the simplest cases, e.g. mean field theories, through the appearance of normalization factors of the form $B_n \sim n^{\rho/2}$, $\rho \neq 1$, still retaining the Gaussian distribution. More generally also new types of distributions can appear. Situations of this kind provide a very important motivation for the search of new regularities or universality properties associated with the limit distributions of "strongly dependent" random variables, i.e. with their collective behavior.

One should mention at this point that phase transitions are not the only situations where strongly dependent random variables are involved. Geophysics seems to offer a wealth of phenomena exhibiting strong correlations.[20] Furthermore as a specially important case we should consider turbulence where some of the concepts that

---

(*) The reader should not confuse a variable like Z with the quantities involved in the thermodynamic limit. For example the expression $\sum_1^n X_i/n$ is expected to have a δ function distribution according to the law of large numbers.

we shall discuss later were introduced for the first time.[18][25] We may remark however that the recent spectacular developments in the theory of phase transitions seem to be entirely independent of the previous history of stochastic processes. The work in the statistical mechanics of phase transitions made originally no special reference to probaility theory and was based on the idea of renormalization group in its various realizations.[24][4] Then it appeared a very technical paper by Bleher and Sinai[2] which discussed a special case, the hierarchical model, in probabilistic terms. This work provided much inspiration for a series of articles by Cassandro, Gallavotti, Knops, Martin Löf[5][9][10] and the author[16] in which the general connection of the renormalization group with the problem of limit theorems in probability theory gradually emerged. In particular in the author's paper was suggested the notion of stable random field (in the terminology adopted in the present lectures) as a generalization of the usual stable distributions in the sense of Levy.[12] This notion corresponds to the fixed point in the renormalization group language. The first mathematically rigorous formulation of this concept for lattice systems has been given independently by Gallavotti and Jona-Lasinio[11] and by Sinai.[22] The case of continuous random fields is more complicated. Using however the language of generalized random fields recently Dobrushin[7] has developed a systematic approach to limit theorems which seems wide enough to include all cases of physical interest, in particular the critical behavior of both continuous and lattice systems. A considerable part of these notes will be devoted in fact to a simplified exposition of Dobrushin's ideas.

It was the translation of renormalization group ideas into probabilistic language to stimulate a more precise characterization of the strong dependence of random variables encountered at the critical point. In an article by the author[17] it was conjectured that violation of strong mixing (see the next section) could provide a useful demarcation line between critical and non critical behavior. This has been substantiated in the work by Cassandro and the author[6] and by Hegerfeldt and Nappi.[13]

On the whole the probabilistic approach to critical behavior seems powerful enough to provide a basis for a comprehensive and rational treatment of the rich heuristics which characterizes much recent work in statistical mechanics and field theory. It seems to constitute also a new way of thinking about several problems, a circumstance which may have quite unexpected consequences. Finally it runs parallel to (and presumably will rejoin) the constructive approach in field theory where probability has already a substantial role.

The plan of the notes is as follows. In Section 2 we make precise the notion of strong dependence of random variables and briefly review some recent results together with some standard facts. In Section 3, we consider the so called Rosenblatt process which provides a method for constructing a rather general class of strongly dependent random processes. We then introduce the notion of stable random field on a multidimensional lattice (Section 4) as a natural generalization of limit distributions to the case of strongly dependent variables. Sections 5, 6, 7, 8 which are based on the recent work by Dobrushin, introduce generalized stable random fields and discuss their relationship with the lattice case together with several other properties. Sections 9 and 10, finally review some constructive rigorous results due to Sinai.

- Acknowledgements

I wish to express my gratitude to Professors Dobrushin and Sinai for communicating to me their recent work and to Prof.B.Tirozzi for several discussions on the results of the "Russian School". During the preparation of these notes I profited from my continuous interaction with Prof.M.Cassandro and from stimulating conversations with Profs. G.Hegerfeldt and C.Nappi.

2. - Weak Dependence and Strong Dependence

There is not a unique natural way to introduce a demarcation line between strong dependence and weak dependence and one has to take into account the nature of the problem considered. For our purposes the interesting concept seems to be "strong mixing" and the reasons will be explained in a moment. To define strong mixing (s.m.) we need some notation. Consider a lattice $\mathbb{Z}^d$ and a variable X_i associated to each site i. In the product space of the variables X_i, the cylinders are sets of the form

$$\{X_{i_1} \in A_1, \ldots, X_{i_M} \in A_M\} \qquad i_1, \ldots, i_M \in \Lambda$$

where Λ is an arbitrary finite set in $\mathbb{Z}^d$ and the A_i are measurable in the space of the variable X_i. We denote with Σ_Λ the σ-algebra generated by such sets. We say that the variables X_i with distribution μ are weakly dependent or that they represent a strong mixing random field if the following condition holds. Given two finite regions Λ_1 and Λ_2 define the distance between Λ_1 and Λ_2 by

$$\delta(\Lambda_1, \Lambda_2) = \min_{i_1 \in \Lambda_1, i_2 \in \Lambda_2} |i_2 - i_1|$$

where $|i_2 - i_1|$ is for example the Euclidean distance. Set

$$\Delta(\Lambda_1, \Lambda_2) = \sup_{A \in \Sigma_{\Lambda_1}, B \in \Sigma_{\Lambda_2}} |\mu(A \cap B) - \mu(A)\mu(B)|$$

Then

$$\Delta(\Lambda_1, \Lambda_2) \leq \alpha(\delta(\Lambda_1, \Lambda_2)) \tag{1}$$

where $\alpha(\delta) \to 0$ as $\delta \to \infty$. α is called the mixing coefficient.

Intuitively the s.m. idea is that you cannot compensate for the weakening of the dependence of the variables due to an increase of their spacial distance, by increasing the size of the sets.

This situation is typical when one has exponential decay of correlations. This has been proved in a recent paper by Hegerfeldt and Nappi(13) for a wide class of random fields which includes ferromagnetic non critical spin systems.

The situation is entirely different at the critical point where one expects that correlations decay according to a power law. In this connection the following result has been proved by Cassandro and Jona-Lasinio.(6)

Theorem 2.1 - A ferromagnetic translational invariant system with pair interactions[*] for which the autocorrelation function

$$E(X_0 X_j) - E(X_0)E(X_j) = R(j)$$

is such that

$$\lim_{L\to\infty} \frac{\sum_{L(s^K-1)\leq j^K < L(s^K+1)} R(j)}{\sum_{0\leq j^K<L} R(j)} \neq 0 \tag{2}$$

does not satisfy the s.m. condition. S is an arbitrary lattice vector with components s^K .

This theorem implies in particular that a critical 2-dimensional Ising model violates s.m. Therefore violation of s.m. seems to provide a reasonable characterization of the type of strong dependence that one encounters in critical phenomena.

So far our discussion has made no reference to the central limit theorem. Actually the s.m. condition was first introduced in probability theory in connection with limit theorems.[21] For one dimensional stochastic processes the following result is well known[14]

Theorem 2.2 - Assume that the sequence X_i is s.m. and $E(X_i)=0$. Let $F_n(z)$ be the distribution function of

$$Z_n = \frac{\sum_1^n X_i}{B_n}$$

If $F_n(z)$ converges weakly to a limit $F(z)$, then $F(z)$ is necessarily stable. If the latter distribution has exponent α , then

$$B_n = n^{1/\alpha} h(n)$$

(*) For a precise definition of ferromagnetic system with pair interactions see[6] and references given there.

where $h(n)$ is a slowly varying function.

We recall that a stable distribution is a distribution which satisfies[12]

$$F(a_1 z) * F(a_2 z) = F(a z)$$

where a_1, a_2, a are positive numbers and * is the convolution .
A slowly varying function is, roughly speaking, a function which increases at most logarithmically.

The generalization of this theorem to multidimensional random fields is straightforward and is discussed in ref.(13). Actually among the stable distributions only the Gaussian seems relevant in physics as all the others have infinite second moment. Therefore in our context s.m. implies the central limit theorem.

The failure of s.m. at the critical point allows for the appearance of new limit distributions. Their description is an extremely important but difficult problem. For ferromagnetic systems it has been shown in[6] that in general, except for the gaussian, the-these distributions cannot be infinitely divisible. [12] Their characteristic functions are entire functions with zeros on the imaginary axis.

3. - The Rosenblatt Process

In this section we discuss a standard procedure to construct examples of strongly dependent random fields. It may be considered also as an introduction to the more general problems discussed in the subsequent sections.

We start from a sequence y_k of independent random variables normally distributed with unit variance. Let

$$X_i = \sum_{k=-\infty}^{-1} |k|^{-a} y_{k+i} \tag{1}$$

with a to be properly chosen. X_i is a stationary Gaussian process characterized by the correlation function $R(i-j) = E(X_i X_j)$ which can be explicitely computed from (1). $R(\ell)$ has the asymptotic scaling form

$$R(\ell) \underset{|\ell| \to \infty}{\longrightarrow} |\ell|^{1-2a} \tag{2}$$

For $\frac{1}{2} < a < 1$ $R(\ell)$ decays at infinity, is not integrable, i.e. $\sum_0^\infty R(\ell) = \infty$, and the sequence (1) is critical in the sense of physicists and strongly dependent in the sense of our previous definition. Violation of strong mixing follows in fact from a well known theorem of Kolmogorov and Rozanov.[14][15] The limit distribution of a block variable of the form (1.1) is Gaussian but the normalization B_M is anomalous

$$B_M^2 = E[(\textstyle\sum_1^M X_i)^2] \underset{M \to \infty}{\longrightarrow} M^{3-2a} \tag{3}$$

Consider now the new process (Rosenblatt process)

$$S_i = X_i^2 - E(X_i^2) \tag{4}$$

This is clearly non Gaussian. From (2) it follows that its correlation function $R^2(\ell)$ is non integrable for $a < 3/4$. Therefore for $\frac{1}{2} < a < 3/4$ the sequence (4) is critical. The limit distribution of a block variable can be computed explicitely and it turns out to be non Gaussian and non stable. Strong mixing is therefore violated. We now outline the calculation. The distribution function of

$$Z_M = \frac{\sum_1^M S_i}{\left(E[(\sum_1^M S_i)^2]\right)^{1/2}} \tag{5}$$

is completely determined by the characteristic function which can be obtained as follow. We first calculate the truncated characteristic function

$$\phi_M(t) = E_M\left(e^{it\sum_1^M X_i^2}\right) =$$

$$= (2\pi)^{-\frac{M}{2}} |R_M|^{-\frac{1}{2}} \int \prod_1^M dx_k \, e^{it\sum_1^M X_i^2} \, e^{-\frac{1}{2}\sum_{i,j=1}^M x_i R^{-1}(i-j) x_j}$$

where $|R_M|$ is the determinant of the $M \times M$ matrix obtained from the correlation function by restricting $1 \le i, j \le M$. This is a Gaussian integral so that

$$\phi_M(t) = |I - 2itR_M|^{-\frac{1}{2}} = \prod_1^M |1 - 2it\mu_M^{(j)}|^{-\frac{1}{2}}$$

where the $\mu_M^{(j)}$ are the eigenvalue of R_M .

Denoting $\sigma_M^2 = E_M[(\sum_1^M S)^2]$ we have

$$\psi_M(t) = E_M\left(e^{it\sum_1^M S_i/\sigma_M}\right) =$$

$$= e^{-it\sum_1^M \mu_M^{(j)}/\sigma_M} \prod_1^M \left|1 - 2it\frac{\mu_M^{(j)}}{\sigma_M}\right|^{-\frac{1}{2}}$$

It can now be shown that(14)

$$b_j = \lim_{M\to\infty} \frac{\mu_M^{(j)}}{\sigma_M}$$

exists. Therefore

$$\psi(t) = \lim_{M\to\infty} \psi_M(t) = \prod_1^\infty |1 - 2itb_j|^{-\frac{1}{2}} e^{-itb_j} \tag{6}$$

Since $\psi(t)$ is analytic at the origin it cannot be stable. As $|t|$ increases it develops singularities on the imaginary axis and presumably it becomes very different from the ferromagnetic characteristic functions mentioned at the end of the previous section.

We may now ask: what happens if we consider more complicated functions of the original Gaussian process? Interesting results in this direction have been obtained by Taqqu but only part of them has been published.[23] In the present context it is worth mentioning a theorem which determines classes of functions leading to the same limit distributions. Let $G(Y_i)$ be an arbitrary function with $E(G(Y_i)) = 0$, $E(G(Y_i)^2) < \infty$. We are interested in the limit distributions of

$$Z_M^G = \frac{\sum_1^M G(Y_i)}{\left[\text{Variance}\left(\sum_1^M G(Y_i)\right)\right]^{1/2}} \tag{7}$$

Define the Hermite rank of G as follows. Let

$$J(q) = E(G H_q) \tag{8}$$

where H_q is the Hermite polynomial of order q. Then

$$\sum_{q=0}^{\infty} \frac{J(q)}{q!} H_q$$

converges to G in the sense of quadratic expectations. Call

$$m = \min_q \left(q : J(q) \neq 0\right) \tag{9}$$

the Hermite rank of G. We have then the following theorem[23]

<u>Theorem 3.1</u> - <u>Suppose G has Hermite rank m and</u> $\sum_0^{\infty} R^m(\ell) = \infty$ <u>Then the limit distribution of (7) is the same as that of</u>

$$\frac{J(m)}{m!} \frac{\sum_1^M H_m(Y_i)}{\left[\text{Variance}\left(\sum_1^M G(Y_i)\right)\right]^{\frac{1}{2}}}$$

For details we refer to the paper of Taqqu.
So far we have limited our considerations to a single block variable of the form (5) or (7).

From now on we generalize our point of view and consider limit distributions also for joint probabilities of 2 or more block variables. Of course in order for this problem to be meaningful one has to let the distances among the blocks to become infinite together with the size of the block. This leads quite naturally to the notion of stable random field or automodel distribution introduced in the next section.

4. - Stable Random Fields - Discrete Case

The rigorous notion of stable random field, as discussed in the present section, was introduced independently by Gallavotti and Jona-Lasinio[11] and by Sinai[22] in the attempt of constructing for the case of strongly dependent random variables, the analog of the stable distributions and a proper setting for limit theorems. The physical motivation was the need of a mathematically precise definition of the renormalization group in the sense of Kadanoff and Wilson and of its connection with the usual field theoretic approach.[16]

As in Section 2 let $\mathbb{Z}^d$ be a lattice in d dimensions and j a generic point of $\mathbb{Z}^d$, $j = (j^1, j^2, \ldots, j^d)$, with j^k integers. X_j is the random variable associated with site j. We define a new random field

$$X_j^M = (R_{\alpha M} X)_j = M^{-\frac{d\alpha}{2}} \sum_{s \in V_j^M} X_s \tag{1}$$

V_j^M is the cube

$$V_j^M = \{ s : j^k M - \tfrac{M}{2} < s^k \leq j^k M + \tfrac{M}{2} \} \tag{2}$$

and $1 \leq \alpha < 2$.
The transformation (1) on X_j induces a transformation on probability measures according to the following equation

$$(R_{\alpha M}^* \mu)(A) = \mu'(A) = \mu(R_{\alpha M}^{-1} A) \tag{3}$$

where A is a measurable set. $R_{\alpha n}$ has the semigroup property

$$R_{\alpha n_1} R_{\alpha n_2} = R_{\alpha n_1 + n_2} \tag{4}$$

A measure μ will be called stable if

$$R^*_{\alpha n} \mu = \mu \tag{5}$$

and the corresponding field will be called a stable random field.(*) We briefly discuss the choice of the parameter α. It is natural to take $\alpha < 2$. In fact the expectation of eq.(1) gives

$$E(|X_j^n|) \leq n^{-\frac{d\alpha}{2}} \sum_{s \in V_j^n} E(|X_s|) = \\ = C n^{d - \frac{d\alpha}{2}}$$

Therefore if $\alpha > 2$ and $E(|X_s|) < \infty$ we have $E(|X_j^n|) \underset{n \to \infty}{\longrightarrow} 0$. This is not surprising as $\alpha = 2$ corresponds to the law of large numbers and the block variable is expected to tend to a sharp value. The condition $\alpha \geqslant 1$ means that we are considering systems which fluctuate stronger than a collection of independent variables ($\alpha = 1$ corresponds to the central limit theorem). As remarked by Dobrushin,[7] mathematically the latter condition is not natural. However it may be considered natural as long as our main concern is critical behavior of ferromagnetic Ising like systems. We shall come back to this later.

We are now ready to introduce the notion of limit distribution for a random field on a lattice. We say that a measure $\tilde{\mu}$ admits a certain stable measure μ (in the sense of (5)) as limit distribution if for some α

(*) Sinai and Dobrushin have suggested the name "automodel" distribution for a measure satisfying (5) in order to avoid the word "stable" which in mathematics has already several meanings.

$$\lim_{n\to\infty} \tilde{\mu}_n = \lim_{n\to\infty} R^*_{\alpha n} \tilde{\mu} = \mu \tag{6}$$

The convergence has to be interpreted as weak convergence of joint distributions i.e. given an arbitrary continuous function of compact support

$$\phi(x_{i_1}, \ldots, x_{i_k}) \qquad 0 < k < \infty \qquad i_\ell \in Z^d$$

$$\lim_{n\to\infty} \int \phi \, d\tilde{\mu}_n = \int \phi \, d\mu$$

The set of $\tilde{\mu}$ satisfying (6) is called the domain of attraction of μ.

Clearly the above definition includes the usual limit theorems as a particular case. It is easy to see that if μ and $\tilde{\mu}$ are products of identical one dimensional distributions (5) reduces to the usual definition of a stable distribution and (6) to the usual notion of limit theorem. The interesting fact is that strongly dependent variables, like those occurring in critical behavior give rise to limit measures which are not products.

So far only special solutions of eq.(5) have been calculated. For gaussian measures satisfying (5) one sees already the beginning of a general theory. Some details will be given in a later section where we shall also consider some solutions which are "close" to a gaussian measure and can be obtained by a kind of perturbation method (bifurcation theory or ϵ -expansion in the physicist's language). Here we point out that a class of non Gaussian solutions of (5) can be constructed along the lines discussed in section 3. The general idea consists in taking functions of a gaussian field, applying the transformation $R^*_{\alpha n}$ with α determined by the variance of the sum of n factors, and then taking the limit $n \to \infty$. What one actually calculates are the moments of the limit process. The possibility of this construction for the Rosenblatt process was indicated in (11). A systematic approach has been developed by Dobrushin[8] within the wider context of generalized random fields.

5. - Stable Random Fields - Continuous Case[7]

A natural approach to limit theorems in the case of continuous random fields is based on the idea of scaling transformation. We need some preliminaries. Let $\mathcal{R}^d$ a linear topological space of real functions over R^d, the d-dimensional Euclidean space. In most cases $\mathcal{R}^d$ will be the Schwartz space $\mathcal{D}(R^d)$ of infinitely differentiable functions with bounded support. A generalized random field over $\mathcal{R}^d$ is a system of random variables $\{\phi_\varphi, \varphi \in \mathcal{R}^d\}$ defined in a probability space $(\Omega, \mathcal{B}, m)$. $\mathcal{B}$ are the Borel sets in Ω and m a measure. A random field must satisfy the linearity condition

$$a_1 \phi_{\varphi_1} + a_2 \phi_{\varphi_2} = \phi_{a_1\varphi_1 + a_2\varphi_2} \tag{1}$$

where $\varphi_1, \varphi_2 \in \mathcal{R}^d$, a_1, a_2 are real numbers.

A probability measure μ can be specified by a collection of one-dimensional distributions

$$\mu \equiv \{\mu_\varphi, \varphi \in \mathcal{R}^d\} \quad \text{defined by}$$

$$\mu_\varphi(B) = m\{\omega \in \Omega : \phi_\varphi(\omega) \in B\} \tag{2}$$

B is a Borel set on the real line. It is easy to see that because of the linearity condition (1), the finite dimensional joint distributions $\mu_{\varphi_1 \varphi_2 \cdots \varphi_k}$ are also determined. Convergence of probability measures is weak convergence of joint distributions.

Define now for any $\varphi \in \mathcal{R}^d$ the scale transformation

$$S_{\alpha\lambda}\varphi(x) = \lambda^{-\frac{d\alpha}{2}} \varphi(\lambda^{-1}x) \tag{3}$$

where $1 \leq \alpha < 2$ as in the previous section. $S_{\alpha\lambda}$ induces a transformation on probability measures according to

$$(S^*_{\alpha\lambda}\mu)_\varphi = \mu_{S_{\alpha\lambda}\varphi} \tag{4}$$

In analogy with (4.5) a measure μ will be called stable if

$$S^{*}_{\alpha\lambda}\,\mu = \mu \tag{5}$$

We shall say that $\tilde{\mu}$ has a limit "in the large" if for some α (*)

$$\lim_{\lambda \to \infty} S^{*}_{\alpha\lambda}\,\tilde{\mu} = \mu \tag{6}$$

If μ exists is clearly stable in the sense of (5). Dobrushin considers also a limit over short distances, $\lim_{\lambda \to 0} S^{*}_{\alpha\lambda}\,\tilde{\mu}$ if this exists for some α . This is essentially a limit "in the large" in momentum space. In the following we shall always refer to limits "in the large".

We now come to an interesting point which explains why in the continuous case one has to consider generalized random fields rather than ordinary random fields. In the latter case ϕ_φ can be represented as an ordinary integral

$$\phi_\varphi = \int_{R^d} \varphi(x)\,X(x)\,dx \tag{7}$$

Then if μ is the distribution of $X(x)$, $S^{*}_{\alpha\lambda}\mu$ will be the distribution of $\lambda^{d(1-\alpha/2)}\,X(\lambda x)$. If μ is stable they coincide. But then we have the following conclusion: if $X(x)$ is a stationary and stable ordinary random field, it is trivial.

In fact from stationarity it follows that the distributions of $X(x)$ and $\lambda^{d(1-\alpha/2)}\,X(\lambda x)$ may coincide only if $X(x) \equiv 0$. Ordinary (non stationary) stable random process were first considered by Lamperti.[19]

(*) We should remark, and this applies also to (4.6), that for the limit (6) to exist, more general multiplicative factors of the form $\lambda^{-\frac{d\alpha}{2}}\,h(\lambda)$, where $h(\lambda)$ is a slowly varying function, may have to be considered. The definition of stable measure however includes only the $\lambda^{-\frac{d\alpha}{2}}$ factor. In the following we shall ignore this complication.

The approach in terms of generalized random fields is also very powerful because, as we shall discuss in the next section, permits to connect explicitely the notions of stable random fields in the discrete and the continuous case.

6. - Relationship between the discrete and the continuous case[7]

We need some auxiliary concepts. We say that the space $\mathcal{R}^d$ can be discretized if the following conditions are fulfilled

i) $\mathcal{R}^d$ contains all the indicators $\chi_{b,c}$ of d-dimensional parallelepipeds

$$\Lambda_{b,c} = \{x : \quad b^k < x^k \leq c^k\}$$

$$\chi_{b,c}(x) = 1 \quad x \in \Lambda_{b,c} \; , \; \chi_{b,c}(x) = 0 \quad x \notin \Lambda_{b,c}$$

ii) the linear space of functions of the form

$$\psi = \sum_{j=1}^{m} a_j \chi_{b_j, c_j}$$

where b_j, c_j have rational components, is dense in $\mathcal{R}^d$.

An important example of such a space in the present context is the following $M(R^d)$. This is given by the finite $\psi \in L_2(R^d)$ such that

$$\|\psi\| = \left[\int_{R^d} |\tilde{\psi}(\lambda)|^2 \prod_1^d (1 + |\lambda^k|) \left(\ln(e + |\lambda^k|) \right)^{-2} d\lambda \right]^{\frac{1}{2}} < \infty \qquad (1)$$

$\tilde{\psi}(\lambda)$ is the Fourier transform of ψ.

The topology in $M(R^d)$ is defined as follows. Let $M(S_r)$ the set of functions in $M(R^d)$ with support in the sphere S_r of radius r contained in R^d. We introduce in $M(S_r)$ the topology of the form (1). The topology in $M(R^d)$ is then the inductive limit topology(*). The following theorem has been proved by

(*) This means the following: a neighborhood of 0 in $M(R^d)$ is any subset V such that $V \cap M(S_r)$ is a neighborhood of 0 in $M(S_r)$ for all r.

Dobrushin.

Theorem 6.1 - $M(R^d)$ can be discretized and $\mathcal{D}(R^d)$ is a dense subspace of $M(R^d)$. The usual topology of $\mathcal{D}(R^d)$ is stronger than that induced by $M(R^d)$.

It is obvious that the functions $\chi_{b,c}$ belong to $M(R^d)$. In fact their Fourier transform is

$$\tilde{\chi}_{b,c}(\lambda) = \prod_{k=1}^{d} \frac{1}{i\lambda^k}\left(e^{i\lambda^k c^k} - e^{i\lambda^k b^k}\right)$$

and satisfies (1). We do not give the proof of the theorem. We only mention that it is done in two steps

a) $\mathcal{D}(R^d)$ is dense in $M(R^d)$

b) $\mathcal{D}(R^d) \subseteq \bar{M}(R^d)$ closure of the space of functions $\psi = \sum_j a_j \chi_{b_j,c_j}$.

Consider now a probability measure μ of a generalized random field in $\mathcal{D}(R^d)$. We shall say that μ admits an extension in $M(R^d)$ if we can find a μ' describing a random field in $M(R^d)$ such that

$$\mu_\varphi = \mu'_\varphi \qquad \forall\, \varphi \in \mathcal{D}(R^d) \tag{2}$$

It is clear that such an extension is unique, although of course it may not exist. We then call discretization of μ (through μ') the distribution $\hat{\mu}$ of the discrete random field

$$X_j = \phi_{\chi_j} \tag{3}$$

where ϕ has distribution μ'. χ_j is the indicator of the cube

$$\Lambda_j = \left\{ x : j^k - \tfrac{1}{2} < x^k \leq j^k + \tfrac{1}{2} \right\} \tag{4}$$

We note that every discrete distribution $\hat{\mu}$ arises from the discretization of some field over $\mathcal{D}(R^d)$. In fact from the discrete field X_j we can construct

$$\phi_\varphi = \sum_{j \in \mathbb{Z}^d} X_j \int_{\Lambda_j} \varphi(x)\,dx \quad , \quad \varphi \in \mathcal{D}(R^d) \tag{5}$$

(5) defines also a random field over M(R^d) whose discretization is X_j with distribution $\hat{\mu}$.

We now have the important theorem

Theorem 6.2 - Given two distributions μ^1 and μ^2 which describe extensions over M(R^d) of random fields over $\mathcal{D}$ (R^d) and such that for some α

$$\mu^2 = S^*_{\alpha M}\, \mu^1 \qquad n = 1,2,\ldots \tag{6}$$

their discretizations satisfy

$$\hat{\mu}^2 = R^*_{\alpha M}\, \hat{\mu}^1 \tag{7}$$

The operation of discretization transforms stable random fields with parameter α into stable random fields with the same parameter

The situation can be visualized with the help of the diagram

$$\begin{array}{ccc} \mu^1 & \xrightarrow{S^*_{\alpha M}} & \mu^2 \\ D \downarrow & & \downarrow D \\ \hat{\mu}^1 & \xrightarrow[R^*_{\alpha M}]{} & \hat{\mu}^2 \end{array}$$

D means discretization.

A final remark. The language of generalized random fields appears as a suitable tool to analyze also the relationship between limit theorems and the multiplicative renormalization groups used with considerable success in field theory(4) (groups à la Gell--Mann and Low). Their special multiplicative structure can be introduced quite naturally following the scheme of ref.(1) based on the notion of multiplicative cocycle.

7. - Structure of the set of stable random fields(7)

We have already mentioned that the problem of constructing stable random fields is a very difficult one and that only a limited number of explicit examples is available so far. It is then natural to examine the possibility of constructing new stable random fields out of the known ones. There are some general properties of stable fields which are interesting in this context.

We denote with A_α^{st} the class of stationary generalized stable random fields over $\mathcal{D}(R^d)$. In the sense explained in the previous section this contains also the random fields over a d-dimensional lattice. We now list some operations which transform A_α^{st} into itself.

a) Multiplication by a constant

Let $T_c\varphi = c\varphi$, then $(T_c^*\mu)_\varphi = \mu_{c\varphi}$ is stable if μ is stable.

b) Convex linear combination

If $\mu^1, \mu^2 \in A_\alpha^{st}$, then for $\lambda^1 \geqslant 0$, $\lambda^2 \geqslant 0$, $\lambda^1 + \lambda^2 = 1$

$$\mu = \lambda^1\mu^1 + \lambda^2\mu^2 \quad \in A_\alpha^{st}$$

c) Convolution

If $\mu^1, \mu^2 \in A_\alpha^{st}$, then $\mu = \mu^1 * \mu^2$ defined by

$$\mu_\varphi = \mu_\varphi^1 * \mu_\varphi^2 \,, \quad \in A_\alpha^{st}$$

d) Closure in the sense of weak convergence

e) Differentiation

We denote by I_d the set of multi indices $i = (i^1, i^2, \ldots, i^d)$,

$i^k = 0,1,2,\dots$, $|i| = i^1 + i^2 + \dots + i^d$. Let $D^i = (D^1)^{i^1} \cdots (D^d)^{i^d}$ where D^k is differentiation with respect to the k-th coordinate. The derivative of order i of the field $\{\phi_\varphi, \varphi \in \mathcal{D}(R^d)\}$ is defined by

$$\phi^i_\varphi = \phi_{(-1)^{|i|} D^i \varphi}$$

If $\mu \in A^{st}_\alpha$, then for $i \in I_d$ $D^{*i}\mu \in A^{st}_{\alpha + \frac{2|i|}{d}}$

Of course operation e) has no-meaning for random fields over a lattice.

We can now formulate in a precise way the problem raised at the beginning of this section. Consider a set $B \in A^{st}_\alpha$. We say that $\bar{B}$ is generated by B if $\bar{B}$ is the smallest closed convex set containing B and such that if $\mu^1, \mu^2 \in \bar{B}$ also $T^*_c\mu$ and $\mu^1 * \mu^2 \in \bar{B}$. Question: is it possible that A^{st}_α be generated by some simple set (possibly finite) of distributions? So far the answer is not known.

8. - Local Mutual Singularity of Stable Random Fields[7]

In this section we discuss a fundamental property of stable random fields which gives additional insight into the difficulties of the subject. We will show following Dobrushin, that any two non coinciding stationary stable random fields have distributions which are locally orthogonal.

For $V \subseteq R^d$ we denote with $\mathcal{D}_V$ the functions $\in \mathcal{D}(R^d)$ which have their support in V and with $\mathcal{B}_V$ the smallest subalgebra of $\mathcal{B}$ with respect to which the functions $\phi_\varphi, \varphi \in \mathcal{D}_V$ are measurable. We say that two distributions μ^1, μ^2 are locally mutually singular or orthogonal if for every $V \subseteq R^d$ the restriction of the corresponding measures m^1, m^2 to $\mathcal{B}_V$ on the common probability space Ω (see section 5 for the relevant definitions) are mutually singular. This means that there exists a $B \in \mathcal{B}_V$ such that

$$m^1(B) = 0 \qquad\qquad m^2(B) = 1 \qquad (1)$$

Theorem 8.1 - Let μ^1, μ^2 be non coinciding stationary ergodic stable random fields. Then they are locally mutually singular.

We sketch the idea of the proof which shows very clearly the role of scale invariance. It is obviously sufficient to show that for every $V \subseteq R^d$ and every $\epsilon > 0$ there exists a $B \in \mathcal{B}_V$ such that

$$m^1(B) < \epsilon \qquad m^2(B) > 1 - \epsilon \tag{2}$$

Since μ^1 and μ^2 are stationary we may take a V containing the origin of the coördinates. Let us assume that μ^1 and μ^2 are characterized by the same α. Since they do not coincide, there will exist functions $\varphi \in \mathcal{D}$ for which $\mu^1_\varphi \neq \mu^2_\varphi$. Then for some continuous function f the averages

$$M^1 = \int_\Omega f(\phi_\varphi(\omega))\, m^1(d\omega)\,, \quad M^2 = \int_\Omega f(\phi_\varphi(\omega))\, m^2(d\omega)$$

will be such that

$$M^1 > M^2 \tag{3}$$

Let us now denote by $L^1(m^i)$, i=1,2 the space of measurable functions integrable with respect to m^i. If V_T is the cube

$$V_T = \{x : \quad -T < x^i \leq T \;, \; i = 1, 2, \ldots d\}$$

by applying the ergodic theorem we have that in the sense of the convergence of $L^1(m^i)$

$$\lim_{T \to \infty} \frac{1}{V_T} \int_{V_T} f(\phi_{E_a \varphi}(\omega))\, da = M^i \tag{4}$$

where E_a is the translation operator defined by $E_a \varphi(x) = \varphi(x-a)$ Then for any $\epsilon > 0$ there exists a T_0 such that for the set

$$B = \left\{ \omega \in \Omega : \frac{1}{V_{T_0}} \int_{V_{T_0}} f\left(\phi_{E_a \varphi}(\omega)\right) da < \frac{M^1 + M^2}{2} \right\} \quad (5)$$

$$m^1(B) < \epsilon \qquad\qquad m^2(B) > 1 - \epsilon \quad (6)$$

In general however B will not be contained in $\mathcal{B}_V$. But now we can use the fact that μ^1 and μ^2 are stable random fields. The set

$$B_\lambda = \left\{ \omega \in \Omega : \frac{1}{V_{T_0}} \int_{V_{T_0}} f\left(\phi_{S_{\alpha\lambda} E_a \varphi}(\omega)\right) da < \frac{M^1 + M^2}{2} \right\} \quad (7)$$

has measures

$$m^1(B_\lambda) = m^1(B) \qquad\qquad m^2(B_\lambda) = m^2(B) \quad (8)$$

Since φ is of bounded support there will exists a cube V_S such that φ vanishes outside it. The support of $S_{\alpha\lambda} E_a \varphi$ will be contained in the cube $V_{(S+|a|)\lambda}$. By chosing λ small enough we can bring any $V_{(S+|a|)\lambda}$ inside V and therefore B_λ inside $\mathcal{B}_V$. The theorem is proved for μ^1, μ^2 having the same exponent α. If $\alpha^1 \neq \alpha^2$ the proof can be carried out along similar lines and it is only slightly more involved.

9. - Gaussian Stable Random Fields(22)

We come again to the crucial question of the explicit construction of stable random fields. In this and the subsequent section we shall review some results of Sinai which provide the first systematic approach to some cases of physical interest. The case of Gaussian processes is very simple as they are completely determined by the correlation function. In physics they appear for example in mean field theories of phase transition. Everything will be restricted to the lattice case. In one dimension the following theorem completely characterizes the unique stationary Gaussian stable random field.

Theorem 9.1 - The stationary Gaussian distribution μ in d=1 is stable if its spectral function has the form

$$\rho(\lambda) = c\,|e^{2\pi i\lambda}-1|^2 \sum_{m=-\infty}^{\infty} \frac{1}{|\lambda+m|^{\alpha+1}} \tag{1}$$

C is a constant > 0

From (1) it follows that

$$G(r) = E(X_0 X_r) = \int_{-\frac{1}{2}}^{\frac{1}{2}} e^{2\pi i\lambda r} \rho(\lambda)\, d\lambda$$

has the asymptotic behavior $G(r) \underset{r\to\infty}{\longrightarrow} |r|^{-(2-\alpha)}$ since $\rho(\lambda) \underset{\lambda\to 0}{\longrightarrow} |\lambda|^{1-\alpha}$.

We make a side remark. As the reader will remember $1 \leq \alpha < 2$ and this restriction was motivated in Section 4. It is clear however that the above formulas have a meaning also for $0 < \alpha < 1$. In this interval $G(r)$ is integrable, actually $\sum_r G(r) = 0$, and the process is not critical in the sense of usual ferromagnetic systems. However it still describes a sequence of strongly dependent variables. Strong mixing is violated as it follows from theorem 2.1 which of course applies to gaussian processes. The difference with respect to a critical Ising model is that in eq.2.2 numerator and denominator now vanish. Violation of strong mixing follows also from general theorems due to the zero of non integral order of the spectral function when $\lambda \to 0$.[15]

For $d > 1$ there are many more possibilities due to the circumstance that we can have different asymptotic behavior in different directions. A complete description of Gaussian stable random fields for $d > 1$ is still missing. A particular class can be constructed as follows. Let $f(\lambda^1, \ldots, \lambda^d)$ a positive homogeneous function of degree $\gamma = d(\alpha+1)$, i.e.

$$f(c\lambda^1, \ldots, c\lambda^d) = |c|^\gamma f(\lambda^1, \ldots, \lambda^d) \tag{2}$$

Then the series

$$g(\lambda) = \sum_{m \in \mathbb{Z}^d} f^{-1}(\lambda + m) \tag{3}$$

converges for $\lambda \notin \mathbb{Z}^d$ and defines a periodic function which can be considered as a function on a d-dimensional torus. Let

$$\rho(\lambda) = \prod_{k=1}^{d} |e^{2\pi i \lambda^k} - 1|^2 g(\lambda) \tag{4}$$

Theorem 9.2 - The Gaussian process with spectral function $\rho(\lambda)$ is stable.

In the isotropic case one may choose

$$f(\lambda^1, \ldots, \lambda^d) = \prod_{k=1}^{d} (\lambda^k)^2 \left(\sum_k (\lambda^k)^2 \right)^{\frac{d(\alpha-1)}{2}}$$

10. - Non Gaussian Stable Random Fields[22]

The construction starts from a study of the linearization of the operator $R^*_{\alpha M}$ around a Gaussian fixed point, i.e. around one of the processes just calculated. Then one studies the spectrum of such a linearization and looks for those values of α for which it contains 1. These are the values for which a non Gaussian solution bifurcates. The new solution is calculated by means of a formal expansion in $\epsilon = \alpha - \alpha_c$ where α_c is the bifurcation point.

Let μ_G^α a gaussian stable measure with exponent α over a d-dimensional lattice. We consider a perturbed measure

$$\mu = \mu_G^\alpha (1 + h) \tag{1}$$

and apply the transformation $R_{\alpha M}^*$. We have

$$R_{\alpha M}^* \mu = \mu_G^\alpha + R_{\alpha M}^* \mu_G^\alpha h \tag{2}$$

From the definition of $R_{\alpha M}^*$ it follows easily that

$$\frac{d R_{\alpha M}^* \mu_G^\alpha h}{d \mu_G^\alpha} = E_G (h | X^M) \tag{3}$$

$E_G(h|X^M)$ is the conditional expectation of h with respect to μ_G^α for fixed values of the block variables X^M defined by eq.(4.1). Eq.(3) defines the differential $D R_{\alpha M}^*$ of $R_{\alpha M}^*$ at the fixed point μ_G^α. Sinai has studied in detail the linear operator $D R_{\alpha M}^*$ and has calculated its eigenvalues and eigenvectors. We summarize the results of his analysis.

He starts by considering polynomial expressions of the form

$$a^{(k)} = \sum_{e_1, \ldots, e_k} a_{e_1 \ldots e_k} X_{e_1} \cdots X_{e_k} \tag{4}$$

From any choice of $a_{e_1 \ldots e_k}$ we can obtain a translational invariant form by taking new coefficients

$$b_{e_1 \ldots e_k} = \sum_{m \in \mathbb{Z}^d} a_{e_1 + m, \ldots, e_k + m} \tag{5}$$

We denote by $b^{(k)}$ the corresponding polynomial form. It is then convenient to decompose the space of forms $b^{(k)}$ into Wick products as follows. Define

$$\begin{aligned} :X_{e_1} \cdots X_{e_k}: = X_{e_1} \cdots X_{e_k} &- \sum_{t_1 \ldots t_{k-2}} c_{t_1 \ldots t_{k-2}} X_{t_1} \cdots X_{t_{k-2}} - \\ &- \sum_{t_1 \ldots t_{k-4}} c_{t_1 \ldots t_{k-4}} X_{t_1} \cdots X_{t_{k-4}} - \cdots \end{aligned} \tag{6}$$

where the coefficients are determined by the condition that expression (6) be orthogonal with respect to the measure μ_G^α to every product $X_{j_1} \cdots X_{j_m}$ $m<k$. The space of forms

$$\sum_{\ell_1,\dots,\ell_k} b_{\ell_1\dots\ell_k} :X_{\ell_1}\cdots X_{\ell_k}: \tag{7}$$

will be called $\mathcal{H}_k$.

Theorem 10.1 - 1) Every space $\mathcal{H}_k$ is invariant under the action of $DR^*_{\alpha n}$, i.e. $DR^*_{\alpha n}\mathcal{H}_k = \mathcal{H}_k$ for $n \gg 1$, $k \geq 1$.

2) In every space $\mathcal{H}_k$ there exists an element h_k which is an eigenvector of $DR^*_{\alpha n}$, i.e. $DR^*_{\alpha n} h_k = n^{\gamma_k} h_k$ with $\gamma_k = d\left(\frac{\alpha k}{2} - k + 1\right)$.

From this theorem there follows a series of consequences. We restrict our considerations to the case of even k. For k=2, $\gamma_2 = \alpha - 1 > 0$ and the space of quadratic forms is always unstable in the sense that $DR^*_{\alpha n}$ drives away from the fixed point μ_G^α. If $1 < \alpha < 3/2$, $\gamma_k < 0$, k=4,6,... and in the spaces $\mathcal{H}_k$ we have a stable spectrum. For $\alpha_k = 2 - \frac{1}{k}$, $\gamma_k = 0$ and the corresponding eigenvalue is equal to 1. At these points, in analogy with the finite dimensional case, it is natural to expect the appearance of new solutions of the equation $R^*_{\alpha n}\mu = \mu$.

For α near 3/2 the new solution is calculated by expanding the perturbation term in (1) in powers of $\epsilon = \alpha - 3/2$ starting with a term $\sim \epsilon h_4$ and solving $R^*_{\alpha n}\mu = \mu$ self-consistently. This has been done by Sinai for n=2. More complete results have been obtained by Bleher and Sinai for the hierarchical model.[3]

References

1) Benettin G., Di Castro C., Jona-Lasinio G., Peliti L., Stella A.L., "On the Equivalence of Different Renormalization Groups" (1976) this volume.

2) Bleher P.M., Sinai Ya.G., Comm.Math.Phys. 33, 23 (1973).

3) Bleher P.M., Sinai Ya.G., Comm.Math.Phys. 45, 247 (1975).

4) Di Castro C., Jona-Lasinio G., in "Phase Transitions and Critical Phenomena", vol. 6, ed. C.Domb and M.S.Green, New York (1976)

5) Cassandro M., Gallavotti G., Nuovo Cimento 25B, 691 (1975).

6) Cassandro M., Jona-Lasinio G., "Asymptotic Behavior of the Autocovariance Function and Violation of Strong Mixing", Preprint ZiF, Bielefeld, (1976).

7) Dobrushin R.L., "Avtomodelnost i Renorm-gruppa Obobshchennyh Sluchainyh Polei" preprint (1976).

8) Dobrushin R.L., "Gaussovskye i Podchinennye Gaussovskim Avtomodelnye Obobshchennye Sluchainye Polia" preprint (1976).

9) Gallavotti G., Knops H., Comm.Math.Phys. 36, 171 (1974).

10) Gallavotti G., Martin-Löf A., Nuovo Cimento 25B, 425 (1975).

11) Gallavotti G., Jona-Lasinio G., Comm.Math.Phys. 41, 301 (1975).

12) Gnedenko B., Kolmogorov A., "Limit Distributions for Sums of Independent Random Variables", (1968).

13) Hegerfeldt G.C., Nappi C.R., "Mixing Properties in Lattice Systems", preprint ZiF Bielefeld (1976).

14) Ibragimov I.A., Linnik Yu.V., "Independent and Stationary Sequences of Random Variables", Groningen (1971).

15) Ibragimov I.A., Rozanov Yu.A., "Processus Aléatoires Gaussiens", Moscow (1973).

16) Jona-Lasinio G., Nuovo Cimento 26B, 99 (1975).

17) Jona-Lasinio G., in "Les Methodes Mathematiques de la Théorie quantique des Champs", Colloques Internationaux C.N.R.S. n.248, p. 207, Marseille (1975).

18) Kolmogorov A., D.A.N. SSSR 26, 115 (1940).

19) Lamperti,J., Tr.Am.Math.Soc. 104, 62 (1962).

20) Mandelbrot B., Van Ness J., SIAM Review 10, 422 (1968).

21) Rosenblatt M., Proc. IVth Berkeley Symp. on Prob. and Math.Stat. 431, Berkeley (1961).

22) Sinai Ya.G., Teor.Ver. i ee Prim 21, 63 (1976). See also Sinai Ya.G., in "Statistical Physics", Proceedings IUPAP Conference, Budapest (1975).

23) Taqqu, M.S., Z. Wahrs. 31, 287 (1975).

24) Wilson K., Kogut J., Phys.Rep. 12C, 76 (1974).

25) Yaglom A.M., Theory of Prob. and its Appl. 2, 273 (1957).

ON THE EQUIVALENCE OF DIFFERENT RENORMALIZATION GROUPS[*]

G. Benettin, C. Di Castro, G. Jona-Lasinio,
L. Peliti, A.L. Stella

Istituto di Fisica dell' Università
and GNSM-CNR Padova

Istituto di Fisica dell' Università
and GNSM-CNR Roma

1. Introduction

If one looks back at the recent developments in the theory of critical phenomena, the Kondo problem and gauge theories, one notices that the same word "renormalization group" appears in connection with approaches which at first sight look very different. In view of the subsequent discussion, it is expedient to introduce some preliminary classification of the various trends.

There are approaches which are called field theoretic in the sense that they use the standard apparatus of quantum field theory based on the formalism of Green's fucntions. However within that context one usually refers to a renormalization group method in the strict sense[1], which is related to certain invariance properties of the theory, or to the so called Callan-Symanzik equations[2]. The two points of view lead to very similar formalisms but their conceptual relationship is not immediately apparent.

Another main line of thought, which in many respects has been dominant in the past few years, is known as the

----

[*] The content of this report was presented in a seminar by A.L. Stella.

Kadanoff-Wilson approach or sometimes, as the "modern version" of the renormalization group[3]. Its relevant feature has been the introduction of a whole series of concepts and calculational schemes which were unknown to the previous tradition of both field theory and statistical mechanics. However, when applied to the same problems, the field theoretic and the Kadanoff-Wilson (K-W) methods give exactly the same results. Actually also within the K-W method several variants are possible, a fact which was recognised rather early.

More recently the zoology of renormalization groups has been enriched by the so-called "non linear" groups [4][5], the possibility of which had been established on the basis of rather general arguments[6][7]. Non linear groups do not appear so simple for a systematic study but they seem to constitute a promising lead.

The existence of so many different "schools" in renormalization group theory, has sometimes created confusion and misunderstandings in the sense that some authors, rather than stressing the very important fact that the different methods, at least as far as they have been developed, give exactly the same results, emphasize their formal difference. In this way one gets easily the impression that the cohincidence of the results, obtained in different ways, is almost accidental and not related to the deep conceptual unity which underlies the various interpretations of the renormalization group idea.

Today, after the success of the different points of view has become an established fact, it is a relevant theoretical and pedagogical problem to make their connections explicit.

At the general level it is not too difficult to extract a leading idea which is common to all renormalization groups. This can be formulated as follows. Given a model of a physical system which is of interest to us for a certain property, try to reduce it by a series of variable transformations to a new model which exhibits this property as the dominant, in such a way, that it can be studied in its simplest possible form. All this can be illustrated with the help of an analogy. In celestial mechanics it is not immediately obvious that a system of planets in interaction exhibits periodic motions. However by performing a series of canonical transformations, one can reduce in some cases the very complicated original equations to a form in which they represent quasi-periodic motions on a torus.

The property which was relevant in the physical systems studied in the recent years, was critical scaling, and in the sense of the example indicated above, scaling may be viewed as corresponding to the classical periodic motions.

It is clear however that the difficult part of the problem consists in making explicit the detailed connection among the different renormalization group methods. This paper represents a contribution in this direction because it makes explicit the analytic connections between some of the schemes mentioned at the beginning.

We think however that the most interesting outcome of the discussion which follows is of a general character. In fact we will show that a multiplicative structure similar to that characteristic of field theoretic renormalization groups can be associated essentially to any renormalization transformation, provided some smoothness conditions of a general nature are satisfied.

2. Field Theoretic Renormalization Groups

The basic ideas of the renormalization group concept in field theory are much older than the recent applications to critical phenomena. What is new is the attempt to construct a systematic theory of scaling starting from the R-G. In view of this circumstance, it is convenient to formulate the problem and introduce the notation having this goal in mind. We begin therefore with one of the main results of phenomenological analysis of critical behaviour: the inverse of the Fourier transform of the two-point correlation function has the form

$$\Gamma^{(2)}_{as}(k,t,\Lambda) = \left(\frac{k}{\Lambda}\right)^{-\eta} k^2 \, f\left(\frac{k\, t^{-\nu}}{\Lambda^{1-2\nu}}\right) \tag{2.1}$$

valid when $k,\ t\to 0$. The meaning of the symbols is as follows: k is the wave vector, $t=T-T_c$ measures the deviation from the critical temperature and Λ is a natural length of the problem which may be taken of the order of the inverse lattice spacing for Ising like systems.

The asymptotic correlations therefore can be characterized by the simple scaling property.

$$\Gamma^{(2)}_{as}(k,t,\Lambda) = s^{\eta}\, \Gamma^{(2)}_{as}(k,t\, s^{\frac{1}{\nu}-2}, \frac{\Lambda}{s}) \qquad (2.2)$$

The field theoretic R-G can be viewed as a generalization of (2) valid for all k and t, which reduces to (2) asymptotically. However one has to be careful because in the exact equations derived from field theory the role of Λ is played by a parameter with a different meaning. If one deals, as it is customary, with a model characterized by an interaction of the form $u\phi^4$ (u is the coupling constant and ϕ the field amplitude), the exact equation satisfied by the renormalized two-point correlation is:

$$\tilde{\Gamma}^{(2)}(k,t,u,\lambda,\Lambda) = Z^{-1}(\lambda',u,\lambda,\Lambda)\tilde{\Gamma}^{(2)}(k,t',u',\lambda',\Lambda) \qquad (2.3)$$

where

$$t' = t\, Z_t^{-1}(\lambda',u,\lambda,\Lambda)\, Z(\lambda',u,\lambda,\Lambda)$$
$$\qquad (2.4)$$
$$u' = u\, Z_u^{-1}(\lambda',u,\lambda,\Lambda)\, Z^2(\lambda',u,\lambda,\Lambda)$$

and Z, Z_t, Z_u are appropriate functions.

We shall discuss later how these functions can be determined when we clarify the meaning of the new variable λ.

The scaling properties emerge from field theory by letting $\lambda' = \lambda/s$, $s\to\infty$ and by assuming that in this limit the dependence of $\tilde{\Gamma}^{(2)}$ on u and Λ disappears, while $Z^{-1}\sim s^{\eta}$, $Z_t^{-1}\sim s^{1/\nu-2+\eta}$.

In this way (2.3) becomes identical with (2.2) provided we identify the roles of Λ and λ. If we had dealt directly with $\Gamma^{(2)}$ and had written an equation of the form

$$\Gamma^{(2)}(k,t,u,\Lambda) = Z^{-1}\, \Gamma^{(2)}(k,t',u',\Lambda')$$

with t' and u', related multiplicatively to t and u, this would not hold with Z's independent of k and t. The usefulness of equation (2.3) resides precisely in the possibility of choosing the λ dependence in such a way that the

Z's do not depend on the variables k and t in which we are interested.

Actually (2.3) can be further simplified by taking from the start the limit $\Lambda=\infty$ when justified by the renormalizability of the theory.

It is now time to recall how (2.3) is constructed. The structure of eq. (2.3) follows from two facts. The first is a trivial invariance property of the diagrammatic expansion of correlation functions. The second, a little less trivial, is connected with the invertibility of $\tilde{\Gamma}^{(2)}$ with respect to some of its arguments.

This will be apparent in a moment. Here we only remark that the parameterization in terms of λ and λ' indicated in (2.3) and (2.4) depends just on this second property. The steps leading to the determination of the functions Z, Z_t, Z_u are the following.

One choses three correlation functions $\frac{\partial\tilde{\Gamma}^{(2)}}{\partial k^2}$, $\frac{\partial\tilde{\Gamma}^{(2)}}{\partial t}$ and $\tilde{\Gamma}^{(4)}$, where $\tilde{\Gamma}^{(4)}$ is the one-particle irreducible four-point function, and requires that the normalization conditions be satisfied

$$\frac{\partial\tilde{\Gamma}^{(2)}}{\partial k^2}(k^2,t',u',\lambda')\Big|_{k^2=0,t'=\lambda'^2} = 1$$

$$\frac{\partial\tilde{\Gamma}^{(2)}}{\partial t}(k^2,t',u',\lambda')\Big|_{k^2=0,t'=\lambda'^2} = 1 \tag{2.5}$$

$$\tilde{\Gamma}^{(4)}(\vec{k}_i,t',u',\lambda')\Big|_{k_i^2=0,\ t'=\lambda'^2} = u'$$

From (2.3) and a similar eq. for $\tilde{\Gamma}^{(4)}$ one then finds the implicit expressions for the Z's

$$Z^{-1} = \frac{\partial\tilde{\Gamma}^{(2)}}{\partial k^2}(k^2,t'Z_tZ^{-1},u,\lambda)\Big|_{k^2=0,t'=\lambda'^2}$$

$$Z_t^{-1} = \frac{\partial\tilde{\Gamma}^{(2)}}{\partial t}(k^2,t'Z_tZ^{-1},u,\lambda)\Big|_{k^2=0,t'=\lambda'^2} \tag{2.6}$$

$$uZ_u^{-1} = \tilde{\Gamma}^{(4)}(\vec{k}_i, t'Z_t Z^{-1}, u, \lambda)\Big|_{k_i^2=0, t'=\lambda'^2}$$

Here one sees clearly where invertibility comes in. In this way, having expressed the Z's in terms of correlation functions, we obtain a set of non linear equations for $\frac{\partial \tilde{\Gamma}^{(2)}}{\partial k^2}$, $\frac{\partial \tilde{\Gamma}^{(2)}}{\partial t}$ and $\tilde{\Gamma}^{(4)}$.

The meaning of λ is clarified by eqs. (2.5) and (2.6). It can also be connected with the subtraction point of standard renormalization theory. For a detailed derivation of the above formulas, we refer the reader for example to (1) where one finds also indications on how to perform practical calculations.

For the purpose of the present paper it is sufficient to explain briefly under which conditions scaling is implied by our scheme.

We first introduce the differential form of eq.(2.3) by taking the derivative with respect to λ' and then setting $\lambda'=\lambda$ (we omit the Λ dependence)

$$\left[\lambda \frac{\partial}{\partial \lambda} - \sigma_t^o\, t \frac{\partial}{\partial t} + \psi_u^o\, u \frac{\partial}{\partial u} - 2\sigma^o\right] \tilde{\Gamma}^{(2)}(k,t,u,\lambda) = 0 \quad (2.7)$$

where

$$\sigma^o = \frac{\partial}{\partial s}(Z^{-\frac{1}{2}})\Big|_{s=1}, \quad \sigma_t^o = \frac{\partial}{\partial s}(Z\, Z_t^{-1})\Big|_{s=1}, \quad \psi_u^o = -\frac{\partial}{\partial s}(Z\, Z_u^{-1})\Big|_{s=1} \quad (2.8)$$

A simple dimensional analysis now shows that (2.7) can be rewritten

$$\left[-k\frac{\partial}{\partial k} - \sigma_t\, t\frac{\partial}{\partial t} + \psi_u(u_\lambda)\frac{\partial}{\partial u_\lambda} - 2\sigma^o + 2\right]\tilde{\Gamma}^{(2)}(k,t,u_\lambda,\lambda) = 0 \quad (2.9)$$

with

$$u_\lambda = u\lambda^{-\varepsilon}, \quad \varepsilon = 4-d, \quad \psi_u(u_\lambda) = u_\lambda \psi_u^o - \varepsilon u_\lambda \quad \sigma_t = 2 + \sigma_t^o \quad (2.10)$$

d is the dimensionality of the system. Similar equations can be written for $\frac{\partial \tilde{\Gamma}^{(2)}}{\partial t}$ and $\tilde{\Gamma}^{(4)}$.

If there exists a u^* for which $\Psi_u(u^*)=0$ we obtain the asymptotic solutions

$$\left.\frac{\partial\tilde{\Gamma}_2^{(2)}}{\partial k}\right|_{k=0} \sim t^{\frac{-2\sigma^o}{2+\sigma_t^o}} = t^{-\eta\nu}$$

$$\left.\frac{\partial\tilde{\Gamma}^{(2)}}{\partial t}\right|_{k=0} \sim t^{\frac{-2\sigma+\sigma_t^o}{2+\sigma_t^o}} = t^{-(\eta+\frac{1}{\nu}-2)\nu} \tag{2.11}$$

with the identification $2\sigma^o(u^*)=\eta$ and $\sigma_t^o(u^*)=\frac{1}{\nu}-2$.

(2.11) implies for Z and Z_t the asymptotic scaling form

$$Z^{-1} \underset{s\to\infty}{\sim} s^{\eta}, \quad Z_t^{-1} \underset{s\to\infty}{\sim} s^{\frac{1}{\nu}-2+\eta} \tag{2.12}$$

Linearizing the function $\Psi_u(u_\lambda)$ around the fixed point, one may also introduce an exponent measuring the speed of approach to the asymptotic region as follows

$$u_{\lambda'} - u^* \sim s^{-\Psi'}(u_\lambda - u^*)$$

with

$$\Psi' = \left.\frac{\partial\Psi_u}{\partial u_\lambda}\right|_{u_\lambda=u^*} \tag{2.13}$$

which allows for the evaluation of the correction terms to the asymptotic expression.

3. Kadanoff-Wilson Linear Transformations

For the purposes of the present paper it is sufficient to describe the usual K-W groups in terms of the following requirements. Consider a family of transformations R_s depending on the real parameter s, acting on the Hamiltonian of a many-body system and satisfying:

a) $R_s(H)=H'$ where H and H' belong to some preassigned space of Hamiltonians I,

b) $R_s(R_{s'}(H))=R_{ss'}(H)$ (composition law).

c) The two particle inverse correlation function transforms in momentum space according to

$$\Gamma^{(2)}(ks,\ R_s(H)) = s^{2-\eta}\ \Gamma^{(2)}(k,H)$$

d) There exists at least one fixed point H^* satisfying for any s

$$R_s(H^*) = H^*$$

e) For τ small enough

$$R_s(H+\tau\Delta H) = R_s(H)+\tau L_s(H)\Delta H$$

where $L_s(H)$ is a linear operator which depends on H. This condition will be called following ref. (8) the smoothness postulate. $L_s(H)$ is the tangent mapping, which transforms vectors in the tangent space at H into vectors of the tangent space at $R_s(H)$.

A condition which is usually imposed on the set of Hamiltonians considered is that they should contain only momenta below a certain cut-off Λ. In such a case c) is meaningful only if $ks<\Lambda$. One can in principle consider much more general cases but for simplicity we shall assume a sharp cut-off Λ which provides a natural unit in terms of which any dimensional quantity can be expressed. We shall call critical any H which satisfies

$$\lim_{s\to\infty} R_s(H) = H^* \tag{3.1}$$

provided the theory corresponding to H^* is not a free theory.

Critical Hamiltonians form a subset I_c of I, which is usually assumed to have the structure of a differentiable manifold.

For $H=H^*$, $L_s(H^*)$ becomes a linear operator acting on the tangent space at H^*.

From b) and e) we obtain easily (9)

$$L_s(R_{s'}(H))L_{s'}(H) = L_{ss'}(H) \tag{3.2}$$

For $H=H^*$ therefore the linear operators $L_s(H^*)$ have

the group property

$$L_s(H^*)L_{s'}(H^*) = L_{ss'}(H^*) \tag{3.3}$$

If we assume, as usual, that $L_s(H^*)$ can be diagonalized, because of (3), its eigenvalues have the form

$$\lambda_i^*(s) = s^{y_i} \qquad i = 1,2..... \tag{3.4}$$

In application one often deals with situations in which $y_1>0$ and $y_i<0$ $(i\neq 1)$. The corresponding set of eigenvectors h_i^* provides a natural basis for the tangent space at H^*: h_i^* $(i\neq 1)$ define the tangent plane to the critical surface I_c at H^*, while h_1^* represents the direction of escape from I_c at the same point.

For a general discussion of scaling properties related to critical hamiltonians H belonging to I_c but different from H^* and in particular for the discussion of the subsequent section, we need further assumptions for $L_s(H)$ which come out as a natural extension of those given above. The rich and interesting structure which emerges is the main justification of our assumptions.

f) For any $H\varepsilon I_c$ we assume the existence of a set $h_i(H)$ of basis vectors normalized in some natural metric in the tangent space at H satisfying:

$$L_s(H)h_i(H) = \lambda_i(H,s)h_i(R_s(H)) \; . \tag{3.5}$$

With the following continuity requirement

$$\lim_{s\to\infty} h_i(R_s(H)) = \lim_{s\to\infty} \frac{L_s(H)h_i(H)}{\lambda_i(H,s)} = h_i^* \tag{3.6}$$

(3.5), for $H=H^*$, reduces to the eigenvalue equation for $L_s(H^*)$, with $\lambda_i(H^*,s)=\lambda_i^*(s)= s^{y_i}$.

From (3.2) and (3.5) it immediately follows

$$\lambda_i(R_s(H),s)\lambda_i(H,s') = \lambda_i(H,ss') \tag{3.7}$$

which has the multiplicative character of the field theoretic transformation for the correlation functions.

If we differentiate equation (3.7) with respect to s' and put s'=1 we obtain:

$$\left[s \frac{\partial}{\partial s} - \sigma_i(H) - \sum_k \Psi_k(H) \frac{\partial}{\partial u_k}\right] \lambda_i(H,s) = 0 \qquad (3.8)$$

where: u_k, k=1,2,... is a set of parameters specifying the Hamiltonian in some preassigned basis,

$$\sigma_i(H) = \frac{\partial}{\partial s'} \lambda_i(H,s') \Big|_{s'=1}$$

and

$$\Psi_k(H) = \frac{\partial}{\partial s'} (R_{s'})_k (H) \Big|_{s'=1}$$

where $(R_s)_k(H)$ represents the component of $R_s(H)$ in the k-th direction.

It is evident that:

$$\sigma_i(H^*) = y_i$$

and

$$\Psi_k(H^*) = 0$$

Equation (3.8) has the same structure as the field-theoretic renormalization group equations with the only difference that the Ψ_k's here depend on a large number of parameters.

In order to give an idea of the physical interpretation of the quantities $\lambda_i(H,s)$, we remark that if the transformation $R_s(H)$ is the Kadanoff block-spin transformation, the λ_i are expressible in terms of conditional correlation functions where the conditioning variable is the block-spin.

Due to the fact that for $H \varepsilon I_c$, H^* is an attractive fixed point, equation (3.8) also implies that asymptotically, for large s, one has

$$\lambda_i(H,s) \underset{s\to\infty}{\sim} \rho_i(H) s^{y_i} \qquad (3.9)$$

The asymptotic form (3.9) will be used in the following section in order to clarify the connection between the K-W and the field theoretic approach.

Here we want to discuss the representation of $L_s(H)$ in the basis $\{h_k^*\}$, which is the one used for practical computations.

Due to the non-orthogonality of the vectors h_k^*, it is convenient to introduce contravariant vectors $h^{\ell *}$ satisfying

$$(h^{\ell *}, h_k^*) = \delta_{\ell k} \tag{3.10}$$

We can then put

$$h_i(H) = \alpha_{ik}(H) h_k^* \tag{3.11}$$

where

$$\alpha_{ik}(H) = (h^{i*}, h_k(H)) \tag{3.12}$$

Thus (5) can be written as

$$\sum_k L_s(H)\alpha_{ik}(H)h_k^* = \lambda_i(H,s) \sum_j \alpha_{ij}(R_s(H))h_j^* \tag{3.13}$$

Defining

$$\lambda_{\ell k}(H,s) = (h^{\ell *}, L_s(H)h_k^*) \tag{3.14}$$

we immediately obtain

$$\sum_k \lambda_{\ell k}\, (H,s)\alpha_{ki}^T\, (H) = \lambda_i(H,s)\alpha_{i\ell}(R_s(H)) \tag{3.15}$$

From (3.15), if we consider the fact that $\alpha_{ik}(R_s(H)) \to \delta_{ik}$ for $s \to \infty$, and take into account the invertibility of $\alpha^T(H)$, which follows from our assumptions, we obtain the following form for λ_{ij}

$$\lim_{s\to\infty} \lambda_{ij}(H,s) \sim s^{y_i}\, C_{ij}(H) \tag{3.16}$$

The previous assumptions are not sufficient to give an explicit determination of the matrix $C_{ij}(H)$, which provides the transformation between the scaling directions at H and the scaling directions at H*.

There is a simple connection between the matrix C_{ij} and the Wegner [10] scaling fields. In fact if we write asymptotically

$$(R_s(H))_i \underset{s\to\infty}{\to} H_i^* + C_i(H)s^{y_i} + \ldots..$$

it is obvious that $C_{ij} = \dfrac{\partial C_i(H)}{\partial H_j}$. Let us check that the $C_i(H)$ are scaling fields. From the above equation we may write:

$$C_i(H) = \lim_{s\to\infty} [R_s(H) - H^*]_i s^{-y_i} .$$

Compute now the total derivative $\dfrac{dC_i}{d\ln s}$:

$$\frac{dC_i}{d\ln s} = \sum_j \frac{\partial C_i}{\partial H_j}\dot{H}_j = \lim_{s'\to\infty}\left[(s')^{-y_i}\frac{\partial(R_{s'}(H))_i}{\partial H_j}G_j(H)\right] \tag{3.17}$$

where $\dot{H}_j = \dfrac{\partial H_j}{\partial \ln s} = G_j(H)$.

On the other hand, from the composition law b)

$$\sum_j \frac{\partial(R_s(H))_i}{\partial H_j} G_j(H) = \frac{\partial(R_s(H))_i}{\partial \ln s} = G_i(R_s(H)) \tag{3.18}$$

From the last two equations it follows

$$\frac{dC_i}{d\ln s} = y_i C_i \tag{3.19}$$

i.e. $C_i(R_s(H)) = s^{y_i}\, C_i(H)$.

As a final remark we notice that the possibility of associating the multiplicative structure (3.7) to a trans formation $R_s(H)$ does not depend on this being multiplicative on the correlation functions. The argument is valid also for any group transformation defined by all the previous conditions except condition C. This has to be replaced by the invariance of the thermodynamic potential[6], which was trivially satisfied in the cases previously considered. Examples of transformations non-multiplicative on the correlation functions will be given in section 6.

4. Formal Equivalence of Field Theoretic and Kadanoff-Wilson Renormalization Groups

In this section we shall compare the field theoretic point of view with the K-W transformations described in section 3.

From the asymptotic expressions (2.12), for the Z's and eq. (2.3), for $\lambda' = \frac{\lambda}{s} \to 0$ and using after differentiation dimensionless quantities, we find

$$\frac{\partial \tilde{\Gamma}^{(2)}}{\partial k^2}\left(\frac{k^2}{\lambda^2}, \frac{t}{\lambda^2}, u_\lambda\right) \cong s^\eta \frac{\partial \tilde{\Gamma}^{(2)}}{\partial k^2}\left(\frac{s^2 k^2}{\lambda^2}, \frac{s^{\frac{1}{\nu}} t}{\lambda^2}, u^*\right) \quad (4.1)$$

Where, in the hypothesis that the theory is renormalizable, the $\Lambda \to \infty$ limit has been taken.

By bringing the s dependence on the left hand side, a consequence of (1) is that the limit

$$\lim_{s\to\infty} s^{-\eta} \frac{\partial \tilde{\Gamma}^{(2)}}{\partial k^2}\left(\frac{k^2 s^{-2}}{\lambda^2}, \frac{s^{-\frac{1}{\nu}} t}{\lambda^2}, u_\lambda\right) \quad (4.2)$$

is finite and independent of u.

On the basis of the results of the previous section, which are not confined to critical Hamiltonians near the fixed point, we shall now derive a similar equation from the Kadanoff-Wilson linear transformations.

We start from a critical Hamiltonian H (e.g. the $u\phi^4$ model with the mass term corresponding to the critical temperature) and we add a small perturbing term of the form

$$\tau\, h_1(H) \qquad (\tau \ll 1)$$

From (3.c), (3.e) and (3.f) we obtain:

$$\begin{aligned} & s^{-2+\eta}\, \Gamma^{(2)}(ks, R_s(H) + \tau L_s(H)\, h_1(H)) = \\ & = s^{-2+\eta}\, \Gamma^{(2)}(ks, R_s(H) + \tau \lambda_1(H,s) h_1(R_s(H)) = \qquad (4.3) \\ & = \Gamma^{(2)}(k, H + \tau h_1(H)) \end{aligned}$$

We now consider the limit $s \to \infty$ in (4.3) with $k' = ks < \Lambda$ and $\tau' = \tau \lambda_1(H,s) << 1$. If we now take all the s dependence on the right hand side and drop for simplicity the primes, we obtain that

$$\lim_{s \to \infty} s^{2-\eta} \Gamma^{(2)} \left(\frac{k}{s}, H + \frac{\tau}{\lambda_1(H,s)} h_1(H) \right) \tag{4.4}$$

is finite and independent of H.

If we take the explicit asymptotic expression (3.9) for $\lambda_1(H,s)$, (4.4) implies that

$$\lim_{s \to \infty} s^{2-\eta} \Gamma^{(2)} \left(\frac{k}{s}, H + \frac{\tau s^{-y_1}}{\rho_1(H)} h_1(H) \right) \tag{4.5}$$

is finite and independent of H.

Let us analyze the meaning of this equation. Formally it is very similar to eq. (4.2), provided we indentify $y_1 = \frac{1}{\nu}$, and in particular it expresses the asymptotic u-independence of the theory if H is the one for $u\phi^4$ model. However the unit of length in (4.5) is the natural unit of the theory, i.e. the cut-off Λ. Formally therefore the roles of λ and Λ are asymptotically the same. This is interesting because it suggests quite naturally a way of looking at the renormalization problem in field theory, different from the usual one based on subtractions. We are referring to what Wilson calls "unconventional renormalization"(11). This consists in giving the parameters of the theory, e.g. the mass term and the coupling u in the ϕ^4 theory, a cut-off dependence which is asymptotically the anomalous dimension of the fixed point scaling behaviour of that parameter. This should produce according to eq.s (4.2) or (4.5) the cancellations necessary for the existence of the theory in the limit of infinite cut-off, a part from overall multiplicative factors.

5. The Callan-Symanzik Equation

One of the most widely used formulations of the field-theoretical renormalization transformations is the Callan-Symanzik equation for the correlation functions.

The equation for the two point vertex function for

instance looks slightly different from eq. (2.7) since an inhomogeneous term appears on the right hand side; it is then argued on the basis of perturbation theory that this inhomogeneous term is irrelevant in the neighbourhood of the critical point.

We shall see here how the C-S equation fits our formulation by simplifying the derivation given in (1). In this context the origin of the inhomogeneous term will become clear.

Any group equation like (2.7) can be transformed into an equation of the Callan-Symanzik type with an inhomogeneous term by specifying it on a submanifold with a reduced number of variables:

$$\lambda = \lambda(t); \; \tilde{\Gamma}_c^{(2)}(k,t,u) = \tilde{\Gamma}^{(2)}(k,t,u,\lambda(t)) \tag{5.1}$$

The derivative of $\tilde{\Gamma}^{(2)}$ with respect to λ is now made of two terms:

$$\frac{\partial\tilde{\Gamma}^{(2)}}{\partial\lambda} = \left(\frac{\partial\tilde{\Gamma}_c^{(2)}}{\partial t} - \frac{\partial\tilde{\Gamma}^{(2)}}{\partial t}\right)\frac{dt}{d\lambda} \tag{5.2}$$

One of them becomes the inhomogeneous term of the new equation for $\tilde{\Gamma}^{(2)}$

$$\left[\lambda\frac{dt}{d\lambda}\frac{\partial}{\partial t} - \sigma_t^o\, t\frac{\partial}{\partial t} + \Psi_u^o\, u\frac{\partial}{\partial u} - 2\sigma^o\right]\tilde{\Gamma}_c^{(2)}(k,t,u) =$$

$$= \lambda\frac{\partial t}{\partial\lambda}\frac{\partial\tilde{\Gamma}^{(2)}}{\partial t}(k,t,u,\lambda=\lambda(t)) \tag{5.3}$$

In the actual C-S equation the internal length scale is given at any temperature by the coherence distance of the fluctuations of the system which is defined as

$$m^2 = \xi(t)^{-2} = \frac{\tilde{\Gamma}^{(2)}(k=0)}{\left.\dfrac{\partial\tilde{\Gamma}^{(2)}}{\partial k^2}\right|_{k^2=0}} \tag{5.4}$$

The model is given as a function of m itself instead of t and satisfies the following normalization conditions at any temperature and therefore for any m

$$\tilde{\Gamma}_c^{(2)}(k,m,u_m)\Big|_{k=0} = m^2 \tag{5.5a}$$

$$\frac{\partial \tilde{\Gamma}_c^{(2)}}{\partial k^2}(k,m,u_m)\Big|_{k=0} = 1 \tag{5.5b}$$

$$\frac{\partial \tilde{\Gamma}_c^{(2)}}{\partial t}(k,m,u_m)\Big|_{k=0} = 1 \tag{5.5c}$$

$$\tilde{\Gamma}_c^{(4)}(\vec{k}_i,m,u_m)\Big|_{k_i^2=0} = u_m m^{\varepsilon} \tag{5.5d}$$

where we have reminded by the subscript c the new normalization conditions imposed on the vertex functions. To avoid complications we maintain in the following the same notation for the vertex functions in terms of m instead of t and use, as it is customary, the renormalized dimensionless coupling constant.

To make the connection with section 2, it is useful to rephrase the field theoretical renormalization group in a slightly different way, so that m appears explicitly as a variable. m is an invariant of the group as it is easily checked from its expression (5.4) and the transformation equation (2.3) for $\tilde{\Gamma}^{(2)}$. Therefore, if the model is parameterized as a function of m instead of t, the same transformation equation for $\tilde{\Gamma}^{(2)}$ now reads

$$\tilde{\Gamma}^{(2)}(k,m,u_\lambda,\lambda)=Z^{-1}\left(\frac{\lambda'}{\lambda},\ u_\lambda\right)\tilde{\Gamma}^{(2)}\left(k,m,\frac{u_\lambda}{\left(\frac{\lambda'}{\lambda}\right)^{\varepsilon}}\ Z^2 Z_u^{-1},\lambda'\right) \tag{5.6}$$

Instead of choosing the N.P. at $t'=\lambda'^2$ as in (2.5), it is now natural to take as normalization point that value $\bar{t}$ of t' for which

$$m(\bar{t},u_\lambda) = \lambda'$$

which leads to

$$Z^{-1}\left(\frac{\lambda'}{\lambda},\ u_\lambda\right) = \frac{\partial \tilde{\Gamma}^{(2)}}{\partial k^2}(k,\lambda',u_\lambda,\lambda)\Big|_{k=0}$$

and

$$\left[\lambda \frac{\partial}{\partial\lambda} + \Psi_u \frac{\partial}{\partial u_\lambda} - 2\sigma^\circ\right] \tilde{\Gamma}^{(2)}(k,m,u_\lambda,\lambda) = 0 \qquad (5.7)$$

instead of eq.s (2.6) and (2.7).

Similarly the equations for $\tilde{\Gamma}^{(4)}$ and $\frac{\partial\tilde{\Gamma}^{(2)}}{\partial t}$ and the related expressions for Z_u^{-1} are obtained.

The present scheme is conceptually entirely equivalent to the one discussed in section 2.

The Callan-Symanzik normalization conditions (5.5) are now recovered by enforcing λ=m for any m.

We are therefore in the situation discussed at the beginning of this section. Since

$$\tilde{\Gamma}_c^{(2)}(k,m,u_m) = \tilde{\Gamma}^{(2)}(k,m,u_\lambda,\lambda=m)$$

the equation for $\tilde{\Gamma}_c^{(2)}$ can now be obtained by specifying eq. (5.7) on the submanifold $\lambda\equiv m$ where it reads

$$\left[m \frac{\partial}{\partial m} + \Psi_u \frac{\partial}{\partial u_m} - 2\sigma^\circ\right] \tilde{\Gamma}_c^{(2)}(k,m,u_m) = m \left.\frac{\partial\tilde{\Gamma}^{(2)}}{\partial m}(k,m,u_\lambda,\lambda)\right|_{\lambda=m}$$

(5.8)

Eq. (5.8) is the same as the Callan-Symanzik equation, with the inhomogeneous term in an unusual form. It is a technical point to reduce this inhomogeneous term to its standard form(1).

The same results are obtained starting from the group equation in its integral form (5.6). This sheds some further light on the origin of this inhomogeneous term.

We may use eq. (5.6) to relate $\tilde{\Gamma}_c^{(2)}(k,m',u_{m'}) = \tilde{\Gamma}^{(2)}(k,m',u_m,m')$ with a given value of the mass m' to $\tilde{\Gamma}^{(2)}(k,m',u_m,m)$ at a fixed normalization point λ=m

$$\tilde{\Gamma}^{(2)}(k,m',u_m,m) = Z^{-1}\left(\frac{m'}{m}, u_m\right) \tilde{\Gamma}_c^{(2)}(k,m',u_{m'}) \qquad (5.9)$$

where

$$u_{m'} = \frac{u_m}{\left(\frac{m'}{m}\right)^{4-d}} Z^2 \left(\frac{m'}{m}, u_m\right) Z_u^{-1} \left(\frac{m'}{m}, u_m\right) \tag{5.10}$$

If we now take the derivative of eq. (5.9) with respect to m' and set m'=m, we once again obtain equation (5.8)

$$\left[m \frac{\partial}{\partial m} + \Psi_u \frac{\partial}{\partial u_m} - 2\sigma^o\right] \tilde{\Gamma}_c^{(2)}(k,m,u_m) = = m' \frac{\partial}{\partial m'} \tilde{\Gamma}^{(2)}(k,m',u_m,m)\Big|_{m'=m} \tag{5.11}$$

with

$$\sigma^o = - m' \frac{\partial}{\partial m'} Z^{-\frac{1}{2}} \left(\frac{m'}{m}, u_m\right)\Big|_{m'=m}$$

$$\Psi_u = m' \frac{\partial}{\partial m'} \left| \frac{u_m}{\left(\frac{m'}{m}\right)^{4-d}} Z^2 Z_u^{-1} \right|\Bigg|_{m'=m}$$

We see that the origin of the inhomogeneous term may be traced back through eq. (5.9) to the necessity of changing the temperature when changing the normalization scale, once normalization conditions such as (5.5) are introduced. In our previous formulation of section 2 the differential equation (2.7) is obtained by differentiation of eq. (5.6) with respect to λ', which actually does not act on the left hand side of the equation.

6. Non-linear transformations

In the Kadanoff-Wilson approach, "non linear" renormalization groups were introduced by releasing one of the conditions specifying the transformation[5]. Specifically the condition c of section 3, which relates the transformed vertex functions to the original ones, is no more valid.

We now discuss some non linear generalizations of the field-theoretic approach described in section 2.

In the group that we shall now introduce the simple multiplicative transformation property of the correlation functions is partially lost.

We may for instance allow for an additional explicit t-dependence of the renormalization factors Z's, so that σ^o, σ_t and Ψ_u are functions of both u and t, not only of u. The group equations for $\tilde{\Gamma}^{(2)}$ and $\tilde{\Gamma}^{(4)}$ then remain of the same form as (2.9), whereas the equation for $\frac{\partial \tilde{\Gamma}^{(2)}}{\partial t}$ is much more complicated.

In order to analyse this case, it is convenient to use the Fourier transform of the three point vertex function $\frac{\delta \tilde{\Gamma}^{(2)}}{\delta t(x)}$, where t(x) is considered as a local source and eventually its constant limit is taken.

As a consequence of the fact that the parameters of the transformation depend in an essential way on t, if we take the functional derivative of the equation for $\tilde{\Gamma}^{(2)}$ with respect to t(x), we obtain the following equation for $\tilde{\Gamma}_t^{(2)}(k,q)=\text{F.T.}\frac{\delta \tilde{\Gamma}^{(2)}}{\delta t(x)}$

$$\left[-k\frac{\partial}{\partial k} - q\frac{\partial}{\partial q} - \sigma_t\, t\frac{\partial}{\partial t} + \Psi_u \frac{\partial}{\partial u_\lambda} + 2 - 2\sigma^o - \sigma_t\right]\tilde{\Gamma}_t^{(2)}(k,q) +$$

$$- \sigma_{tt}(q)t\,\tilde{\Gamma}_t^{(2)}(k,q) + \Psi_{ut}(q)\tilde{\Gamma}_u^{(2)}(k,q) + \tag{6.1}$$

$$- 2\sigma^o{}_t(q)\tilde{\Gamma}^{(2)}(k) = 0$$

where

$$\sigma_{tt}(q)=\text{F.T.}\left(\frac{\delta\sigma_t}{\delta t(x)}\right), \sigma^o_t(q)=\text{F.T.}\left(\frac{\delta\sigma^o}{\delta t(x)}\right), \Psi_{ut}(q)=\text{F.T.}\left(\frac{\delta\Psi_u}{\delta t(x)}\right)$$

$$\Gamma_u^{(2)}(k,q)=\text{F.T.}\left(\frac{\delta\Gamma^{(2)}}{\delta u_\lambda}\right) \tag{6.2}$$

and in the limit $q\to 0$

$$\sigma_{tt}(0) = \frac{\partial\sigma_t}{\partial t},\ \sigma^o{}_t(0) = \frac{\partial\sigma^o}{\partial t},\ \Psi_{ut}(0) = \frac{\partial\Psi_u}{\partial t},\ \tilde{\Gamma}_u^{(2)}(k,0)=$$

$$= \frac{\partial\tilde{\Gamma}^{(2)}(k)}{\partial u_\lambda} \tag{6.3}$$

The second line of equation (6.1) reflects the non multiplicative character of the transformation.

Our transformation cannot be anymore completely specified by simple normalization conditions similar to (2.5).

However we can still construct satisfactory field-theoretic schemes where the transformation maintains a certain degree of arbitrariness in its definition.

This arbitrariness could be for example exploited to simplify the study of cross-over effects by varying the flow diagrams in the Hamiltonian parameter space.

We now indicate two possible examples asymptotically equivalent, i.e. in the scaling region, to the multiplicative case.

If we look at the structure of (6.1) a natural kind of imposition is the requirement that σ°, σ_t and Ψ_u be chosen in such a way that we obtain the standard multiplicative structure for some fixed temperature, e.g. the critical temperature. To this purpose, it is sufficien to impose

$$\sigma_t^{\circ}(q)\Big|_{t=0} = \Psi_{ut}(q)\Big|_{t=0} \equiv 0 \qquad (6.4)$$

so that the second line of eq. (6.1) vanishes at t=0.

We are now in a position to determine σ°, σ_t and Ψ_u at t=0 by requiring in analogy with the multiplicative case that for $k=q=\lambda$ and $t=0$

$$\tilde{\Gamma}_t^{(2)}\Big|_{k=q=\lambda, t=0} = 1, \quad \frac{\partial\tilde{\Gamma}^{(2)}}{\partial k^2}\Big|_{k=\lambda, t=0} = 1, \quad \Gamma^{(4)}\Big|_{k_i=\lambda \text{ s.p.}, t=0} = u_\lambda \qquad (6.5)$$

In this way σ°, σ_t and Ψ_t at t=0 coincide with those appropriate to a multiplicative group, while their t-dependence is still compatible with various specifications. The critical indices therefore coincide with those of section 2, while flow diagrams in the Hamiltonian space can be very different.

We may realize a different version of our non linear groups by trying to impose the normalization conditions

$$\left.\frac{\partial\tilde{\Gamma}^{(2)}}{\partial k}\right|_{k=\lambda} = 1, \quad \left.\tilde{\Gamma}^{(4)}\right|_{k_i=\lambda \text{ s.p.}} = u_\lambda, \; \forall \; t \qquad (6.6)$$

which are of the same type as those originally used by Gell-Mann and Low.

For a constant $t(x)=t$ indipendent of x, Ψ_u and σ^o are fully specified and depend explicitely on t. Equation (6.1) is therefore in order also in this case in its full generality. It can be shown that infrared divergences arise in σ_{tt}, σ_t^o and Ψ_{ut} at t=0. We can take care of them by imposing that the second line of eq. (6.1) vanishes at q=k=0 for any t.

If we now assume that this happens also for $q=k=\lambda$ the equation for $\tilde{\Gamma}_t^{(2)}$ at this point reduces to the simple form of the multiplicative case, and we can therefore determine $\sigma_t(u,t=0)$ by imposing the normalization condition

$$\left.\tilde{\Gamma}_t^{(2)}\right|_{k=q=\lambda,t=0} = 1$$

$\sigma_t(u,t)$ is also determined since its derivative $\sigma_{tt}(0)=\frac{\partial\sigma_t}{\partial t}$ is known for all t. $\sigma_{tt}(q)$ is still arbitrary as a function of q except at q=0 and $q=\lambda$, where it is determined by the vanishing of the second line of eq. (6.1). This makes clear that the extension of the original Gell-Mann and Low transformation to include the temperature behaviour is not a trivial problem.

References

1) Di Castro C., Jona-Lasinio G., Peliti L.: Ann.of Phys. 87, 327 (1974).

   Di Castro C., Jona-Lasinio G.: "Phase Transitions and Critical Phenomena", Vol. 6, edited by C. Domb and M.S. Green - Academic Press London, New York 1976.

2) Brezin E., Le Guillou J.C., Zinn-Justin J.: Phys.Rev. D8, 2418 (1973).

3) Wilson K.G., Kogut J.: Phys.Rep. 12C, 76 (1974). This and the two previous articles contain a wide list of references. One may also consult the lucid review by Ma S.K., Rev. Mod. Phys. 45, 589 (1973).

4) Niemejer T.H., Van Leeuwen J.M.J.: Physica 71, 17 (1974).

5) Bell T.L., Wilson K.G.: Phys. Rev. B10, 3935 (1975).

6) Jona-Lasinio G.: Proc.Nobel Symp. 24, 38 Academic Press (N.Y.) 1973.

7) Wegner F.: Journ.of Phys. C7, 2098 (1974).

8) Cassandro M., Gallavotti G.: Nuovo Cim. 25B, 691 (1975)

9) An object satisfying a condition like (3.2) is known in the mathematical literature as a multiplicative cocycle. See for example Oseledec V.I., Trans.Moscow Math. Soc. 19, 197 (1968).

10) Wegner F.J.: Phys. Rev. B5, 4529 (1972).

11) Wilson K.G.: Phys. Rev. D7, 2911 (1973).

LIST OF PARTICIPANTS

MM. T. Balaban (Warsaw Univ., Poland)
G. Benzi (Roma, Italy)
O. Bratteli (Bielefeld, RFA)
E. Brézin (C.E.N. de Saclay, France)
J. Bricmont (Louvain la Neuve, Belgium)
Mrs. M. Combescure (Orsay, France)
MM. G. De Francesci (Frascati, Italy)
G. Dell'Antonio (Napoli, Italy)
C. Di Castro (Roma, Italy)
J. Dimock (Genève Univ., Switzerland)
Mrs. M. Dumont (Mons, Belgium)
MM. F. Dunlop (I.H.E.S., Bures sur Yvette, France)
V. Enss (Bielefeld, R.F.A.)
H. Epstein (I.H.E.S., Bures sur Yvette, France)
J. Feldman (M.I.T., USA)
R. Figari (Nàpoli, Italy)
R. Flume (CERN, Genève)
J. Frolich (Princeton, N.J., USA)
C. Gawedzki (Warsaw Univ. Poland)
J. Glimm (The Rockefeller Univ., New York, USA)
M. Gourdin (Univ. Pierre et Marie Curie, Paris, France)
B. Hasslacher (C.I.T., Pasadena, USA)
K. Hepp (E.T.H. Zurich, Switzerland)
L. Jacobs (M.I.T., USA)
A. Jaffe (Harvard Univ., USA)
G.E. Johnson (Maryland Univ. USA)
G. Jona Lasinio (Roma, Italy)
J. José (Brown Univ. Providence, USA)
B. Julia (E.N.S. Paris, France)
L. Kadanoff (Brown Univ. Providence, USA)
H. Lehmann (Hamburg Univ., RFA)
M. Lévy (Univ. Pierre et Marie Curie, Paris, France)
J.C. Le Guillou (Univ. Pierre et Marie Curie, Paris, France)
M. Lüscher (Hamburg Univ., RFA)
A. Luther (Nordita, Copenhagen, Dk)
G. Mack (Hamburg Univ., RFA)
J. Magnen (Ecole Polytechnique, Palaiseau, France)

G. Martinelli (Roma, Italy)
O. McBryan (Rockefeller Univ. New York, USA)
P.K. Mitter (Univ. Pierre et Marie Curie, Paris, France)
Mrs. C. Nappi (Napoli, Itlay)
MM. H. Neuberger (Tel Aviv Univ., Israel)
A. Neuveu (E.N.S., Paris, France)
K. Osterwalder (Harvard Univ., USA)
G. Parisi (Roma, Italy)
Y.M. Park (Bielefeld, RFA)
G. Passarino (Torino, Italy)
K. Pohlmeyer (Hamburg Univ. RFA)
MM. O. Ragnisco (Roma, Italy)
P. Renouard (Ecole Polytechnique, Palaiseau France)
M.W. Roth (FNAL, Batavia, USA)
A. Rouet (CNRS,Marseille, France)
B. Schroer (Freie Univ. Berlin, RFA)
R. Sénéor (Ecole Polytechnique, Palaiseau, France)
T. Spencer (The Rockefeller Univ., New York, USA)
A. Stella (Padova, Italy)
R. Stora (CNRS, Marseille, France)
G. Sylvester (Yeshiva Univ. New York, USA)
K. Symanzik (DESY, Hamburg, RFA)
H.J. Thun (Berlin, RFA)
P.C. West (Imperial College, London, GB)
K. Wilson (Cornell Univ. Ithaca N.Y. USA)
J. Zinn-Justin (CEN de Saclay, France)
F. Zirilli (Roma, Italy)
Y. Demay (Nice, France)

INDEX